P9-BJU-707

The Gulf of California

Arizona-Sonora Desert Museum Studies in Natural History

SERIES EDITORS

Richard C. Brusca, Ph.D.
Christine Conte, Ph.D.
Mark A. Dimmit, Ph.D.

The Gulf of California
Biodiversity and Conservation

Edited by
Richard C. Brusca

The University of Arizona Press and
The Arizona-Sonora Desert Museum | Tucson

The University of Arizona Press

www.uapress.arizona.edu

Library of Congress Cataloging-in-Publication Data
The Gulf of California : biodiversity and conservation / Richard C. Brusca, editor.
p. cm.
Includes bibliographical references and index.
1. Biodiversity—Mexico—California, Gulf of. 2. Wildlife conservation—Mexico—California, Gulf of. 3. Natural history—Mexico—California, Gulf of. I. Brusca, Richard C.
QH95.4.G85 2010
508.3164'1—dc22

2009046445

Manufactured in the United States of America on acid-free, archival-quality paper containing a minimum of 30 percent post-consumer waste and processed chlorine free.

15 14 13 12 11 10 6 5 4 3 2 1

To Carlene and Elizabeth, who were always nearby during my earliest years in the Gulf; and to the memory of our beloved Natalie, who grew up on the shores of the Sea of Cortez to became Queen of the Tec.

Contents

Foreword ix
RODRIGO A. MEDELLÍN

Introduction 1
RICHARD C. BRUSCA

1 Origin, Age, and Geological Evolution of the Gulf of California 7
JORGE LEDESMA-VÁZQUEZ AND ANA LUISA CARREÑO

2 Physical, Chemical, and Biological Oceanography of the Gulf of California 24
SAÚL ALVAREZ-BORREGO

3 Reefs That Rock and Roll: Biology and Conservation of Rhodolith Beds in the Gulf of California 49
RAFAEL RIOSMENA-RODRÍGUEZ, DIANA L. STELLER, GUSTAVO HINOJOSA-ARANGO, AND MICHAEL S. FOSTER

4 Invertebrate Biodiversity and Conservation in the Gulf of California 72
RICHARD C. BRUSCA AND MICHEL E. HENDRICKX

5 Fishes of the Gulf of California 96
PHILIP A. HASTINGS, LLOYD T. FINDLEY, AND ALBERT M. VAN DER HEIDEN

6 The Importance of Fisheries in the Gulf of California and Ecosystem-Based Sustainable Co-Management for Conservation 119
MIGUEL Á. CISNEROS-MATA

7 Sea Turtles of the Gulf of California: Biology, Culture, and Conservation 135
JEFFREY A. SEMINOFF

8 Ospreys of the Gulf of California: Ecology and Conservation Status 168
JEAN-LUC E. CARTRON, DANIEL W. ANDERSON, CHARLES J. HENNY, AND ROBERTO CARMONA

9 Marine Mammals of the Gulf of California: An Overview of Diversity and Conservation Status 188
JORGE URBÁN R.

10 A Brief Natural History of Algae in the Gulf of California 210
RICHARD MCCOURT

11 Ecological Conservation in the Gulf of California 219
MARÍA DE LOS ÁNGELES CARVAJAL, ALEJANDRO ROBLES, AND EXEQUIEL EZCURRA

Bibliography 251

About the Contributors 331

Index 337

Foreword

Every person should spend at least one night a month out in the open under the stars, to clean his soul of all arrogance. The Silesian poet and novelist J. von Eichendorff wrote these words nearly two hundred years ago, and today—when arrogance seems to rule every aspect of the human universe—they are more true than ever. Our arrogance ranges from delusions of ruling the world, to ignoring what our own alarming findings are clearly screaming at us, to directly destroying that which sustains our life and that of all the planet's inhabitants—all the while thinking that our deeds will go unchecked and fantasizing that we are doing it to live in a "better" world. We are pushing the world's ecosystems to collapse, impoverishing our own life-support systems and earth's rich biological diversity. We are attacking the source of our livelihoods from all directions.

The Gulf of California is one of those places that continually remind us of our indelible and intimate connection to the natural world. Whether we are strolling along its rocky coasts or sandy beaches, peering into tidal pools, enjoying its shallow waters and finding "almejas chocolatas" for supper, on board a ship, panga, or dinghy in a research or leisure quest, measuring trees in its rapidly vanishing mangrove forests, diving along its shores or in deeper waters, or simply observing one of those spectacular, unforgettable sunsets endemic to the Gulf, our omnipresent links to biological diversity are so obvious that we could ignore it only by conscious effort. The Gulf's fisheries resources reach the farthest extremes of Mexico, and many of them—such as spiny lobsters, shrimp, sea cucumbers, abalone, sharks, and others—end up on dinner plates around the world. Thus, it is not too far a stretch to say that the Gulf of California is one with the world; it is a true universal treasure that benefits humans across the planet. The entire world should be concerned about its future; the benefits the Gulf provides are of great value and a constant source of comfort and subsistence to millions of people on all continents and in many countries. Surely, conservation of the Gulf is a much higher priority than building a wall between two countries that benefits nobody but threatens ecosystems, biodiversity, and a good-neighbor policy. History has proven that much. In the words of Spanish-American poet George Santayana: "Those who do not learn from history are doomed to repeat it."

I welcome this opportunity to write a few paragraphs as a foreword for this timely and thought-provoking book. Its multidimensional approach is especially welcome and makes it a useful reference for future work and for the decision-making processes needed to guide the sustainable use and conservation of its resources. From the origins of the Gulf of California to its physical and chemical characteristics, and from urgently needed conservation alternatives for fisheries and the entire Gulf ecosystem to the knowledge of its invertebrates, fishes, cetaceans, and sea turtles, this wide-ranging book provides new insights and offers clear paths for continuing the long battle to achieve sustainable use that is solidly based on robust conservation science. A very welcome chapter is the one on the ecology and conservation of osprey, an iconic bird that provides one more unforgettable sight for visitors to the region; anyone who has seen an osprey nest anywhere along the peninsula has been branded a true "son of the Gulf." Rick Brusca is the perfect individual to guide this project and see it through. His forty-year investment in knowledge of the Gulf, together with his boundless enthusiasm and commitment to its conservation, are an unstoppable combination that has produced scores of outstanding pieces. Rick's work has already made a significant difference for the benefit of the Gulf. *The Gulf of California: Biodiversity and Conservation* is one more link in this very long chain.

This book is also a remarkable example of international, multi-institutional, interdisciplinary collaboration. The careful balance between Mexican and non-Mexican authors provides a clear example of the power of collaboration. Only through collaborative projects such as this one will Mexico (and, indeed, any other country) achieve success in answering large-scale, critically important questions. Much more collaboration is required to drive the conservation messages home. Dozens of courses, workshops, programs, projects, lectures, publications, and educational materials are still required in a variety of settings, from indigenous villages to schools and universities to isolated towns near oases to large resorts and tourist towns, new developments, and even the giant cruise ships that increasingly affect the Gulf. All of the people in these population hubs need the information contained in this book, even if they don't yet realize it.

If only our bureaucrats and politicians had spent but a single night under the skies of the Gulf of California—a single one of the thousands of treasured nights experienced by all the authors and, no doubt, also by the readers of this book—then our future, and that of the Gulf and its creatures, would be brighter and the loss of its resources could be curbed. It is not yet too late.

—Rodrigo A. Medellín

The Gulf of California

Introduction

RICHARD C. BRUSCA

One cannot visit the Gulf of California (Sea of Cortez) without recognizing its remarkable and singular nature (color plate 1).[1] The accumulation of species diversity since the Gulf's opening ~5.6 million years ago has produced one of the most biologically rich marine regions on earth. The coastlines, offshore benthic regions, and pelagic waters of the Gulf are famous for supporting not only an extraordinary biological diversity but also exceptionally high productivity and large populations in all marine taxa: invertebrates, fishes, cetaceans, sea turtles, and birds. Its 6,000 *recorded* animal species is estimated to represent about 70 percent of the *actual* (total) faunal diversity lurking in its rich waters. So productive are the Gulf's waters that about half of Mexico's total fisheries production comes from the region. Small pelagic fish constitute 25–40 percent of the fisheries landings in Mexico, and over 70 percent of those landings (predominantly the Pacific sardine, *Sardinops caeruleus*) are taken in the Gulf of California. More than 500,000 tons of seafood are taken from the Gulf annually, and this figure does not include the wasted by-catch (which would probably double, triple, or quadruple that tonnage). This book provides a benchmark for understanding the nature of this extraordinary productivity and diversity, how it is threatened, and in what ways it is being protected.

Continental Mexico itself is regarded as one of the world's "megadiverse" countries, containing over 10 percent of the living species on earth on only 1.7 percent of the world's land. But all of Mexico's biological diversity is at risk, both on land and in its oceanic waters—including the Gulf of California. In the Gulf, areas targeted for bottom trawling (i.e., in waters shallower than 150 m) have had their benthic ecosystems decimated over the past fifty years as a result of excessive disturbance from shrimp trawlers (and other bottom trawlers). However, we have almost no knowledge regarding current or past community composition and food web structure for the Gulf's offshore benthic *or* pelagic communities. Data suggest that pelagic oceanic food webs are quite resilient, and most can probably recover from anthropogenic insults once we change our ways. However, it is yet to be seen if badly damaged benthic communities can recover from severe disturbance, such as years of bottom trawling, and studies elsewhere suggest that a return to previous community structure might not necessarily occur when disturbances have been significant.

One of the most pressing research needs in the Gulf is to achieve an integrated understanding of community and ecosystem structure in this large marine ecosystem and of how profound have been the effects of bottom trawling on this system.

Much of the mainland coast has been devastated by "development" (urbanization, marina construction, aquaculture, and other forms of habitat disturbance and destruction) and by overcollecting of marine life by both residents and tourists. These coastlines now harbor but a pale shadow of their former diversity. It is human nature for tourists to pick up creatures found along the shore and to take them as souvenirs, but every time this happens a piece of the food web disappears, and today millions of souvenir seekers carry away tidepool animals every year. Hardly a habitat in the Gulf has been left undisturbed by humankind.

Since the mid-1980s, a growing conservation movement has emerged in northwestern Mexico that has had a powerful and positive influence on conservation in the Gulf. These emerging conservation efforts have been crucial to establishing protected areas and conservation priorities for the Gulf and its islands, working with artisanal fishers to help develop sustainable fisheries, and working with state and federal government agencies to encourage better protection of marine and coastal resources. As a result of such efforts, fisheries laws over the past decade have tightened up—gillnetting may soon become illegal, and bottom trawling may become better regulated (and, it is hoped, soon banned).

If the conservation movement in the Sea of Cortez continues with its present momentum then new areas will receive protection, and better enforcement of protected regions should follow. Fortunately, one still can find islands and remote coastal refuges, areas not easily accessible by road or large fishing boats, that serve as important shelters for species extirpated elsewhere. As more areas are protected, these species refugia will become vitally important source areas for recruitment and repopulation to areas that had previously lost their diversity. It remains to be seen, however, if the vast offshore benthic ecosystem that has been so devastated by bottom trawlers will ever regain its original biological community structure.

As Mexico continues to move forward in protecting the Gulf of California, some areas loom large as critical habitats in need of urgent protection. These habitat conservation priorities include: the sea around all of the Gulf's islands, which are essential population refugia for the region's littoral and shallow-water biodiversity; esteros and estuaries, which harbor unique assemblages of marine life and also play critical roles in the life histories of most commercial species, both finfish and shellfish; the Canal de Infiernillo, which is the only sizable seagrass bed in the Gulf; the four unique beachrock (*coquina*) outcroppings of the Northern Gulf (Puerto Peñasco, Punta Borrascoso, San Felipe, and Coloradito); offshore hard bottoms and seamounts where large "reef fish" (e.g., groupers, snappers, jacks, creolefish, giant hawkfish) spawning aggregations occur; and rocky shores throughout the Gulf.

A keystone among Mexico's environmental laws is its "endangered species act": Norma Oficial Mexicana NOM-059-ECOL-2001, Protección ambiental-especies nativas de México de flora y fauna silvestres—Categorías de riesgo y especificaciones para su inclusión, exclusión o cambio-lista de especies en riesgo. Other key federal laws for biodiversity protection are: Ley Orgánica de Administración Pública Federal, Reglamento Interior de la Secretaría de Medio Ambiente y Recursos Naturales (SEMARNAT), Ley General del Equilibrio Ecológico y la Protección al Ambiente, Ley de la Vida Silvestre, and Ley Federal sobre Metrología y Normalización. In addition, Mexico is a member of CITES—the Convention on International Trade in Endangered Species—which protects many species including all of the whales and sea turtles in Mexican waters. (One of the best examples of species protection is the group of sanctuaries that Mexico established on the Pacific coast of Baja California Sur for gray whales.) Together, these laws constitute a solid framework for protecting Mexico's natural resources. However, they are frequently undermined by lack of political will and lack of funding for enforcement. Of critical concern today is the continued use of traditional shrimp trawls that destroy the benthos and, in the upper Gulf, capture an estimated 120,000 endangered juvenile totoaba annually. Also of critical concern is the impact of poorly regulated shrimp farm development in or adjacent to wetlands throughout the Gulf.

One of the most important conservation steps taken for the Gulf was the protection of its island archipelago. The Gulf of California encompasses more than 900 islands and islets. Because their sizes range from tiny to huge, they effectively side-step the so-called SLOSS (single large reserve vs. several small ones) conservation biogeography debate: both the "bigger is better" and "more is better" arguments are satisfied by protecting the entire archipelago. CONABIO's (Comisión Nacional para el Conocimiento y Uso de la Biodiversidad) 2002 "Programa de Regiones Marinas Prioritarias de México" was also an important recent step forward, and the Gulf figured prominently in the recognition of areas requiring priority attention.

This book contains 11 chapters by 24 Gulf of California experts. The goal is to provide a benchmark volume on the status of major biodiversity and conservation issues in the region. The decision to incorporate all of the references cited in the chapters into a single, expanded Gulf bibliography was made in order to provide a more useful source list for readers. It is hoped that the book will be useful not only to scholars but also to anyone who has a passion for this extraordinary body of water and a concern for its future.

I wish to express my deep gratitude to the authors of this volume. They are all very busy people who set aside time to develop chapters specifically for this book, and they put up with my countless e-mails badgering them for this and that. I also thank the organizations that supported the 2004 Gulf of California Conference, which was organized by the Arizona-Sonora Desert Museum and served as the foundation of this volume: The David and Lucile Packard Foundation, World Wildlife Fund Gulf of California Program, Conservation International Gulf of California Program, Sonoran Institute, Nature Conservancy's Tucson office, the Sonoran Desert Waters Institute (formerly, the Sonoran Sea Aquarium), the George A. Binney Foundation, CEDO Intercultural, and Priscilla Baldwin. There are too many people who helped make the Gulf Conference a success, and bring this book to fruition, to be listed here, but very special thanks go to Linda Brewer, Allyson Carter, Yajaira Gray, Sergio Knaebel, Blair Kuropatkin, Shannan Marty, and Barry McCormick. Over my past four decades of working in the Gulf, many good friends have emerged to offer intellectual and spiritual fellowship, support, and some darn good tequila-drinking nights. They helped me in innumerable ways, put up with my lopsided sense of humor, and in some cases traveled the road with me from our formative and halcyon years in the '60s. Together, we have seen the emergence of a strong and growing conservation ethic in the Gulf of California. My life would have been impoverished were it not for their trustworthy friendship: Luís Bourillón, Kathy Boyer and Stan Gregory, Jose Luis Carballo, Machángeles Carvajal, Christine Conte, Richard and Tiffney Cudney, Richard Felger, Ana Luisa Figueroa, Lloyd Findley, Sonnie and Elizabeth Findley, Jesús García, Todd and Lisa Haney, Jeff Hartman, Lindsey Haskin, Phil Hastings and Marty Eberhardt, Michel Hendrickx and Mercedes Cordero, Roy Houston, Jonathan Mabry, Larry Marshall, Tim Means, Rodrigo Medellín, Francisco Molina, Ed Pfeiler, Alejandro Robles, Steve Schuster, Jeff Seminoff, George Staley, Don Thomson, Jorge Torre and Jaqui Garcia, Sandra Trautwein, Peggy Turk and Rick Boyer, Albert van der Heiden and Sandra Guido, Omar Vidal, Linda Yvonne and Samantha Maluf, and many others. Finally, I don't have words to thank the wonderful Wendy Moore, whose daily counsel and unqualified love help me reach my potential—with her, I have found the time of my life.

NOTE

1. In this book, the names Sea of Cortez and Gulf of California (or simply "the Gulf") are used interchangeably. Over the past 450 years, four names (primarily) have been used for this region. Mar Vermejo (Vermillion Sea) is the original name given by Francisco Ulloa in 1539, the first person to sail to the head of the Gulf and prove to Europeans that Baja California is a peninsula. Mar de Cortés (Sea of Cortez, Sea of Cortés) is the name adopted by the Spaniards shortly after Ulloa's journey and original naming opportunity. The "z" spelling is a U.S. anglicization in use as early as 1870; its popularity and use have been increasing in recent years. Mar Rojo (Red Sea) is a name adopted by some Spaniards in the late 1500s, but it never really caught on. Golfo de California (Gulf of California) is a name adopted by the Spaniards sometime in the late 1500s and that gradually grew in popularity over Mar de Cortés. Contemporary usage is split between Mar de Cortés (Sea of Cortez) and Golfo de California (Gulf of California). Modern Mexican cartography uses both names (in the Spanish version, of course), whereas modern U.S. maps use Gulf of California (although maps in the popular press often use Sea of Cortez).

The terms "Baja" and "Baja California" are often used loosely. The technically correct usages are as follows. "Baja California peninsula" is a peninsula of North America that includes two modern Mexican states: "Baja California" is the 29th Mexican state (known as "Northern Territory of Baja California" before its statehood in 1952); "Baja California Sur" is the 31st Mexican state (known as "Southern Territory of Baja California" before its statehood in 1974). The term "Baja California" may also refer to a former Mexican territory (1804 to 1931) that comprised the entire Baja California peninsula.

CHAPTER 1

Origin, Age, and Geological Evolution of the Gulf of California

JORGE LEDESMA-VÁZQUEZ AND
ANA LUISA CARREÑO

Summary

The present-day topography of the Baja California peninsula evolved through a series of tumultuous geological events that began many millions of years ago (Ma). Between 130 and 90 Ma, during the Cretaceous Period, a sea-floor crustal plate (the Farallon Plate) in the eastern Pacific began to subduct beneath the western edge of North America/Mexico. The plate was being driven northeastward by a highly active rift or sea-floor spreading center, the East Pacific Rise; to the west of the spreading center, the great Pacific Plate was slowly being pushed northwestward by the East Pacific Rise. The Rise itself was also shifting slowly eastward, following along "behind" the Farallon Plate. The subduction of the Farallon Plate caused mountain building along the northwestern coast of Mexico. Today, much of this now weathered mountain range is on the Baja California peninsula, recognized as the "Alta California–Baja California Peninsula Range Batholith"—originally on the continental mainland, but today extending from southern California to nearly the 28th parallel on the Baja peninsula. Studies of the batholithic rocks in this mountain chain (e.g., radiometric dating, geomagnetic analyses, etc.) reveal that when it formed in the Cretaceous, what is now southern California and the Baja California peninsula were located 330–400 km south of their current geographic positions, with the tip of the peninsula-to-be positioned at Bahía Banderas, at the present-day location of Cabo Corrientes (near Puerto Vallarta), Jalisco. One can envision this ancient configuration by mentally sliding the peninsula to the southeast and positioning it against the mainland.

By the middle Oligocene (~30 Ma), the Farallon Plate had largely completed its protracted descent beneath the North American Plate, and at that point in time the North American Plate and the East Pacific Rise came to intersect on the coast of continental Mexico. As the spreading center sank beneath continental North America, a narrow slice of crust shifted west of the rift and accreted, or "stuck" itself onto the Pacific Plate. The final subduction of the Farallon Plate drove the development of a terrestrial volcanic arc in the early to middle Miocene (24–11 Ma) along what is now roughly the axis of the modern Gulf of California. The rocks generated from that volcanism today constitute the basement of the "Gulf Extensional Province." These rocks can be seen in the uplifted mountains of the easternmost Baja California peninsula, exposed as thick andesite flows in the Sierra La Giganta and other ranges. By the late Miocene, 12–6 Ma, the entire margin of northwestern Mexico had coupled itself to the northwesterly moving Pacific Plate and began to be gradually pulled away from the North American Plate (i.e., from Mexico) as it was pushed northwestward by the continuing sea-floor spreading at the East Pacific Rise.

The rifting process continued unabated throughout most of the Miocene. In the late Miocene, 12–6 Ma, subsidence resulting from Pacific–North American plate-boundary motion and associated rifting in the region created a depressed subcoastal trough on the west coast of Mexico (the so-called Proto-Gulf Extension). Paleontological evidence suggests that beginning 9–6.3 Ma, this trough was filled by one (or more) marine incursions, creating what has been called the "Proto-Gulf." By ~5.6 Ma the rifting of the modern Baja California peninsula and the development of the San Andreas Fault system were underway and the Gulf had begun to open. These geotectonic activities led to the development of the "Gulf of California Geological Province," which includes the Salton Trough in its north, the only part of the province that is not covered by ocean waters today (but has sporadically been so in the past).

Spreading center sea-floor basins in the Gulf opened sequentially from south to north, from 5.6 Ma for María Magdalena Rise, to 3.7–3.4 Ma for the Alarcón Basin, and 2.1 Ma for the Guaymas Basin. Evidence suggests that at 1.5 Ma the San Andreas Fault system had an average displacement rate of ~35 mm/year, decreasing to ~9 mm/year about 90 thousand years ago. Today, the rates have again picked up, and the displacement rate is ~27 mm/year.

Prior to the opening of the Gulf a series of marine incursions flooded the newly formed Proto-Gulf Extensional Trough on the west coast of Mexico. These marine incursions formed seaways that have been collectively (and loosely) referred to as the "Proto-Gulf." Miocene marine sediments and paleontological data suggest either a single seaway, opening in the south from a location near the mouth of the modern Gulf, and extending north all the way to the Salton Trough, or perhaps as far north as San Gorgonio Pass (e.g., Oskin and Stock 2003), or separate marine incursions (e.g., Helenes and Carreño 1999), in the south and in the north, fed by narrow inlets (sometimes confusingly referred to as "transpeninsular" seaways, although the peninsula had not yet formed). The northern seaway would have probably been located south of the Sierra San Pedro Mártir, perhaps around the present-day location of Bahía San Ignacio, on the Pacific coast of what is today the Baja California peninsula. In either case, the northern basin was probably isolated from the southern basin for much of the time. Contrary to what is sometimes written, it is not known if a marine incursion persisted until the opening of the modern Gulf, around 5.6 Ma, when the peninsula separated from the mainland, and there is no physical evidence of such.

The modern Gulf of California was thus formed from a tectonic structural rift, and today comprises a 1,200-km-long peripheral extension of the eastern Pacific Ocean varying from ~200 km wide at its mouth to 85 km at its narrowest point. The modern Baja California peninsula is about 1450 km in length. Thinning of the earth's crust below the waters of the Gulf due to this spreading process also resulted in the development of hydrothermal vents in the Gulf, active since late Miocene times, and many of these vents persist, such as the Guaymas Basin, where geothermal liquids and gases discharge through fractures in sea-floor rock and accumulated sediments.

Since the Gulf began to make its appearance in the late Miocene, numerous islands have broken off the peninsula and the mainland as a consequence of the extension process, or as products of faulting and uplift associated with tectonic activity along the many faults that arise off the East

Pacific Rise that today bisects the Gulf (e.g., islas Salsipuedes, San Lorenzo Norte and Sur, Carmen, Danzante, Coronado, Monserrat, Santa Catalina, San José, Espíritu Santo). Other Gulf islands are the result of recent volcanism (e.g., islas Coronado, San Luis, and Tortuga off Santa Rosalía). Over 900 islands, islets, and emergent rocks have been identified in the Gulf's waters, making it one of the world's largest island archipelagos. The sizes and locations of more than 230 of these are given in Table 1.1-1 of Case et al.'s (2002) *A New Island Biogeography of the Sea of Cortés.*

By the early or mid-Pliocene, the Colorado River had begun depositing its outflow and sediments into the rift-derived Salton Trough. Over time, the filling of the trough developed a lacustrine basin, and as the basin's lake grew in size the river eventually overflowed and reached the upper Gulf, thus initiating the Colorado River delta ecosystem.

Introduction

The Gulf of California and Baja California peninsula possess an impressive marine and terrestrial biological diversity that was ultimately derived through historical geological processes that facilitated species dispersals, isolation events, speciation events (originations), and extinctions. The modern fauna of the region, both terrestrial and marine, exhibits a broad variety of distribution patterns, much of which is directly associated with the separation of the peninsula from mainland Mexico, with ancient configurations of the peninsula, and with the formation of the islands of the Gulf.

The discipline of phylogeography investigates the effects of geological events on the evolution of organisms; and phylogeographic inference benefits from a well-established knowledge of the sequence and timing of geological events in a particular region. Biogeographic affinities of fossils encountered in Neogene rocks, as well as phylogeographic studies of recent marine and terrestrial biota inhabiting the area, have been used to test hypotheses concerning origin, migration, evolution, and extinction of species on the Baja California peninsula and in the Gulf of California and its islands. In order to explain patterns of biological diversity, paleontologists and biologists search the geological record of an area for events that can help explain their hypotheses and interpretations. In spite of a large body of information on the origin and geological evolution of the Gulf of California and surrounding areas, several significant questions and conflicts concerning its Neogene history still exist.

Two recent and interesting controversies began when Smith et al. (1985) assigned a date of 12.9 ± 0.4 Ma to marine deposits intercalated with volcanic breccia on Tiburón Island, and when Neuhaus (1989) assigned a date of 21–15 Ma for a volcanic unit beneath the marine rocks in the Gulf, predating by almost 8 million years the conventional dating of seafloor extension in the region. Both of these studies seem to contravene not only the geophysical data (Oskin and Stock 2003) but also some of the available geological information (Escalona-Alcazar et al. 2001; Pacheco et al. 2006).

Some recent hypotheses have proposed Miocene seaways across the peninsular region that created biogeographic barriers that were vicariant events leading to genetic separation of populations of terrestrial vertebrates (Grismer 1994; Helenes and Carreño 1999; Riddle et al. 2000; Murphy and Aguirre-León 2002; Carreño and Helenes 2002; Lawlor et al. 2002; Ledesma-Vázquez 2002; Zink 2002; Montanucci 2004; Leaché and Mulcahy 2007; Leaché et al. 2007). Some of these proposals have pointed to pre-peninsular (i.e., mainland) coastal seaways as the barriers, in agreement with marine sediment and paleontological data suggesting Miocene marine transgressions. Others, however, have suggested that the peninsula was already in place when the seaways formed, in direct conflict with the geological data, which date the opening of the Gulf at around 5.6 Ma.

The Geological Evolution of the Gulf of California

The Gulf of California Geological Province can be divided into five regions: (1) the modern subaerial Salton Trough, (2) the northern Gulf of California, (3) the central Gulf region, (4) the southern Gulf, and (5) the mouth region (fig. 1.1). Nowadays, only the Salton Trough is not covered by marine waters, although it was connected to marine waters of the Gulf in the early history of the region.

Wegener (1928) was the first to hypothesize—based on the shape of the coastlines of the Baja California peninsula and western Mexico—that the peninsula was previously attached to the Mexican mainland and that the separation was driven by activity of the San Andreas Fault. Later, other workers expanded on this idea, noting that both strike-slip (pull-apart) faulting and

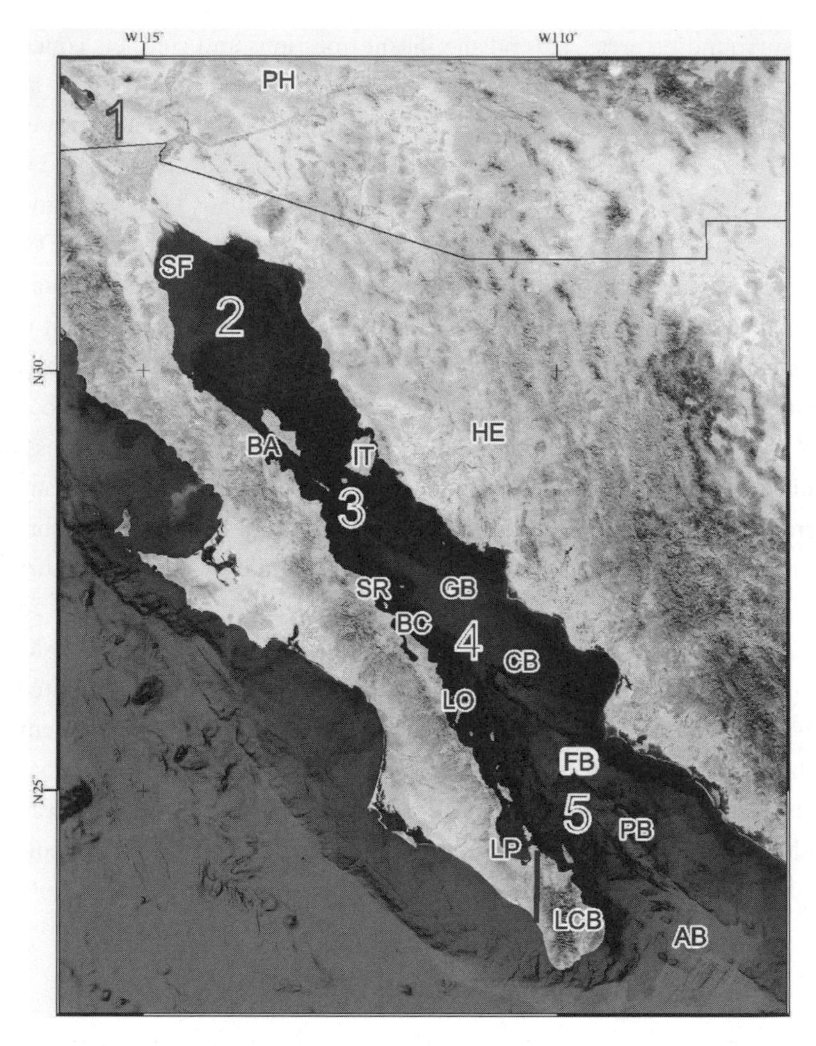

1.1. Merger of NASA images captured by Aqua/Modis sensor on November 30, 2003, with seafloor information from a collection of surveys by DANAO7RR, DANAO8RR (*SIO*), and other bathymetric data, showing the locations of: the modern subaerial Salton Trough (1), the northern Gulf of California (2), the central Gulf region (3), the southern Gulf of California (4), and the mouth region (5). Localities on land, from north to south: Phoenix (PH), San Felipe (SF), Hermosillo (HE), Tiburón Island (IT), Bahía de Los Ángeles (BA), Santa Rosalía (SR), Bahía Concepción (BC), Loreto (LO), La Paz (LP), and Los Cabos block (LCB). Oceanic basins within the gulf, from north to south: Guaymas Basin (GB), Carmen Basin (CB), Farallón Basin (FB), Pescadero Basin (PB), and Alarcón Basin (AB). La Paz fault, solid dark line. Image merger prepared by Alejandro Hinojosa-Corona (CICESE).

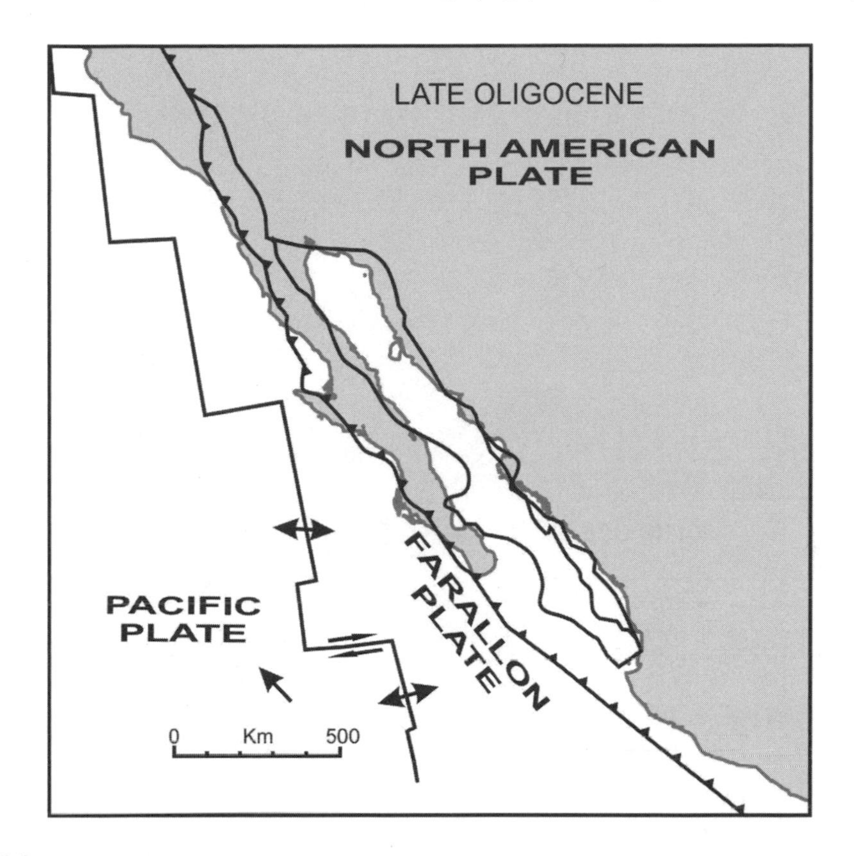

1.2. Late Oligocene. Simplified paleogeographic reconstruction for the margin of North America at the time of arrival of the Farallon Plate and ridges. Solid arrows indicate the presence of a subduction zone. Modified from Nicholson et al. (1994).

rotation probably contributed to the formation of the peninsula and the Gulf (Hamilton 1961).

By the 1960s, studies of seismic refraction, gravity, heat-flow, and bathymetrics in the Gulf began to accumulate, and several hypotheses of Gulf opening were put forth (Rusnak and Fisher 1964). Some of these hypotheses proposed a total displacement of 475 km (Phillips 1964), while others proposed 260 km of Gulf dilation over 4 to 6 million years (Larson et al. 1968). Since the 1960s, considerable research has been done, and today it is generally believed that the opening of the Gulf was driven by two sequential extensional events (Stock and Hodges 1989).

During the late Oligocene (~30 Ma) the peninsula of Baja California was still attached to the Mexican mainland (fig. 1.2), and continental

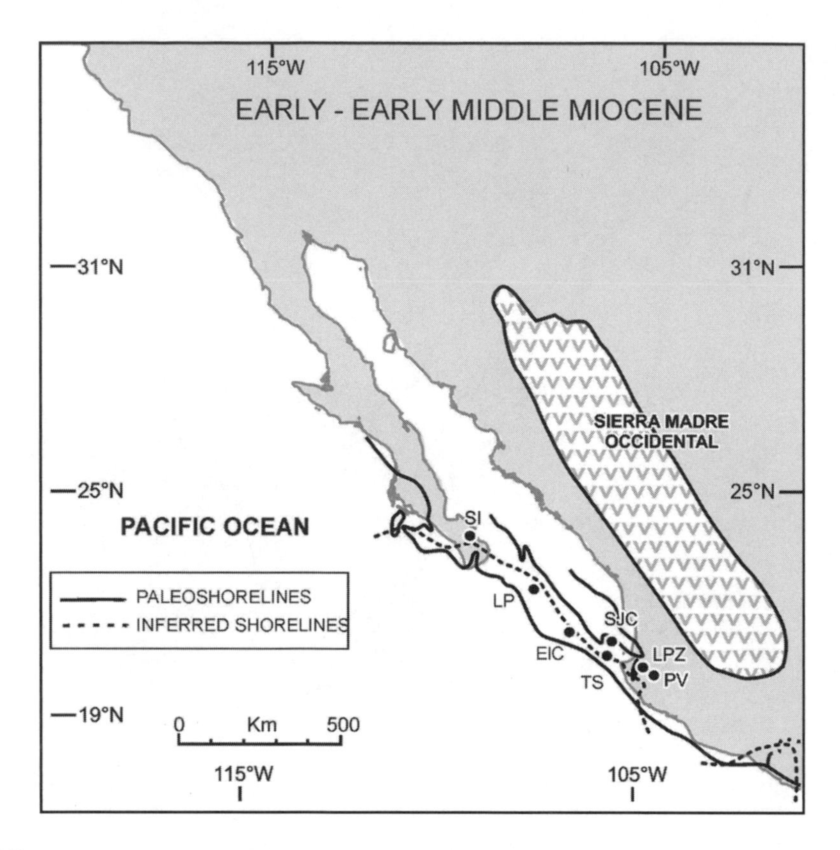

1.3. Early–early middle Miocene. Approximate extent of shallow marine areas on the western margin of North America (see text for details). ElC = El Cien; LP = La Purísima; LPZ = La Paz; SJC = San Juan de la Costa; SI = San Ignacio; TS = Todos Santos; PV = Puerto Vallarta. After Smith (1991).

sedimentation dominated the Mexican Pacific coastal region. Around 22 Ma (early Miocene) an eastward strike-slip faulting occurred on the Pacific coast, west of the peninsular batholith, primarily due to interactions between the Pacific and North American Plates. This strike-slip faulting drove an accelerated subduction regime of the Continental Borderland and neighboring areas, which was coincident with a global high sea-level stand. This same phenomenon produced shallow to outer shelf marine deposits of the San Joaquin Valley (Addicott 1970; Barnes and Mitchell 1984; Olson 1990; Huddleston and Takeuchi 2007), the El Cien–San Juan de la Costa region (Kim and Barron 1986; Carreño 1992b), and the La Purísima (Kim and Barron 1986) and La Paz (Schwennicke et al. 1996) formations. The rest of the peninsula does not contain evidence of marine sedimentation (fig. 1.3).

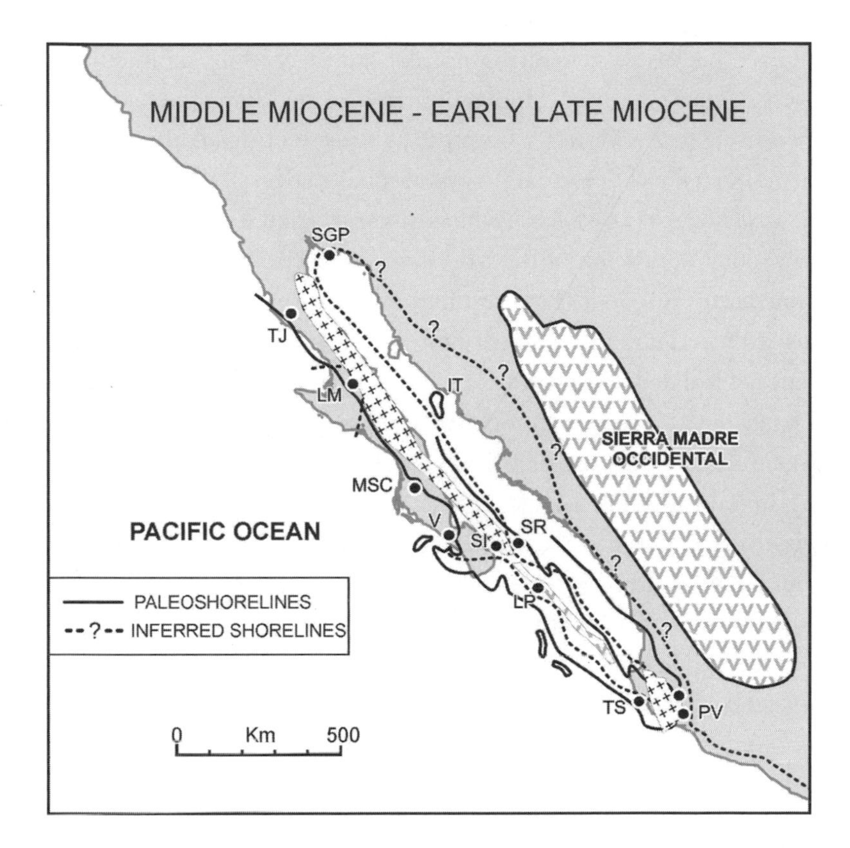

1.4. Middle Miocene to early late Miocene. Shallow marine sedimentation occurs in the areas of San Gorgonio Pass (SGP), Tijuana (TJ), La Misión (LM), Mesa San Carlos (MSC), Vizcaíno (V), San Ignacio (SI), Todos Santos (TS), Puerto Vallarta (PV), and La Purísima (LP). A volcanic arc installed in what is now roughly the axis of the Gulf of California, which constitutes the basement of the Gulf Extensional Province. The peninsula remains attached to the mainland and contains only volcanic, volcaniclastic, and continental sedimentary rocks (see text for details). After Smith (1991).

By the middle Miocene (fig. 1.4), shallow marine sedimentation had occurred in the Tijuana and La Misión areas (Minch 1967; Minch et al. 1984; Ashby and Minch 1988; Ashby 1989; Ledesma-Vázquez and Kasper 1989) as well as in the Mesa San Carlos (Novacek et al. 1991), Vizcaíno (Helenes-Escamilla 1980), and La Purísima (Hertlein and Jordan 1927) regions. A volcanic arc was established in what is now roughly the axis of the Gulf of California, constituting the basement for the Gulf Extensional Province (Gastil et al. 1979). At this time, extension initiated at the eastern side of the peninsula generating normal faulting, while subduction along the western margin ceased. High mountains formed along the eastern side of the peninsula (e.g., Sierra La Giganta) as a result of this faulting process.

Extensional deformation began to take effect in the late middle and early late Miocene (15–8 Ma), which was associated with a drop in volcanic activity due to cessation of the Miocene volcanic arc in the Salton Trough and northern Gulf area. This resulted in the formation of several shallow basins with periodic influence of marine waters (fig. 1.4), inferred by the presence of isolated fossil records reported from the Yuma area (Dean 1996; McDougall et al. 1999), Sonora (Gómez-Ponce 1971; Smith in Gastil et al. 1999), on Ángel de la Guarda Island (Escalona-Alcázar and Delgado-Argote 1998), in Bahías de los Ángeles (Delgado-Argote et al. 2000) and Las Ánimas in the San Lorenzo Archipelago (Escalona-Alcázar et al. 2001), and in the northern Gulf region (Helenes et al. 2005). Some authors (Oskin and Stock 2003) have rejected the idea of an early (prior to 8 Ma) marine incursion in the region to become the northern Gulf on grounds of the time and transgressive characteristics of the benthic organisms, arguing instead for a reworked presence for planktonic species encountered at those localities.

Nevertheless, structural studies and igneous rock records, including dates for the Gulf region, also clearly indicate an earlier marine extension for the northern region than for the southern Gulf, one that ranges in age from middle to late Miocene and ranges in extent from the southern Sierra Juárez to Santa Rosalía on the peninsula. In addition, there are post–late Miocene age data from Bahía Concepción all the way to the La Paz area (Hausback 1984; Sawlan and Smith 1984; Lee et al. 1996; Nagy 2000). Lee et al. (1996) indicated that normal faulting extension for the northern Gulf region is bracketed between 15.98 and 10.96 Ma. Sawlan and Smith (1984) recognized that extension began for the southern Gulf shortly after 10 Ma but prior to the latest Miocene. With the aforementioned in mind, there seems to be good evidence that a northern basin predated a southern one, which would allow for the older marine deposition evidenced in the stratigraphic record.

Helenes and Carreño (1999) hypothesized the presence of a middle Miocene seaway connecting to a large marine embayment to explain these late Miocene deposits. A marine incursion in the region of the present-day northern Gulf had been ruled out due to the presence, since the Cretaceous, of the Peninsular Range batholithic belt (Bohannon and Parsons 1995). The presence of large amounts of volcanic, volcaniclastic, and nonmarine sedimentary rocks covering the central and southern parts of the peninsula also prevents a marine incursion from the south in a similar way as occurs today. In Baja California Sur, only the La Purísima and El Cien regions contain marine sediments. As a possibility, Helenes and Carreño (1999) proposed that a marine seaway, leading to a large embayment, in the San Ignacio area could have facilitated the marine incursion into the area now represented by the northern Gulf (fig. 1.5), based on the presence of middle Miocene marine sedimentary rocks capped by 10 Ma tholeiitic basalt flows together with the presence of vents that might indicate a zone of weakness in the crust.

In late Miocene times, the western margin of the peninsula-to-be shifted to a minor compressive tectonic regime while in the north a marine basin developed. The central "Gulf" region continued to trap volcanic deposits, whereas the Los Cabos block was catching marine sedimentation similar to the western margin of Nayarit, Jalisco, and Michoacán. Deposits located in the north, as well as those located to the east, generally reflect shallow water depths in the early late Miocene and upper bathyal depths deposits in the latest Miocene, whereas those located on the west coast of the mainland were shelf deposits. Along the western margin of the Baja peninsula, marine deposits have been recorded at Vizcaíno (slope deposits) and in the Santa Rita area (shelf deposits). According to Winker and Kidwell (1996), by ~6.5 Ma marine inundation had reached San Gorgonio Pass (California), within 50 km of the Pacific Ocean near the present-day Los Angeles Basin (but never reached or connected back to the Pacific Ocean). Therefore, the late Miocene (8.3–6.5 Ma) configuration shows shallow to deep marine basins developed in the Salton Trough (McDougall et al. 1999), Sierra Cucupá east of Laguna Salada (Vázquez-Hernández et al. 1996), San Felipe (Boehm 1984), Tiburón Island (Ingle in Gastil et al. 1999), San Esteban Island (Carreño in Calmus et al. 2008), Santa Rosalía (Carreño 1983), and Bahía Concepción (Carreño in Ledesma-Vázquez et al. 1999) as well as in the southernmost Gulf (McCloy 1984; Carreño 1992a) at María Madre Island (Carreño 1985; McCloy et al. 1988), Punta Mita, Nayarit (Ingle in Gastil et al. 1999), and Punta Maldonado, Guerrero (Juárez-Arriaga et al. 2005). This has been interpreted as the first great incursion of the Pacific Ocean,

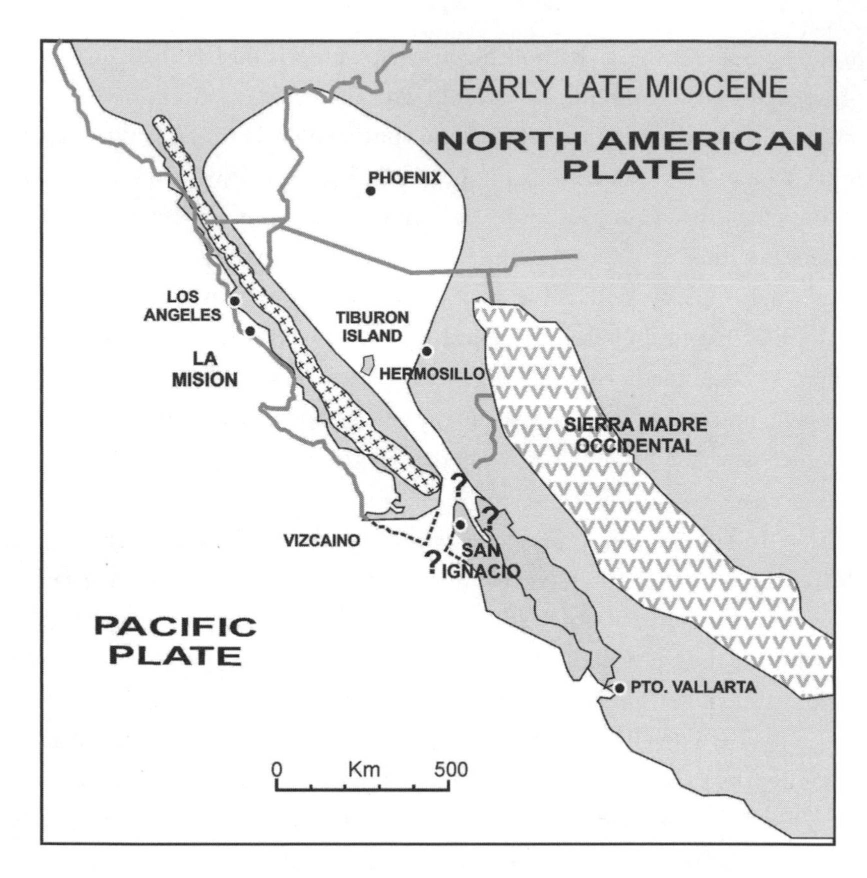

1.5. Early late Miocene. The peninsula of Baja California remains attached to the mainland and an early proto–Gulf of California marine incursion into the northern Gulf of California occurred by means of a transpenisular seaway located at San Ignacio (originating during the late middle–early late Miocene). After Helenes and Carreño (1999) and Ledesma-Vázquez (2002).

which initiated about 8.2–7.5 Ma and began at the tip of Baja (still attached to the mainland). Configuration of the area has produced a scenario (fig. 1.6) in which the Baja California peninsula must have been exhumed at its east side with a deeper marine basin to the west, whereas Baja California Sur contained shelf deposits on the west and east side but was restricted by the intensive volcanisms.

According to Nagy and Stock (2000), spreading began in the early Pliocene around the mouth of the Gulf, and sequentially appearing spreading axes and basins opened in the southern Gulf region from southeast to northwest from 5.6 Ma for the Maria Magdalena Rise, 3.7–3.4 Ma for

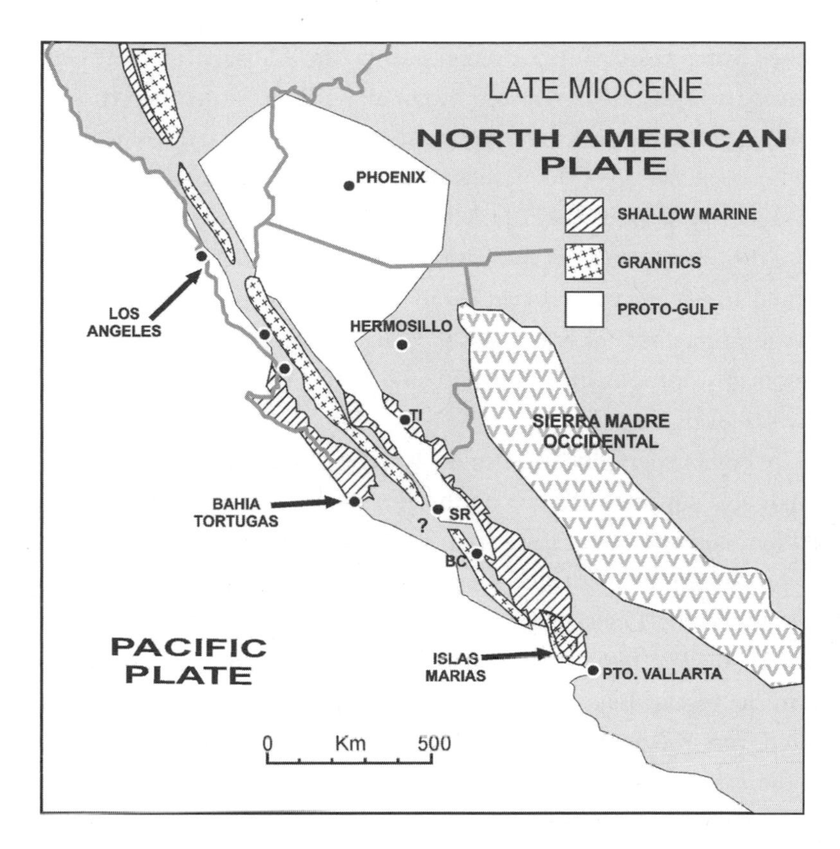

1.6. Late Miocene. Proposed reconstruction of first greatest incursion of the Pacific Ocean underway at the tip of the Baja peninsula (which is still attached to the mainland) in the late Miocene–earliest Pliocene (see text for details). The Los Cabos block remained connected to the rest of the peninsula and the eastern margin received marine deposits, which reached bathyal depths. TI = Tiburón Island; SR = Santa Rosalía; BC = Bahía Concepción; ? = site of probable seaway.

Alarcón, and 2.1 Ma for the Guaymas Basin. Spreading took ~3 million years to move from the mouth to the central Gulf, and it is only now beginning to arrive in the northern Gulf. At the time of the Miocene–Pliocene transition, the Gulf Extensional Province varied in width from 400 km in the northern Gulf to 250 km in the south (Stock and Hodges 1989). The Gulf of California accommodates a maximum of 255 ± 10 km of post–6 Ma dextral slip (Oskin et al. 2001).

The Cabo Trough is an excellent place to test hypotheses concerning origin, migration, evolution, and extinction of species. Geological evidence indicates that the southern part of the Baja California peninsula was attached to mainland Mexico at least until 5 Ma. Thus, rifting between Baja California and mainland Mexico began along the continental crust before the actual Gulf of California began to form—and before the oldest oceanic crust between the tip of the peninsula and the East Pacific Rise was formed (De Mets 1995). However, late Miocene (10–7 Ma) subsidence in the region of the future mouth of the Gulf appears to have created relatively deep, coastal marine embayments that led to marine sediment deposits in the area predating the actual opening of the Gulf (Carreño 1992a). The area was subsequently lifted in the Pliocene, and these Miocene deposits are now above sea level.

Phylogeographic studies on terrestrial fauna inhabiting the Cabo Trough also suggest that the southernmost peninsula has not been separated for a long period of time from mainland Mexico. Two scenarios have been proposed for this part of the peninsula. One scenario (Anderson 1971) suggests that the Los Cabos block migrated independently northeastward along the La Paz fault approximately 50 km to its present location relative to the rest of Baja California, which implies that the block remained isolated from the peninsula for a long time. The other hypothesis suggests that the Los Cabos block did not separate from the mainland as an isolated landmass but instead remained connected to the rest of the peninsula (fig. 1.7). According to Fletcher et al. (2000) and others, evidence points toward the second scenario, but under a particular geological history that provides a clearer understanding of the evolutionary history of the species occurring in the region. During early middle Miocene (16–10 Ma), along the western margin of the Los Cabos block, shallow seaways were present that promoted marine deposition. This could have divided the geographic ranges of terrestrial populations into vicariant Cape region populations and northern populations. Afterwards, the region was uplifted and the eastern margin received marine deposits in the early Miocene (Carreño 1992a), which reached bathyal depths in the late Miocene–Pliocene. Late Pliocene to early Pleistocene shallow marine rocks (Carreño and Smith 2007) and terrestrial rocks (Ferrusquía-Villafranca and Torres-Roldán 1980; Miller 1980) indicate a regressive event, which would have allowed colonization of terrestrial species throughout the area.

Separation of terrestrial faunas began with the opening of the southern region (mouth) of the Gulf 4–5 Ma. Nevertheless, interchange between

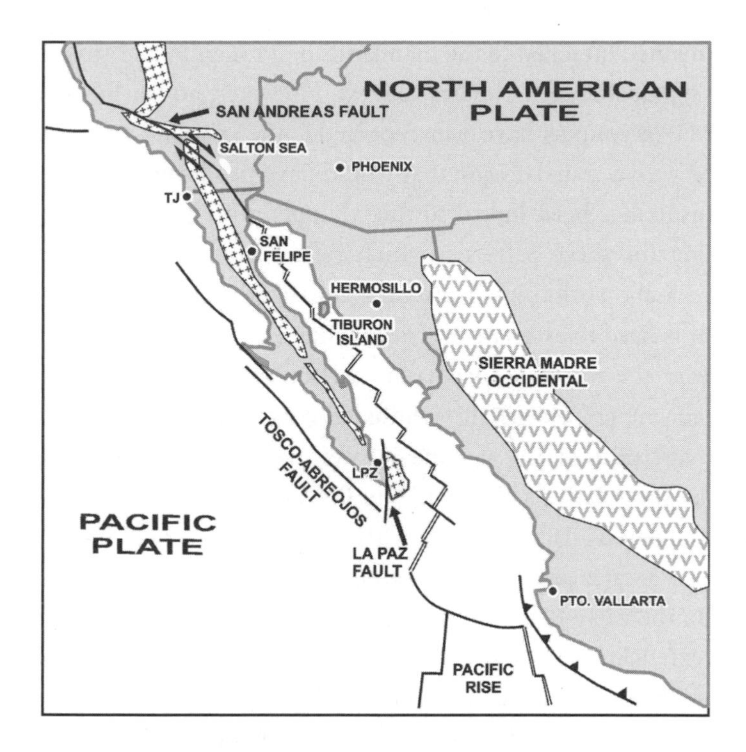

1.7. Proposed migration of Los Cabos block along the La Paz Fault; see text for details.

mainland Mexico, the islands, and the tip of the peninsula (or the rest of the peninsula) may have occurred later via island hopping, depending on sea-level stands and the geographic distribution of various islands during detachment and isolation. Espíritu Santo and Partida (sur) islands have Cretaceous basement and thick sequences of Miocene volcanic and volcanoclastic rocks similar to those cropping out at nearby La Paz. Also, Islas Santa Catalina, Santa Cruz, San Diego, San José, and Cerralvo all contain rocks that are related to those forming the southern tip of the peninsula, interpreted as rapid uplift during the strike-slip movement at the mouth of the Gulf. A Pleistocene (~1 Ma) age has been given for the probably first isolation of these islands (Carreño and Helenes 2002), which could have formed land bridges or "stepping-stones" for dispersal of terrestrial organisms.

In another context, studies on mammals (e.g., Lawlor et al. 2002) have supported the existence of a young, mid-transpeninsular seaway across the Vizcaíno Desert or, alternatively, that the tip of the Baja California peninsula remained attached to the mainland for a much longer time than geophysical and geological estimates suggest. However, no marine rocks of this age (~1 Ma) or younger have been reported in the area, and no regional event associated with a transgression that would have produced a seaway through the peninsula has been linked to this younger time period. One plausible explanation for these patterns, which reflect a probable vicariant event, is that the Cabo region was isolated as an island for some period of time.

It is hard to estimate precisely when effective isolation of the Cabo region from the mainland occurred, particularly considering that during the Pleistocene–Holocene the peninsula experienced relatively slow and uniform vertical motions at a mean rate of 100 ± 50 mm/10^3 years and, judging by the relative stability of late Pleistocene terraces in the southern parts of the peninsula (ca. 125,000 BP), uplift rates seem to have decreased (Ledesma-Vázquez and Johnson 2001).

In the early Pliocene (5 Ma), the initiation of a right-lateral strike-slip and extension regime began development of the modern tectonic configuration of the Gulf of California (Zanchi 1994; Mayer and Vincent 1999) and the complete transfer of the Baja California peninsula from the North American Plate to the Pacific Plate. Other than the late Miocene embayment that persisted until Pliocene times in the northern Gulf, sedimentary deposits are recorded only south of Bahía Concepción, and their age shows an interesting position compared to the age of marine deposits to the north. A key locality is at San Nicolás Basin (southeast of Mulegé), were the unit includes large-scale sand wave deposits up to 60 m thick, which denote extreme tidal conditions during their deposition (Ledesma-Vázquez and Johnson 2001). Another locality with a clear tidal signature for the Pliocene is at the southern tip of Monserrat Island (Ledesma-Vázquez et al. 2007). Perhaps even more interesting is the oldest unit in the San Nicolás Basin, Toba San Antonio, which X-ray diffraction (XRD) analysis suggests was deposited in a dry environment; this implies that, during the initial deposition (about 3.3 Ma), the basin was not under water at all (Berry and Ledesma-Vázquez 1998).

Several distinctive units extend 8 km across Ensenada El Mangle north of Loreto along the shoreline. This sequence correlates with the middle Pliocene Piacenzian Stage based on tuffs near the base of the succession that yield an age of 3.3 ± 0.5 Ma. At Carmen and Monserrat islands, marine deposits not older than about 3.1–3.5 Ma have been reported, and it has been suggested that Pliocene deposition at the Loreto embayment took place not earlier than 3.3 Ma. However, work in progress by the present authors at Carmen and Monserrat indicates an age of 5.3 to 6.0 Ma for the upper bathyal marine sedimentation, which is coincident with Gulf spreading.

In the early Pliocene, the Colorado River was depositing its waters and sediments into the Salton Trough, and over time the filling of the rift in that region developed a lacustrine basin (Winker and Kidwell 1996). Sometime near the beginning of the Pliocene, waters of this basin/lake reached the northernmost Gulf of California, as indicated by the upper sedimentary record of the Altar Basin (Pacheco et al. 2006), although this event is unclear in the Wagner, Consag, and Tiburón Basins.

At this time and during the Pliocene, when the Gulf's configuration was close to that of the present day, some of the islands within the Gulf were detached from the Baja California peninsula and the Mexican mainland as a consequence of the extension and right-lateral slip faulting that was taking place. At some point in the process, basins with oceanlike floors developed in the central and southern parts of the Gulf, while smaller and fault-bounded basins formed along the eastern margin of the peninsula. Other islands (e.g., San Luis Island) are a result of recent volcanism.

ACKNOWLEDGMENTS

Ledesma-Vázquez thanks CONACyT, PRF, and UABC under project "Atlas de Ecosistemas Costeros en el Golfo de California: Pasado y presente," for partial support for his research. Both authors acknowledge support from grant PAPIIT IN116308.

CHAPTER 2

Physical, Chemical, and Biological Oceanography of the Gulf of California

SAÚL ALVAREZ-BORREGO

Summary

The Gulf of California is a dynamic marginal sea of the eastern Pacific Ocean. Today, nutrient input to the Gulf from rivers is very small and has only local coastal effects. The Gulf has three main natural fertilization mechanisms: wind-induced upwelling, tidal mixing, and water exchange between the Gulf and the Pacific Ocean. These natural fertilization mechanisms have made the Gulf of California more resistant to anthropogenic effects (e.g., those due to construction of dams) than are other ecosystems, such as the eastern Mediterranean Sea. Upwelling occurs off the mainland (eastern Gulf) coast with northwesterly winds ("winter" conditions from December through May) and off the Baja California peninsula coast with southeasterly winds ("summer" conditions from July through October), with June and November as transition periods. Upwelling off the mainland coast is strong and has a marked effect on phytoplankton communities (chlorophyll concentration values can exceed 10 mg m^{-3}). When combined with eddy circulation, upwelling increases the phytoplankton biomass across the Gulf. However, because of strong stratification during summer upwelling off the Baja coast, the southeasterly winds have a weak effect on phytoplankton biomass, causing an increase in chlorophyll *a* only to values around 0.5 mg m^{-3} in spite of wind magnitudes similar to those of winter.

Tides in the Gulf of California are produced by co-oscillation with the adjacent Pacific. In the Gulf, the tidal wave is progressive: the time of high or low water is progressively later traveling northward in the Gulf. The time difference between the entrance and the vicinity of the Colorado River is approximately 5.5 hr for high water and 6 hr for low water. The result is that low water at one end of the Gulf occurs at about the same time as high water at the other end. The northern Gulf exhibits spectacular tidal phenomena, with a range of >7 m during spring tides in the uppermost Gulf and >4 m in the Midriff Islands region. Tidal mixing in the latter region produces a vigorous stirring of the water column down to >500 m depth, with the net effect of carrying colder, nutrient-rich water to the surface and creating an ecological situation similar to constant upwelling. This also has the effect of making the areas around the islands of the Gulf a source of CO_2 to the atmosphere. Satellite ocean color data show that most of the Gulf has low photosynthetic pigment concentrations during the summer, but they remain high in the waters around the Midriff Islands because of tidal mixing.

The lowest offshore surface temperatures and the highest surface nutrient and total dissolved inorganic carbon concentrations in the entire Gulf are persistently found in the Midriff Islands region. As a result, primary productivity is high year-round, and this area supports large numbers of sea birds and marine mammals. South of the Ángel de la Guarda and Tiburón Islands, the Gulf has basically the same thermohaline structure as that of the Eastern Tropical Pacific, with modifications at the surface due to excess evaporation. At intermediate depths, the concentration of oxygen in some places is undetectable by the Winkler method. This oxygen minimum plays an important role in the ecology and geology of the Gulf, as it influences the distribution of pelagic and benthic organisms and sedimentation patterns. Laminated diatomaceous sediments are formed where the basin slopes intersect the oxygen minimum in the water column. These laminated sediments are paleo-ecological records of seasonal, interannual, and decadal changes of hydrographic conditions. Satellite-derived data revealed a dramatic reduction in chlorophyll at the entrance to the Gulf during El Niño 1982–1983, with values down to ~20 percent of those for non–El Niño years, but there was a relatively small impact in the Midriff Islands and northern regions.

Introduction

The first physical oceanographic data for the Gulf were collected in 1859–1861, when John Xantus measured tides at Cabo San Lucas for the U.S. Coastal Survey. The first physical oceanographic work off the west coast of mainland Mexico was during the *C. & I.S.S. Hassler* cruise with Alexander Agassiz on board. The *Hassler* expedition worked at Cabo Corrientes in 1872 (Schwartzlose and Alvarez-Borrego 2002). The first major scientific program was a series of cruises in 1888, 1889, 1904, and 1911 on the U.S. Fish Commission vessel *Albatross* (Hedgpeth 1945). The March 1889 *Albatross* sea-surface temperature (SST) data show the effect of an El Niño event in the Gulf of California, with positive anomalies as large as those of 1983 (Alvarez-Borrego 1990). McGee (1900) first reported on the powerful tides at the Midriff Islands region of the Gulf and observed that strong tidal currents, aided by frequent gales, caused vigorous marine erosion that resulted in submarine terraces up to a mile wide, covered with shallow water. Thorade (1909) was the first to describe the Cabo San Lucas thermal front, and he used ship-drifting data to determine that the surface currents of the southern Gulf generally agreed with the direction of the wind. Thorade postulated that surface outflow (and compensating deep inflow) occurred in winter whereas surface inflow (deep outflow) occurred in summer. Modern investigations of the Gulf of California began with the 1939 and 1940 E. W. Scripps cruises, where data were collected on the bathymetry and geology of the ocean along with its chemical and physical features. Using data from the 1939 cruise, Sverdrup (1941) described the unique oceanographic characteristics of the Midriff Islands region of the Gulf, which are due to strong mixing by phenomena associated with tides and upwelling of deep waters.

The Gulf of California is a dynamic marginal sea of the eastern Pacific Ocean, and it has been described as an area of great fertility since the time of early explorers. Lying between the arid peninsula of Baja California and the almost equally arid states of Sonora and Sinaloa, it comprises a large evaporation basin that is open to the Pacific at its southern end (the mouth of the Gulf). Oceanographically (and paleotectonically), the southernmost limit of the Gulf of California is a line connecting Cabo San Lucas and Cabo Corrientes (Roden 1964). Topographically, it is divided into a series of basins and trenches, deepening to the south and separated from each other by transverse ridges (fig. 2.1a) (see Ledesma-Vázquez and Carreño, chapter 1 in this volume). Maximum basin depths at the entrance to the Gulf may be >3,000 m but as shallow as ~200 m in the northernmost region. This topography is primarily a result of strike-slip faulting (Shepard 1950;

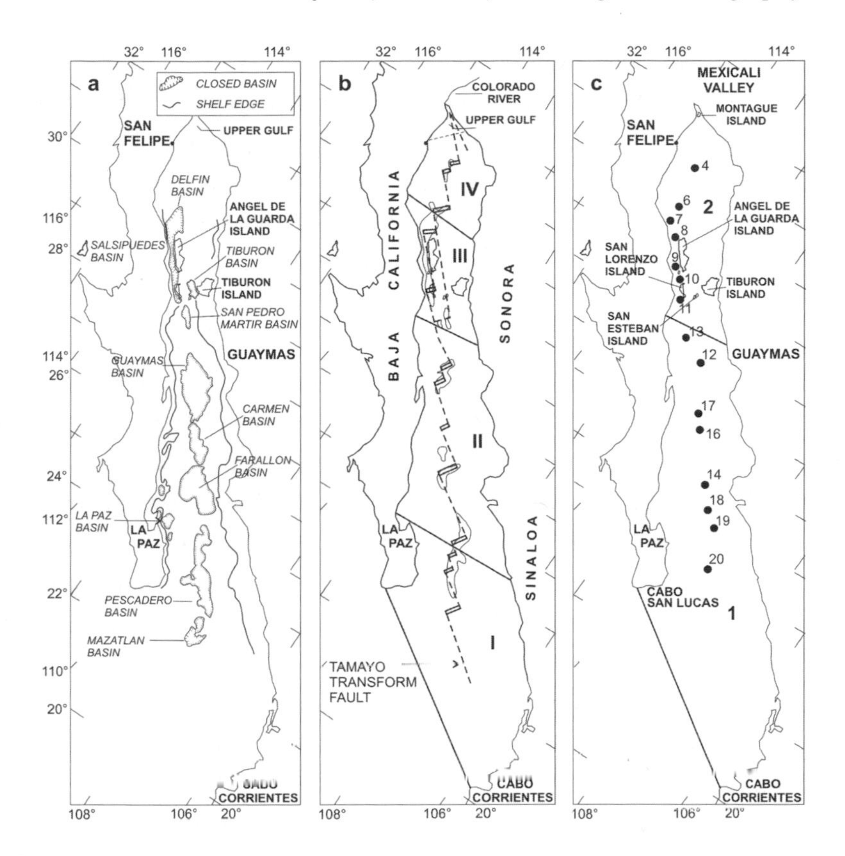

2.1. (a) Shelves and basins in the Gulf of California. Basins are defined by depth contour at sill depth (after van Andel 1964). (b) Schematic map of Gulf of California tectonics. Dashed lines show approximate positions of transform faults; double solid lines show locations of spreading centers as suggested by the bathymetry (after Reichle 1975). Roman numerals are the Gulf's regions defined by Hidalgo-González and Alvarez-Borrego (2004) for calculating primary production from satellite data for the "cool season." (c) Location of hydrographic stations of the April–May 1974 R / V *Alexander Agassiz* cruise that generated the data shown in figure 2.2. Large Arabic numerals are the Gulf's regions defined by Hidalgo-González and Alvarez-Borrego (2004) for calculating primary production from satellite data for the "warm season."

Ledesma-Vázquez and Carreño, this volume). The tensional faults separating the strike-slip faults are loci of seafloor spreading (Sykes 1968) (fig. 2.1b).

The Gulf is separated into two rather distinct areas by the islands of Ángel de la Guarda and Tiburón, which—together with several smaller islands—are known as the Midriff Islands. Much of the area north of the Midriff Islands is at shelf depth, with few shelf breaks. South of the Ángel de la Guarda and Tiburón Islands, the Gulf has well-developed continental shelves. On the western side of the Gulf, the shelf is generally rocky and narrow, with a sharp shelf break between 80 and 100 m; the eastern shelf is wider (van Andel 1964). The eastern coast has many more coastal lagoons than the western coast, a consequence of the number of rivers that drain into the Gulf from the Sierra Madre Occidental range on the mainland.

Today, nutrient input to the Gulf from rivers is very small and has only local coastal effects, most notably on the coasts of southern Sinaloa and Nayarit. The Gulf mainly has three natural fertilization mechanisms: wind-induced upwelling, tidal mixing, and the thermohaline circulation. The nutrient contribution from mangrove lagoons along the mainland coast, south of the Midriff Islands region, is probably significant but has not been studied. Accumulation of biogenic silica in the sediments of the central Gulf (mostly composed of diatoms) was about 27 times the dissolved silica provided by rivers draining into the Gulf (on a per annum basis) before the completion of the Hoover Dam in 1935 (Calvert 1966), and the amount of other nutrients (such as PO_4 and NO_3) needed for phytoplankton photosynthesis was proportionally even greater compared to those input by rivers. River input is greatly diminished today, since dams have been built upstream to divert the water for agricultural and urban uses. Thus, water exchange between the Gulf and the Pacific Ocean supplies sufficient nutrients via a mechanism—upwelling and tidal mixing—that continuously supplies dissolved nutrients to the euphotic zone. These natural fertilization mechanisms have made the Gulf of California more resistant to anthropogenic effects (such as those due to dam construction) than other ecosystems, such as the eastern Mediterranean.

For example, before the High Aswan Dam was fully operational in 1965, the Nile River plume had been characterized by strong blooms of diatoms and copepods, and an extensive fishery had operated based on sardines that migrated into the area to exploit the plankton. In addition, a strong shrimp fishery was based on stocks that were adapted to the pattern of river runoff. After the dam was built, the catch of sardines declined by 90 percent and that of shrimp by 75 percent, and the levels have remained well below pre-dam values (Drinkwater and Frank 1994).

The drastic extent of this reduction, caused by collapse of the pelagic ecosystem, has not been observed in the Gulf of California but only in the estuaries themselves and coastal areas adjacent to river mouths. For example, *Mulinia coloradoensis,* a bivalve mollusc endemic to the upper Gulf, had a much larger population before the Colorado River dams were built; now this species is in danger of extinction, possibly because of the increase in salinity (Flessa 2001).

Meteorological Aspects: Winds and Upwelling

The moderating effect of the Pacific Ocean upon the climate of the Gulf is greatly reduced by an almost uninterrupted chain of mountains, 1 to 3 km high, in Baja California. To the east and north, elevations in the Sierra Madre Occidental typically exceed 1,500 m. Significant air flow below 800 m is usually channeled along the Gulf, and it is open to direct oceanic influence only from the south. The Gulf of California is thus a semi-enclosed basin in both a meteorological sense and an oceanic sense (Badan-Dangon et al. 1991). The climate of the Gulf is therefore more continental than oceanic, and this contributes to the large annual and diurnal temperature ranges observed there (Hernández 1923, cited by Roden 1964). There is more precipitation on the east than on the west side of the Gulf. The northern half of the Gulf is dry and desertlike, with annual rainfall of less than 100 mm. In the southeast, rainfall along the coast increases to about 1,000 mm per year. South of the Ángel de la Guarda and Tiburón Islands, most of the rain falls between June and October; in the north, however, most of the rain falls during winter (Page 1930 and Ward and Brooks 1936, cited by Roden 1964). Skies are generally clear, which makes the Gulf an ideal place for calibration of satellite color sensors.

The number of rainy days per year decreases from about 60 at Cabo Corrientes to about 5 along the central Baja California coast and the northern Gulf region in general. However, the year-to-year variation in rainfall in the Gulf is large and depends strongly on the incidence of tropical storms and hurricanes. Summer tropical storms along the west coast of Mexico seldom enter the Gulf region, and those that do usually dissipate in the states of Sinaloa and Sonora, although a few travel north all the way to Arizona (Harris 1969). During the summer, tropical Pacific air flows northward up the Gulf and over the United States (drawn by low pressure at the higher latitudes), and a portion of the large pool of tropical moisture present off western Mexico is channeled up the Gulf and over the southwestern United States, where it contributes to the summer monsoon season in the Sonoran Desert (Reyes and Cadet 1986). Often these airflows take the form of relatively strong pulses of southeasterly winds, 10 m s^{-1}, that last as long as 5 days (Badan-Dangon et al. 1991).

The wind field over the Gulf is essentially monsoonal in nature, from the NW during winter and from the SE during summer, and it is markedly coherent both along and across the Gulf (Bray and Robles 1991). Marinone et al. (2004) have shown that satellite-derived data (QuikScat) confirm this monsoonal nature of winds in the Gulf. The seasonal signal of the monsoon-type winds tends to be modified by the nearshore sea breeze. Moderate northwest gales lasting 2–3 days at a time are frequently experienced in the northern Gulf between December and February. These winds are particularly strong in the Ballenas Channel (also called the Salsipuedes Basin), between Ángel de la Guarda Island and the peninsula; they may occasionally raise such a heavy sea that navigation becomes impossible (Roden 1964).

Upwelling occurs off the eastern coast with northwesterly winds (winter conditions from December through May) and off the Baja California coast with southeasterly winds (summer conditions from July through October), with June and November as transition periods. Upwelling off the eastern coast is strong, has a marked effect on phytoplankton communities (chlorophyll concentration values can exceed 10 mg m^{-3}), and—because of eddy circulation—increases the phytoplankton biomass across the Gulf (Santamaría-del-Ángel et al. 1994a). Upwelling events last for a few days, after which relaxation allows for stabilization of the water column and blooming of phytoplankton communities. Yet because of strong stratification during summer, upwelling off the Baja coast with southeasterly winds has a weak effect on phytoplankton biomass. As a result, chlorophyll *a* increases only to values around 0.5 mg m^{-3} in spite of wind magnitudes similar to those of winter (Santamaría-del-Ángel et al. 1999). Another important reason for the weak effect of summer upwelling is that, on the western side of the Gulf, winds and surface sea currents often flow in the opposite direction (Lluch-Cota 2000).

Tides and Tidal Mixing

Tides in the Gulf of California are produced by co-oscillation with the adjacent Pacific; that is, they are induced by tides coming in from the Pacific. They are due mainly to variations of sea level at the entrance and not to gravitational attraction of the moon and sun on the Gulf's waters per se (Ripa and Velázquez 1993). The effect of direct gravitational driving by sun and moon is barely 5 percent of oceanic driving, and it is usually neglected when calculating tidal energy dissipation (Filloux 1973) or when constructing models to represent the semidiurnal and diurnal components of the surface tide in the Gulf of California (Ripa and Velázquez 1993). In the Gulf, the tidal wave is progressive: the time of high or low water is progressively later traveling north up the Gulf. The time difference between the entrance and the vicinity of the Colorado River is approximately 5.5 hr for high water and 6 hr for low water. The result is that low water at one end of the Gulf occurs at about the same time as high water at the other end (Roden 1964).

The northern Gulf exhibits spectacular tidal phenomena, with a range of >7 m during spring tides in the uppermost Gulf and >4 m in the Midriff Islands region. The rates of tidal energy dissipation are high. Tidal mixing between the islands of San Lorenzo and San Esteban produces a vigorous stirring of the water column down to >500 m depth, with the net effect of carrying colder, nutrient-rich water to the surface (Simpson et al. 1994) and creating an ecological situation similar to constant upwelling. This also has the effect of making the areas around the islands of the Gulf a source of CO_2 to the atmosphere (Hidalgo-González et al. 1997). Satellite ocean color data show that, during summer, most of the Gulf has low photosynthetic pigment concentrations, but these concentrations remain high in the waters around the Midriff Islands and in the upper Gulf because of tidal mixing (Alvarez-Borrego 2002). In the entire Gulf, the lowest surface temperatures (Robinson 1973; Soto-Mardones et al. 1999) and lowest surface pH, as well as the highest surface nutrient and total dissolved inorganic carbon concentrations (Alvarez-Borrego et al. 1978; Gaxiola-Castro et al. 1978), are persistently found in the Midriff Islands region (fig. 2.2). Thus, primary productivity is high year-round, and this area supports large numbers of sea birds and marine mammals

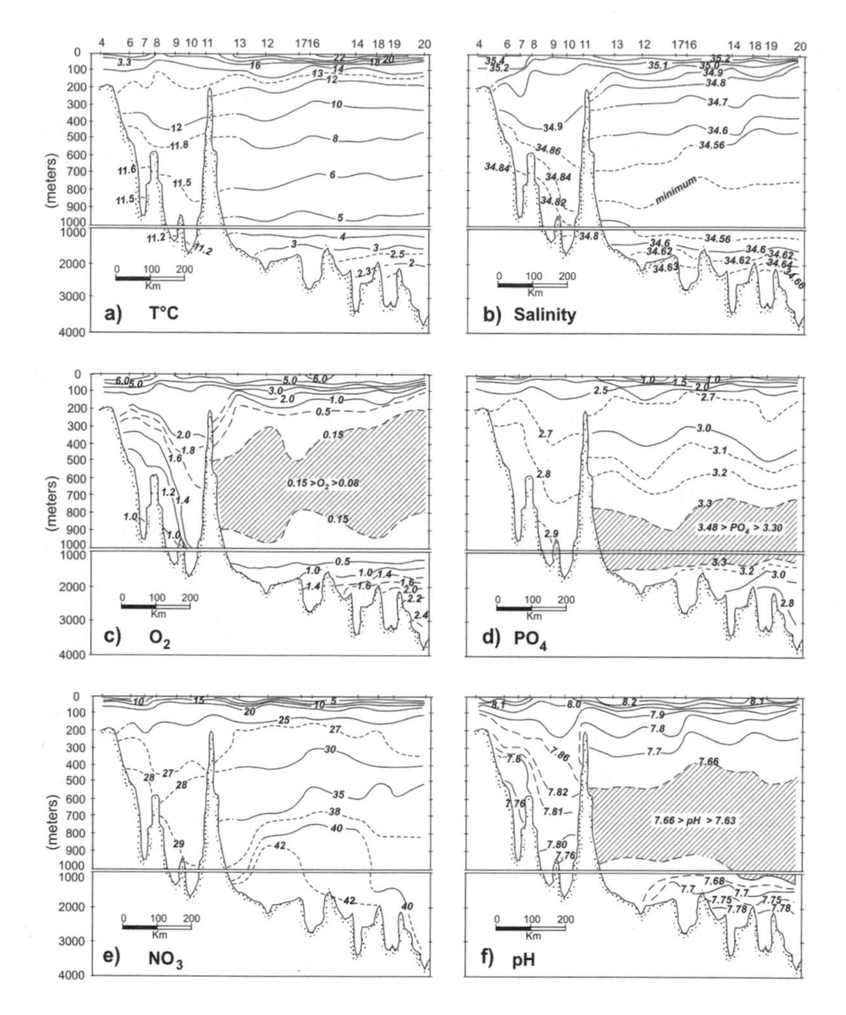

2.2. Vertical distribution of: (a) temperature (°C); (b) salinity; (c) dissolved oxygen (mℓ ℓ$^{-1}$); (d) phosphate (μM); (e) nitrate (μM); and (f) pH. Panels a, b, c, and f are after Gaxiola-Castro et al. (1978); panels d and e are after Alvarez-Borrego et al. (1978). Upper numbers are the hydrographic stations as shown in figure 2.1c.

(Tershy et al. 1991). Tidal currents in the narrows between the islands and the coast, and also in the passages connecting the Gulf with semi-enclosed lagoons, are strong. The speed of these currents is variable and depends on the stage of the moon and the prevailing winds, but they have been reported up to 3 m s^{-1} (6 knots) for Ballenas Channel (Alvarez-Sánchez et al. 1984).

The Effect of Air–Water Heat Exchange on Circulation

In the Gulf of California there is, in spite of evaporation, an annual mean net water–atmosphere heat flux into the sea of >100 W m^{-2} (watts per square meter). This heat must be exported to the Pacific somehow, for otherwise the Gulf's temperature would be increasing (Lavín et al. 1997a). Less dense, warmer surface water has a net flow out from the Gulf into the Pacific. To balance this flow, relatively deep colder water has a net flow into the Gulf. This net flux is the inverse of that observed between the Mediterranean and the Atlantic (through the Strait of Gibraltar), where nutrient-depleted surface waters flow into the Mediterranean.

Circulation is the system of currents that does not include the tidal currents; in other words, circulation consists of the residual currents. Tidal currents produce a forward and backward motion. Circulation is of interest because it is responsible for transporting materials (e.g., pollutants, animal eggs) from one place to another. Bray (1988) proposed an oversimplified three-layer circulation for the Gulf of California, in which water was envisioned as generally inflowing between 250 and 500 m, outflowing between 50 and 250 m, and having a surface layer (the top 50 m) with a tendency to reverse direction with the seasonal winds. Marinone (2003) used a three-dimensional model to conclude that heat and salt flowed out of the Gulf within the top 200 m and into the Gulf at depths of 200 to 600 m, and most of the Gulf's "heat budget" was defined as occurring in the upper 350 m. The net heat flux is an export of 17×10^{12} W to the Pacific. This circulation pattern has profound ecological implications, because inflowing deep water has higher inorganic nutrient concentrations than outflowing surface water (Alvarez-Borrego and Lara-Lara 1991). However, the thermohaline circulation is very slow and does not have a significant effect on the general circulation system (Roden 1958). Marinone (2003) concluded that the three-layer circulation pattern proposed by Bray (1988) does not represent the thermohaline circulation but rather the general circulation of the Gulf caused by all the forcing agents (including winds, tides, and the Pacific Ocean).

Circulation and the Transport of Passive Particles

Gilbert and Allen (1943) studied the phytoplankton communities of the central Gulf with samples mainly collected during the February–

March E. W. Scripps cruise in 1939. Based on the dynamics of these communities, the researchers postulated that a net inflow of deep water and a net outflow of surface water characterize the general circulation of the Gulf. As a consequence of this circulation, deep water that is rich in plant nutrients is brought to the euphotic zone, where the nutrients can be utilized by phytoplankton. According to Gilbert and Allen, the main factor affecting this circulation is the wind, which brings about upwelling and an outward transport of surface water. The vertical distribution of water density in the central and southern Gulf at the time of the 1939 cruise showed alternating upward and downward displacements of the isopycnal surfaces (surfaces that join points of equal water density), and this led Sverdrup (1941) to postulate the existence of a standing internal wave with three nodes. Gilbert and Allen (1943) speculated that this internal wave would be associated with three circulation cells or horizontal eddies, with maximum speeds up to 50 cm s^{-1}, capable of carrying phytoplankton communities from one side of the Gulf to the other. Roden (1958) used 1947 ship-drifting charts prepared by the U.S. Hydrographic Office. He concluded that, in winter, southward currents north of Cabo Corrientes characterize the surface circulation of the Gulf. During the summer, a current flows northward along the coast of Mexico and enters the Gulf at the eastern and central regions of the mouth, with a southward flow near Baja California. This is in agreement with the results obtained by Granados-Gallegos and Schwartzlose (1974), who used drift bottles in their research. Lepley et al. (1975) used satellite photographs (infrared and multispectral imagery) to detect what they interpreted as a counterclockwise (cyclonic) surface gyre during summer in the northern Gulf, which reversed during winter to form an anticyclonic gyre.

Beier (1997) developed a numerical model that was linear, bidimensional, and baroclinic (where, unlike barotropic models, velocity changes with depth); it incorporated two layers and bathymetry. This model is driven by the effects of the Pacific Ocean (PO), the wind (W), and the air–water heat flux (Q). The results of Beier's model for the surface layer (0–70 m) show anticyclonic circulation during winter (water enters from the Pacific Ocean at the Baja side and leaves the Gulf at the mainland side) and cyclonic circulation during summer (figs. 2.3a and 2.3b). However, there are continuous changes throughout the year. Circulation is not always clearly cyclonic or anticyclonic (Beier 1997). In general, the results shown in figure 2.3

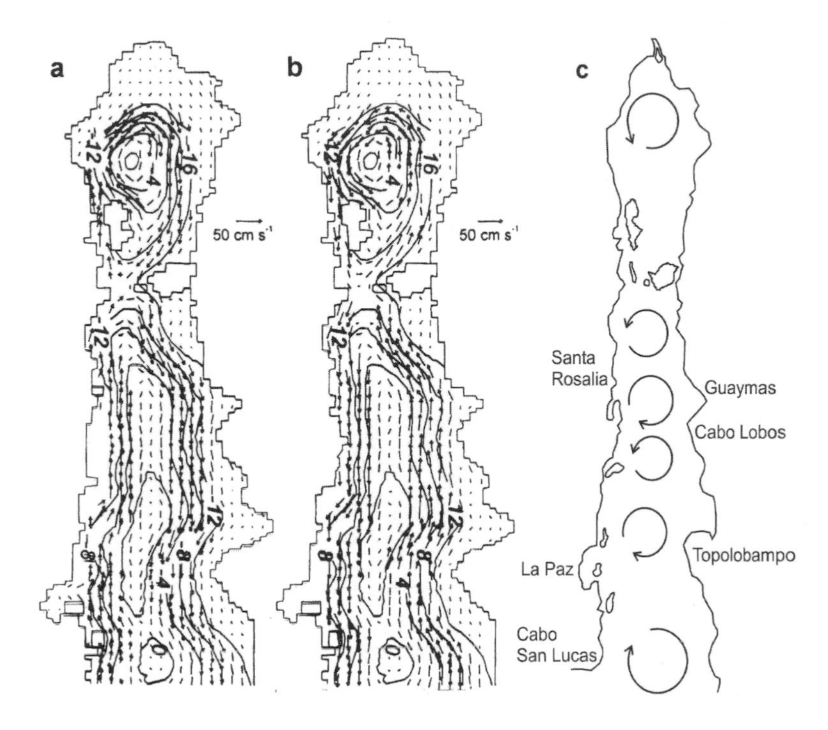

2.3. Contours of sea level (cm) and circulation of the Gulf of California for (a) February and (b) August; after Beier (1997). Notice that sea-level values are negative for February and positive for August. (c) Summer eddies deduced from satellite data on ocean color (after Pegau et al. 2002).

(a and b) are in agreement with observations reported by previous authors. One striking result is the gyre that forms in the northern Gulf. This gyre is in agreement with the one postulated by Lepley et al. (1975) and with the one described by Lavín et al. (1997b) using direct observations with ARGOS drifters. Lavín et al. (1997b) described the presence in the northern Gulf of a cyclonic gyre during summer and an anticyclonic gyre during winter. Both gyres had mean speeds of ~30 cm s^{-1} near the edge. This kind of circulation suggests that neutrally buoyant substances and passive organisms may become trapped for extended periods in the northern Gulf.

Marinone (2003) simulated the mean and seasonal circulation of the Gulf with a three-dimensional numerical model driven by the Pacific Ocean, tides, winds, and the air–water heat exchange. In general, Marinone's (2003) and Beier's (1997) results for the surface layer are similar, but there are some important differences. Marinone (2003) confirmed the

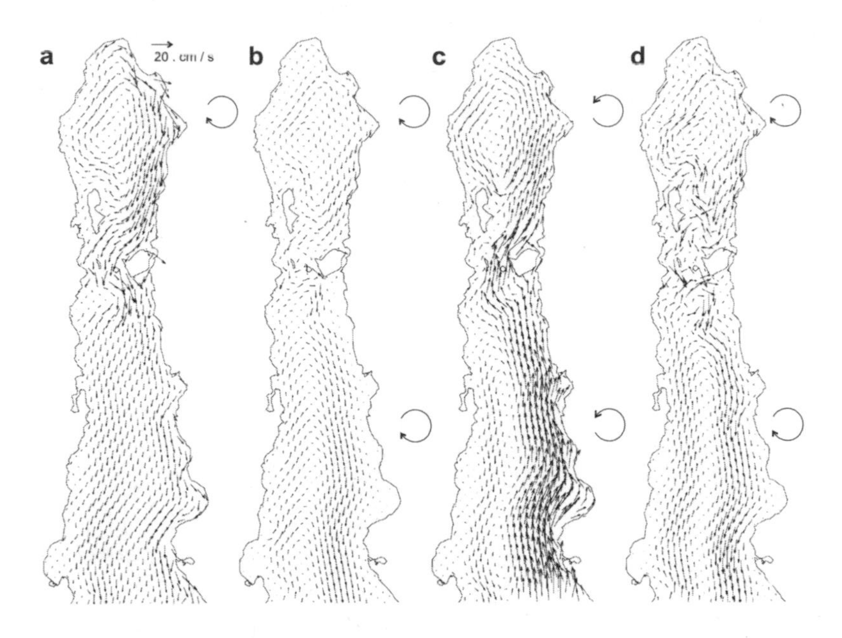

2.4. Modeled surface circulation (0 – 10 m). The curved arrows are drawn to help visualize the sense of circulation. (a) February, (b) May, (c) August, and (d) November (after Marinone 2003).

northern Gulf's anticyclonic gyre with winter conditions, reversing direction during summer; he also concluded that tides in this region of the Gulf play an important role in producing residual currents and that both tides and winds compete against the ocean effects to produce the circulation. However, with Beier's (1997) model there are long periods of no communication between the northern and southern Gulf whereas with Marinone's (2003) model there is strong exchange between the two portions of the Gulf: relatively large surface speeds between the two big Midriff Islands most of the time, to the south in winter and to the north in summer (fig. 2.4). With Marinone's (2003) model, the general circulation south of the Midriff Islands is due to winds and the Pacific effect, and two periods of anticyclonic circulation (fig. 2.4b, April–May, and fig. 2.4d, November–December) and one cyclonic period (fig. 2.4c, July–September) occur each year. From December through February there is only southward surface flow throughout the whole southern Gulf (fig. 2.4a) because the wind-induced currents overcome those produced by the Pacific Ocean; this result is closer to the circulation proposed by Gilbert and Allen (1943) and described by Roden

(1958) and Granados-Gallegos and Schwartzlose (1974). Also, with Beier's (1997) model the speeds of circulation are the same on both sides of the Gulf (figs. 2.3a and 2.3b), whereas with Marinone's (2003) model the speeds are much greater at the eastern side of the Gulf, resulting in very little water exchange between the northern and southern parts of the Gulf through Ballenas Channel. Finally, Marinone's model shows currents being much stronger during the summer than during the rest of the year.

There are few direct measurements to test these theoretical results. Merrifield et al. (1986) reported on the results from current-meter moorings in the central and northern Gulf. Their most completely instrumented section was one in the Guaymas Basin. The mooring at the eastern side of Guaymas Basin, with bottom depth of 1000 m, had instruments located at depths of 50, 300, 500, and 850 m, and it produced data from May to November 1984. The unfiltered currents show a whole spectrum of variability, including the effect of propagating coastally trapped waves. Sometimes the currents reversed direction for about 2 weeks or less. At 50 m the flow was generally northward until late August, when it reversed and became weaker until the end of the sampling period. Current speeds were 10–50 cm s^{-1} northward and ~20 cm s^{-1} southward. The deeper time series show little northward tendency until November, with the exception of an ~2-week period of northward direction at the beginning of July, and they generally show weak southward flow through the summer. Badan-Dangon et al. (1991) used an Acoustic Doppler Current Profiler (ADCP) in January 1990 to study the currents in the Midriff Islands region and confirmed a deep flow toward the northern Gulf and outflow in the upper layers. López and García (2003) generated a 9-month time series and found a persistent current flowing close to the bottom, at a depth of 340 m, at the sill that separates Tiburón Basin and Delfín Basin. The current flowed northward at an average speed of 27 cm s^{-1}. López and García reported that, at depths above 260 m, mean currents were much more variable and weaker but veered abruptly toward the NW close to the bottom; furthermore, the currents over the shelf were in opposition to those over the sill, which is consistent with a compensating mechanism.

There are two ways to study currents and circulation: in the Lagrangian manner, particles are followed as they move from place to place (e.g., with drifters); in the Eulerian manner, particle movement is studied at fixed points (e.g., with moored current meters). Gutiérrez et al. (2004)

used the results from Marinone's (2003) model to describe the Lagrangian surface circulation in the Gulf. They performed a computer experiment to study the advection of 710 particles for annual and monthly periods. During the cyclonic period (July–August), the particles from the central and southern regions travel to the northern region, where they are trapped by a gyre for more than a month. During the anticyclonic period, the particles flow to the southern part of the Gulf. Some particles escape to the Pacific Ocean; others remain in the Gulf and then return to the northern part during the cyclonic period. In the central and southern regions, a trapping zone was found on the peninsula side, where the particles travel less than 100 km in 30 days. In contrast, there is a rapid transit zone over the continental coast. The particle escape time from the Gulf into the Pacific is shortest during December–February: less than 2 months for particles near the mainland coast in the central and southern Gulf but more than 12 months for particles in the northern Gulf or near the peninsula. The escape time is longer for summer (July–August): 6 to 8 months for particles near the peninsula in the central and southern Gulf, and more than 12 months for particles in the northern Gulf or near the mainland coast (Gutiérrez et al. 2004). These estimates of particle escape times are minima because eddies at the central and southern Gulf were not taken into account in the numerical simulations.

There is evidence of gyres in the central and southern Gulf that are not predicted by Beier's (1997) or Marinone's (2003) model. Badan-Dangon et al. (1985) used March–April satellite SST imagery to describe the evolution of an upwelling plume, which originated north of Guaymas and eventually reached most of the way across the Gulf. The movement of this plume was associated with a basinwide anticyclonic circulation. Sokolov and Wong-Rios (1973) postulated that sardine eggs and larvae are transported from the spawning grounds near the east coast to the west coast of the Gulf by the surface circulation, and Hammann et al. (1988) indicated that the specific mechanism for this east–west transport is the general eddy circulation, with jets of cool water extending from the Guaymas area to the west coast of the Gulf. Santamaría-del-Ángel et al. (1994a) used satellite-derived photosynthetic pigment concentration data to describe the temporal and spatial variability within the Gulf. Their results show a clear seasonal pattern. Pigment maxima occur on both sides of the Gulf during winter, which they attributed to surface eddy circulation that brings upwelled water from the eastern to the western coast.

Emilsson and Alatorre (1997) used drifters to describe a summer baroclinic cyclonic gyre at the entrance to the Gulf, with maximum speeds of >55 cm s^{-1} at its western edge. Using SeaWIFS color imagery for the summers of 1997–2001, Pegau et al. (2002) described a series of jets, squirts, and eddies throughout the whole Gulf. These eddies are not clearly seen in the summertime SST images. According to these latter authors, this series of eddies along the Gulf exhibits an alternating sense of rotation (fig. 2.3c). It is important to recognize that sea-surface temperature derived from infrared satellite imagery originates only from a thin film (~0.02 mm) and may not represent the temperature, or thus the flow, at greater depths (Stewart 1985). Nevertheless, sea-surface color satellite data (chlorophyll concentration) results from processes that are integrated over approximately the first optical depth (~8 m) (Gordon and Morel 1983). Navarro-Olache et al. (2004) used closely spaced expendable bathythermograph (XBT) and conductivity-temperature-depth (CTD) data to study the subsurface structure of features—such as fronts, jets, and eddies—observed in a series of SST images at the end of October. They described the evolution and subsurface structure of a cool filament that had advanced due south at ~50 cm s^{-1} along the edge of a pool of warm water ~110 km in diameter in the Guaymas Basin. The density distribution at the edge of the warm pool produced a geostrophic jet extending down to ~80 m and with a volume transport of ~0.7 Sv (1 Sv = 10^6 m^3 s^{-1}). The cool filament detached from the mainland off Guaymas, crossed the basin, hit the peninsula, and then curled clockwise to the north and east—clearly suggesting a gyre.

It is well known that sea level changes with the circulation. With anticyclonic circulation the gyre's edges have low sea level, and the reverse happens when it is cyclonic. For example, Beier's (1997) results show that, near the coasts, sea level is as much as 32 cm lower during winter than during summer (figs. 2.3a and 2.3b). This has significant ecological effects on the estuaries and coastal lagoons of the Gulf. During winter, which is characterized by anticyclonic circulation or southward flow and upwelling, coastal lagoons on the eastern shore have much lower sea levels. Consequently, larger extensions of mudflats are exposed to the air, and for longer periods of time, during low tides. This is advantageous for migratory birds that use these coastal lagoons as feeding grounds, because the infauna is more available to them. With cyclonic circulation during summer, sea level increases at the eastern shore and coastal lagoons are more inundated, which could make them better as nursery grounds for oceanic species.

Vertical Distribution of Physical and Chemical Water Properties

South of the Ángel de la Guarda and Tiburón Islands, the Gulf has basically the same thermohaline structure as that of the Eastern Tropical Pacific, with modifications at the surface due to excess evaporation (Sverdrup 1941; Roden 1964) (figs. 2.2a and 2.2b). The Gulf consists of six water masses as follows: Gulf of California Water (GCW, with salinity $S \geq 35$ and temperature $T > 12°C$); California Current Water (CCW, $S \leq 34.5$ and $12 \leq T°C < 18$), which is present only at the mouth of the Gulf; Equatorial Surface Water (ESW, $S < 35$ and $T \geq 18°C$); Subtropical Subsurface Water (SSW, $34.5 < S < 35$ and $9 \leq T°C < 18$), which invades most of the Gulf; Pacific Intermediate Water (PIW, $34.5 \leq S < 34.6$ and $4 \leq T°C < 9$); and Pacific Bottom Water (PBW, $S > 34.5$ and $T < 4°C$). These latter two water masses are present only south of the Ángel de la Guarda and Tiburón Islands. Beneath the surface water masses (ESW, CCW, and GCW), in order of increasing depth, are: SSW, characterized by a salinity maximum (maximum in the sense of passing from relatively low to higher values and then lower values again, with depth; this is not maximum salinity, which occurs in GCW where $S > 35$); PIW, characterized by a salinity minimum (at 750–900 m; fig. 2.2b); and PBW, below the intermediate water, characterized by an increase in salinity to about 34.66 close to the bottom of the entrance basins. The GCW is either ESW or SSW that has been transformed by evaporation (Roden and Groves 1959). The ESW has a clear seasonal pulse, penetrating the Gulf up to the Guaymas Basin during summer but only near the entrance during winter, except during El Niño years (Alvarez-Borrego and Schwartzlose 1979). This summer invasion of the ESW, a warm and oligotrophic water mass, has an effect similar to that of an annual ENSO (El Niño Southern Oscillation) event, suppressing phytoplankton production drastically. In general, summer phytoplankton production is about one quarter of that for winter (Hidalgo-González and Alvarez-Borrego 2004).

South from the Ángel de la Guarda and Tiburón Islands there are salinity, oxygen, and pH minima, as well as a phosphate maximum, at intermediate depths (500–1,100 m), whereas in Ballenas Channel, between Ángel de la Guarda and the peninsula, these properties vary monotonically with depth (fig. 2.2). South of the Ángel de la Guarda and Tiburón Islands, near the bottom of the basins, temperature increases and salinity and oxygen decrease northward from basin to basin. Waters within these basins tend to have the characteristics of those at the corresponding sill depths. Near the bottom, salinity changes from 34.66 at Pescadero Basin (stations 19 and 20), to 34.64 at Farallon Basin (station 14), to 34.63 at Carmen Basin (station 16), and to 34.62 at Guaymas basin (station 12)—compared with 34.68 in the open Pacific. At 2,500 m, oxygen content ranges from 2.3 $m\ell \; \ell^{-1}$ at Pescadero to 1.7 at Farallon and to 1.4 at Carmen (Gaxiola-Castro et al. 1978).

At intermediate depths (500 to 1,100 m), the concentration of oxygen in some places is undetectable by the Winkler method. This oxygen minimum plays an important role in the ecology and geology of the Gulf, as it influences the distribution of pelagic and benthic organisms and sedimentation patterns. Laminated diatomaceous sediments are formed where the basin slopes intersect the oxygen minimum in the water column (Calvert 1964). Burrowing organisms do not live in this poorly oxygenated zone, and the absence of biogenic disturbance allows the laminations to become finely developed. These laminated sediments are paleoecological records of seasonal, interannual, and decadal changes of hydrographic conditions. For example, Baumgartner et al. (1985) studied phytoplankton of size > 24 μm preserved in the laminated sediments of Guaymas Basin, which are composed of heavily silicified diatoms and silicoflagellates, and found that their year-to-year variability is strongly correlated with interannual sea-level anomalies. El Niño periods are generally marked by greater numbers of individuals within species whose distribution is limited to tropical and subtropical waters. With respect to reconstruction of primary production time series, Gilbert and Allen (1943) cautioned: "It should be borne in mind . . . that the diatoms whose frustules make up the diatomaceous deposits are not the same as those that make up the bulk of the phytoplankton pulses." Also, Holmgren-Urba and Baumgartner (1993) studied these laminated sediments to reconstruct time series of scale deposition rates for Pacific sardine, northern anchovy, Pacific mackerel, Pacific hake, and an undifferentiated group of myctophids. The time series were resolved into 10-year sample blocks and extend from approximately 1730 to nearly 1980. This reconstruction shows a strongly negative association between the presence of sardines and anchovies, with anchovies dominating throughout the nineteenth century and with only two important peaks of sardine scale deposition: one in the twentieth century and one at the end of the eighteenth century.

The Ballenas Channel is isolated from the central part of the Gulf by a submarine ridge. The sill depth of this ridge is about 450 m (Rusnak and Fisher 1964). The topography of the channel is very irregular, with maximum depths exceeding 1,500 m. The water in the basin comes from mixing between the surface and sill depths by strong tidal currents and breaking internal waves. The outstanding hydrographic features of the Ballenas Channel are high temperatures, salinities, and oxygen at great depths, and the absence of oxygen and salinity minima (Sverdrup 1941). Compared with conditions elsewhere in the Gulf, the temperature, salinity, and dissolved oxygen in the channel at 1,000 m are higher by 6°C, 0.4, and 1 $m\ell\ \ell^{-1}$, respectively (fig. 2.2). North from Ángel de la Guarda Island there are also no oxygen, pH, or salinity minima. Hydrographic conditions at Tiburón Basin, between the Ángel de la Guarda and Tiburón Islands, are very similar to those at Ballenas Channel (Sverdrup 1941). However, maximum bottom depths for Tiburón Basin are only about 550 m, and it has more open communication with the southern Gulf than does Ballenas Channel. Reported surface nutrient values for the Midriff Islands region are as high as 2.0, 13.0, and 29.0 μM for (respectively) PO_4, NO_3, and SiO_2 under winter conditions (Alvarez-Borrego et al. 1978) and as high as 2.7, 13.1, and 17.8, respectively, under summer conditions (Cortés-Lara et al. 1999).

There are almost no data on trace metals in the Gulf's waters except for the intensely studied Guaymas Basin hydrothermal region (e.g., Gieskes et al. 1991). Delgadillo-Hinojosa et al. (2001) reported on the dissolved cadmium (Cd) distributions for the upper 1,000 m of the Gulf's area between the mouth and the Midriff Islands. Cadmium is a micronutrient that typically behaves much like PO_4 (Libes 1992). As with the other nutrients, the highest Cd surface concentrations were found at the Midriff Islands region (0.21–0.35 nmol ℓ^{-1}) and the lowest were found at the Gulf's mouth (0.08–0.16 nmol ℓ^{-1}). The Gulf's surface waters are enriched with

Cd compared to the adjacent Pacific Ocean (0.002–0.003 nmol ℓ^{-1}), and this is due to the same processes that increase the macronutrient's surface concentrations. Concentrations of Cd in the deep portions of the water column of the Gulf are not different from those of the open Pacific Ocean.

Nutrient Fluxes and Phytoplankton Production

Using the estimates of water exchange at the Gulf's mouth from Roden and Grove (1959), ~1.6 Sv in or out of the Gulf, Calvert (1966) calculated that the Pacific Ocean net supply is approximately 10^{14} g of dissolved silica per annum, which is about 7 times the silica deposited in the sediments of the central Gulf every year. In contrast, Bray (1988) estimated a total annual average transport in or out of the Gulf of only 0.9 Sv, with a standard error of 0.11 Sv. Bray estimated that roughly half of the total transport is attributable to horizontal recirculation, yielding a *net* inflow of deep water, or outflow of near-surface water, of about 0.45 Sv. This would yield an annual net input of silica to the Gulf of only 0.28×10^{14} g, about twice the amount deposited in the sediments of the central Gulf.

Given Bray's (1988) value for the water exchange together with data from figure 2.2, a rough estimate of the PO_4 annual net inflow from the Pacific is about 4×10^{12} g and that for nitrate is $\sim 26.4 \times 10^{12}$ g; the near-surface annual net outflows are $\sim 2 \times 10^{12}$ g and $\sim 6.9 \times 10^{12}$ g for PO_4 and NO_3, respectively. Thus, the annual net inputs of PO_4 and NO_3 to the Gulf are $\sim 2 \times 10^{12}$ g and $\sim 19.5 \times 10^{12}$ g, respectively. Using data in table 2 of Delgadillo-Hinojosa et al. (2001) and Bray's (1988) value for water exchange at the Gulf's mouth, we can estimate the annual net input of Cd into the Gulf to be $\sim 7.8 \times 10^{6}$ mols, or $\sim 877 \times 10^{6}$ g.

If Redfield's ratio for the carbon–nitrogen relationship applies to the Gulf of California (Alvarez-Borrego et al. 1975a, 1978), then the total annual phytoplankton new production (i.e., that part of total phytoplankton production based on new nutrients and not on recycled nutrients) of the whole Gulf would be $\sim 25 \times 10^{12}$ g (or $\sim 25 \times 10^{6}$ tons) of fixed carbon.

Hidalgo-González and Alvarez-Borrego (2004) used satellite imagery of ocean color to estimate an average new production value for the whole Gulf of $\sim 29.2 \times 10^{6}$ tons of fixed carbon per year for the period 1999–2002 (non–El Niño years). In light of the difference in methodologies and the uncertainties associated with the numbers involved in the calculations, the agreement between these figures and the estimate based on nitrate net input to the Gulf is quite good, with a difference of only about 17 percent. The greatest uncertainties of Hidalgo-González and Alvarez-Borrego's (2004) estimates concern the photosynthetic parameters that describe the relationship between photosynthesis and light. Changing these parameters by a single standard error changes the estimate for new phytoplankton production by nearly 18 percent. Adjusting the net water flux so that the two new primary production values are equal would yield a value of 0.53 Sv, which is within the 95 percent confidence interval of Bray's (1988) estimate. Thus, the two estimates for new primary production—based on net nitrate input to the Gulf and color satellite imagery—are not significantly different. This result supports Bray's (1988) value for the water exchange between the Gulf and the Pacific Ocean, and it indicates that Roden and Groves (1959) overestimated it.

If we assume a steady state, then this new primary production value must be compensated by export of reduced forms of inorganic nitrogen after respiration (mainly bacterial), export of particulate organic matter (POM) from the pelagic ecosystem to the sediments of the Gulf, and export of dissolved organic carbon (DOC) and POM to the Pacific Ocean in the surface layer. Extraction due to fisheries is negligible and accounts for less than 1 percent of new production. There are no available ammonia data for Gulf waters. However, if we assume that data for the Eastern Tropical Pacific (McCarthy, 1980) can be used as a first approximation to estimate the net flux, then ammonia concentration for Gulf waters deeper than 200 m is negligible and the mean concentration for the first 200 m is less than 0.5 mmols m^{-3}, which would produce a net export to the Pacific Ocean of only about 2 percent of the net input of nitrogen in the form of NO_3 into the Gulf. If we assume a mean deposition rate of particulate organic matter for the whole Gulf that is equal to that for the Carmen and Guaymas Basins (~0.02 gC m^{-2} day^{-1}) (Thunell et al. 1993), then the export to the sediments would be only about 3 percent of the new phytoplankton production of the Gulf. Thus, most (>94 percent) of the export must be in the form of DOC and POM. In general, though, POM is only about 3 percent of DOC (Libes 1992) and so most of the export must be in the form of DOC. For equilibrium to obtain between new production and export, the mean DOC concentration difference between the first 200 m and that of the 200 – 600 m layer would be ~130 μM, which is within the range reported in the literature.

Hidalgo-González and Alvarez-Borrego (2004) divided the Gulf into four regions during winter (fig. 2.1b, roman numerals) and into two regions during summer (fig. 2.1c, arabic numerals) for estimating average integrated total (P_{Tint}) and new (P_{newint}) primary production (gC m^{-2} day^{-1}) for the period 1997–2002. Values for P_{Tint} had a large seasonal variation (e.g., 1.16–1.91 for winter versus 0.39–0.49 for summer; see table 2.1). Values for P_{newint} for winter increased from the entrance region to the Midriff Islands region and then remained the same from there to the northern Gulf, with values up to twice as large in the two latter regions (≤1.33 gC m^{-2} day^{-1}) as in the entrance (≤0.48). As shown in table 2.2, the values of P_{newint} for summer were less than half those for winter. In spite of the limited data, a clear interannual P_{Tint} variability is evident for the entrance region, with lowest values for the 1997–1999 El Niño event. This kind of variation was also present for the central Gulf, but the effect was much weaker (see table 2.1) because of its strong dynamics. Santamaría-del-Ángel et al. (1994b) described a similar El Niño effect on satellite-derived chlorophyll concentrations (Chl_{sat}) for the winters of 1982/83 and 1983/84 using data from the Coastal Zone Color Scanner. These latter authors reported a dramatic suppression of Chl_{sat} at the entrance to the Gulf during El Niño 1982–1983, with values down to ~20 percent of those for non–El Niño years but with a relatively small impact in the Midriff Islands and northern regions.

The Upper Gulf and the Colorado River

The upper Gulf of California, the triangle north of Puerto Peñasco and San Felipe and near the Colorado River Delta, is a shallow area with depths largely <30 m. This region has the greatest seasonal hydrographic changes of the entire Gulf. After construction of Glen Canyon Dam was completed in 1963, regular input of Colorado River water to the upper Gulf ceased. During 1979–1987, however, large water releases became necessary because of higher than average snowmelts in the upper basin of the river. In the upper Gulf, surface temperatures increase from the southeast to the northwest in summer; the opposite occurs in winter. Minimum and maximum surface temperatures have been recorded west of Montague Island:

TABLE 2.1. Satellite-derived integrated total primary production (P_{Tint}) per unit area (gC m^{-2} day^{-1} or TonC km^{-2} day^{-1}) and for whole regions in the Gulf of California (TonC day^{-1}). The regions are shown in figures 1b and 1c (after Hidalgo-González and Alvarez-Borrego 2004).

Year	I		II		III		IV	
	P_{Tint} (gC m^{-2} day^{-1})	P_{Tint} (TonC day^{-1})	P_{Tint} (gC m^{-2} day^{-1})	P_{Tint} (TonC day^{-1})	P_{Tint} (gC m^{-2} day^{-1})	P_{Tint} (TonC day^{-1})	P_{Tint} (gC m^{-2} day^{-1})	P_{Tint} (TonC day^{-1})
1997–1998	1.16	5.8×10^4	1.52	9.0×10^4	1.45	3.1×10^4	1.52	2.6×10^4
1998–1999	1.20	6.0×10^4	1.67	9.8×10^4	1.60	3.5×10^4	1.60	2.7×10^4
1999–2000	1.85	9.3×10^4	1.74	10.3×10^4	1.73	3.7×10^4	1.60	2.7×10^4
2000–2001	1.80	9.0×10^4	1.91	11.3×10^4	1.62	3.5×10^4	1.68	2.8×10^4
2001–2002	1.78	8.9×10^5	1.87	11.0×10^4	1.66	3.6×10^4	1.65	2.8×10^4

Year	1		2	
	P_{Tint} (gC m^{-2} day^{-1})	P_{Tint} (TonC day^{-1})	P_{Tint} (gC m^{-2} day^{-1})	P_{Tint} (TonC day^{-1})
1997	0.42	4.6×10^4	0.43	0.93×10^4
1998	0.39	4.3×10^4	0.42	0.91×10^4
1999	0.43	4.7×10^4	0.45	0.97×10^4
2000	0.49	5.3×10^4	0.44	0.95×10^4
2001	0.48	5.2×10^4	0.43	0.93×10^4
2002	0.47	5.1×10^4	0.43	0.93×10^4

TABLE 2.2. Satellite-derived integrated new primary production (P_{newint}) per unit area (gC m^{-2} day^{-1} or TonC km^{-2} day^{-1}) and for whole regions in the Gulf of California (TonC day^{-1}). The regions are shown in figures 1b and 1c (after Hidalgo-González and Alvarez-Borrego 2004).

Year	I		II		III		IV	
	P_{newint} (gC m^{-2} day^{-1})	P_{newint} (TonC day^{-1})	P_{newint} (gC m^{-2} day^{-1})	P_{newint} (TonC day^{-1})	P_{newint} (gC m^{-2} day^{-1})	P_{newint} (TonC day^{-1})	P_{newint} (gC m^{-2} day^{-1})	P_{newint} (TonC day^{-1})
1997–1998	0.38	2.2×10^4	0.71	4.2×10^4	1.11	2.4×10^4	1.16	1.95×10^4
1998–1999	0.39	2.3×10^4	0.75	4.4×10^4	1.22	2.6×10^4	1.22	2.05×10^4
1999–2000	0.48	2.9×10^4	0.75	4.4×10^4	1.33	2.9×10^4	1.26	2.07×10^4
2000–2001	0.46	2.7×10^4	0.81	4.8×10^4	1.29	2.8×10^4	1.29	2.17×10^4
2001–2002	0.38	2.2×10^4	0.79	4.7×10^4	1.31	2.8×10^4	1.25	2.10×10^4

Year	1		2	
	P_{newint} (gC m^{-2} day^{-1})	P_{newint} (TonC day^{-1})	P_{newint} (gC m^{-2} day^{-1})	P_{newint} (TonC day^{-1})
1997	0.27	3.2×10^4	0.28	0.62×10^4
1998	0.25	2.8×10^4	0.28	0.62×10^4
1999	0.27	3.0×10^4	0.30	0.66×10^4
2000	0.31	3.4×10^4	0.28	0.62×10^4
2001	0.29	3.2×10^4	0.27	0.60×10^4
2002	0.27	3.0×10^4	0.27	0.60×10^4

8.25°C in December and 32.58°C in August (Alvarez-Borrego et al. 1973). A monthly hydrographic sampling carried out during 1972–1973 (years of no Colorado River water release) showed that salinity in general maintained a surface gradient with values increasing northwestward. Salinity ranged from a minimum of 35.28 in October to a maximum of more than 38.50 in July (Alvarez-Borrego et al. 1973). With data obtained between December 1993 and June 1996, this hydrographic behavior has been shown to persist so long as there is no Colorado River freshwater release (Lavín et al. 1998). The Colorado River freshwater release of March–April 1993 provided an opportunity to observe the effect of the positive-estuary behavior on the upper Gulf. In contrast to the "normal" inverse estuarine situation, surface salinity decreased from 35.4 near San Felipe to 32 at ~10 km south of Montague Island. The surface temperature was normal for the time of the year (Lavín and Sánchez 1999).

For the northern Gulf of California, Alvarez-Borrego et al. (1978) reported PO_4, NO_3, and SiO_2 surface concentrations in the ranges of 0.7–1.0 μM, 0.0–4.0 μM, and 6.1–18 μM, respectively. Their northernmost station was a little south of latitude 31° N. In the area of the upper Gulf adjacent to Montague Island, values are in the ranges 0.3–3.1 μM and 3.3–18.3 μM for PO_4 and NO_3, respectively (Hernández-Ayón et al. 1993). Agricultural drainage water is carried from the Mexicali Valley to the Colorado River estuary, and the effects of this flux include relatively low salinity and high NO_3 concentrations at the internal extreme of the inverse estuary (Alvarez-Borrego 2003). Nieto-García (1998) compared nutrient concentrations in the upper Gulf for the springs of 1993 (wet year) and 1996 (dry year), finding that NO_2, NO_3, and PO_4 were lower in the spring of 1993 (with freshwater input) than during 1996. Only dissolved silicate was significantly higher during 1993, indicating a clear influence of river water. Alvarez-Borrego (2003) indicated that a possible explanation for the relatively low NO_2, NO_3, and PO_4 values in 1993 is that the river flow diluted the agricultural drainage water carried from the Mexicali Valley and also "cleaned" the surface sediments of the estuary.

ACKNOWLEDGMENTS

Thanks to J. M. Dominguez and F. Ponce for the artwork.

CHAPTER 3

Reefs That Rock and Roll

Biology and Conservation of Rhodolith Beds in the Gulf of California

RAFAEL RIOSMENA-RODRÍGUEZ, DIANA L. STELLER, GUSTAVO HINOJOSA-ARANGO, AND MICHAEL S. FOSTER

Introduction

Imagine an underwater field of closely packed, purple-pink spheres about the size of golf balls, each composed of numerous calcareous branches radiating out from the center of the sphere. The spheres and hundreds of species of animals and seaweeds living on and in them move with the motion of waves and currents. These spheres are rhodoliths, and beds of these calcareous red algal spheres comprise a rarely mentioned but common habitat in global nearshore environments. Rhodoliths are not generally included in natural history and diving guides. The rhodoliths themselves are attractive because of their color and range of morphologies and because of the variety of organisms associated with the beds. Local people view rhodolith beds in terms of the associated rich fisheries or as recruitment sites for harvestable species.

Rhodoliths are morphologically diverse, free-living, nongeniculate, coralline red algae (Rhodophyta: Corallinales) that form extensive living beds worldwide (color plate 2A) and are abundant in fossil deposits from the early Cretaceous to the Pleistocene (Aguirre et al. 2000; Foster 2001). As with many attached reef-forming coralline species (Steneck and Adey 1976), rhodolith size, shape, and branching vary among species and—primarily as a result of variation in water motion—within species (Bosence 1976; Foster et al. 1997). Much like the maerl beds common to the northeastern Atlantic, rhodolith beds are benthic communities dominated by rhodoliths that collectively create a fragile, structured biogenic matrix (color plate 2B). This matrix provides habitat for diverse assemblages of invertebrates and algae (Cabioch 1969; Keegan 1974; Bosence 1979; Steller et al. 2003). Rare, unusual, and endemic species in rhodolith beds have been reported from geographically diverse locations including the northeastern Atlantic (De Grave 1999), Norwegian fjords (Freiwald et al. 1991), the Mediterranean (Ballesteros 1988), the tropical Atlantic (Ballantine et al. 2000), the Indian Ocean (Weber-Van Bosse and Foslie 1904; Scoffin et al. 1985), and the tropical and subtropical Pacific, including the Gulf of California (Reyes-Bonilla et al. 1997; Clark 2000; James et al. 2006; Foster et al. 2007).

Living rhodoliths occur as an upper layer of pigmented, generally rounded thalli that overlay carbonate sediments (color plate 2C). The latter are often the result of long-term accumulation of dead rhodoliths plus hard parts from other calcified organisms such as corals and molluscs (Minoura and Nakamori 1982; Steller and Foster 1995; Boscence and Wilson 2003). In the northeastern Atlantic, unconsolidated rhodolith deposits have long been harvested for a variety of human uses including as a soil amendment, and in many other locations the beds are harvested for clams and scallops (Blunden et al. 1977; Briand 1991; review in Steller et al. 2003). Recent studies in the northeastern Atlantic and Mediterranean have shown that these common benthic environments are not resilient to direct disturbances caused by harvesting (Blunden et al. 1981) and fisheries trawling (Hall-Spencer and Moore 2000; Bordehore et al. 2003) or to indirect impacts from activities that reduce water quality through siltation and eutrophication (Blunden et al. 1981; Hily et al. 1992; Grall and Glemarec 1997). Slow growth rates (review in Foster 2001; Steller et al. 2007a) combined with the negative impacts of burial make rhodoliths particularly vulnerable to both disturbance and extraction (Adey and McKibben 1970; Potin et al. 1990). These community attributes have led to the recent listing of rhodolith beds as threatened and protected in New Zealand (Department of Conservation 1998), Australian (Director of National Parks 2005), and European (Birkett et al. 1998) coastal habitats.

Rhodoliths in the Gulf of California have been known to science since the late 1880s (Hariot 1895), but the exceptional abundance of fossil and living beds (fig. 3.1) dominated by these algae, their high diversity and importance to nearshore ecology, and the need for their conservation in the

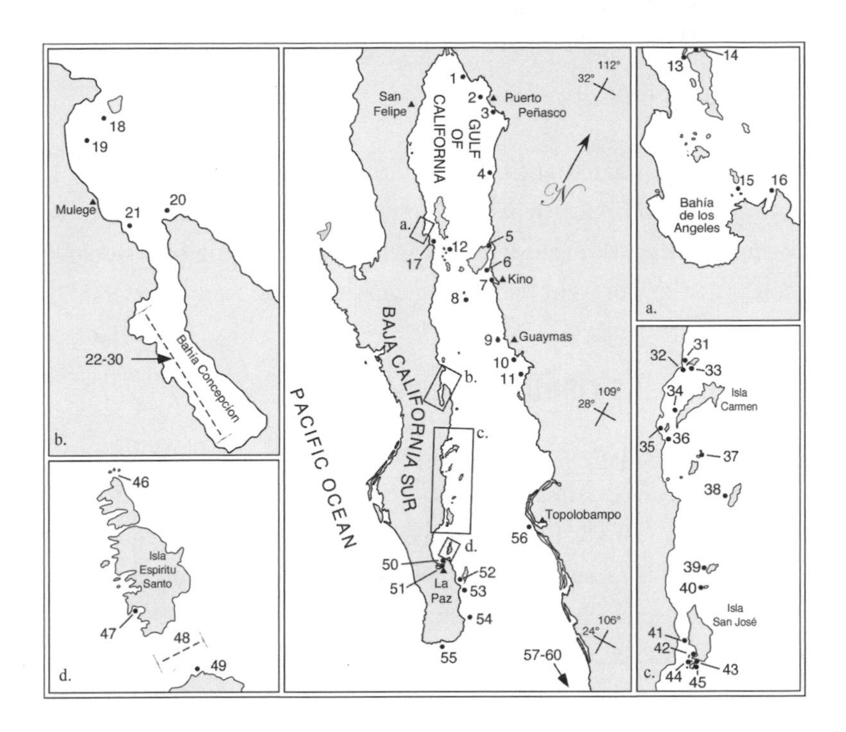

3.1. Known locations of living rhodolith beds in the Gulf of California. A rhodolith bed consists of an area of 10% or greater cover of rhodoliths. All beds except those designated E.Y.D (E. Yale Dawson; Dawson 1960b) or O (other) were observed by one or more of the authors (contact the authors for details). Dawson's observations were based on dredged material or were not specified. Numbers correspond to following sites by region. *Puerto Peñasco/Guaymas Region:* (1) Punta Borrascosa (O); (2) Manto Peñasco (O); (3) Estero Morua (O); (4) Cabo Tepoca; (5) Estero Arenas (O); (6) Estero Santa Rosa; (7) Isla Alcatraz; (8) Isla San Pedro Martir (O); (9) Isla San Pedro de Nolasco; (10) Las Gringas; and (11) Bahía Bocochibampo (E.Y.D.). *Bahía de los Ángeles Region:* (12) Isla Partida (E.Y.D.); (13) Puerto Refugio (O); (14) Canal Mejia (O); (15) Isla Cabeza de Caballo; (16) Bahía de San Francisco Pond Island (E.Y.D., O); (17) Bahía de las Palomas (O); (18) Punta Chivato I; and (19) Punta Chivato II. *Bahía Concepción Region:* (20) Punta Aguja (Concepción, E.Y.D.); (21) Los Machos; (22) Santispac; (23) Isla Blanca; (24) Isla Coyote; (25) Morro Tecomates; (26) El Cardón; (27) Isla El Requeson; (28) La Cueva (Correcaminos); (29) Los Pocitos; and (30) El Coloradito. *Loreto Region:* (31) Punta Bajo; (32) Isla Coronado Channel; (33) Isla Coronado; (34) Isla Carmen (in Bahia Salinas, E.Y.D., O); (35) Isla Danzante (O); (36) Puerto Escondido (E.Y.D.); (37) Islas Las Galleras; (38) Isla Catalina (O); (39) Isla Santa Cruz (E.Y.D.); (40) Isla San Diego; (41) Isla San José; (42) Isla San José Estero; (43) Isla El Pardito; (44) La Lobera; and (45) Isla San Francisquito. *La Paz Region:* (46) Los Islotes (O); (47) Bahía San Gabriel; (48) Canal de San Lorenzo; (49) Punta Galeras; (50) Isla Gaviota; (51) Bahia de La Paz (Malecon/historic site, E.Y.D., O); (52) Isla Cerralvo (O); (53) Punta Perico; (54) Cabo Pulmo (O); and (55) Bajo La Gorda (E.Y.D.). *Sinaloa Region:* (56) Topolobampo (E.Y.D.) and (57) Isla Venado. *Nayarit Region:* (58) Isla Isabel (O); (59) Isla Maria (O); and (60) Isla Marietas. Note that beds 57–60 are off the map (to the south).

Gulf was not appreciated until the early 1990s (review in Foster et al. 1997). Our purpose is to review what is known about rhodoliths in the Gulf of California and then to discuss this information, from a global perspective, as it concerns their present and future conservation.

Spatial Distribution of Living Beds

When compared with kelp forests, sea grass meadows, and other major communities dominated by macrophytes, rhodolith bed distribution is poorly known. This is because beds are often not visible from the ocean surface, commonly occur in areas thought to be simply soft bottoms, and can shift in location. The world distribution map in Foster (2001), updated from that of Bosence (1983), shows that currently known living beds are widespread from polar regions to the tropics, with large concentrations in the northwestern Atlantic, Caribbean, Gulf of California, eastern Brazilian Shelf and Western Australia. Such a distribution indicates that the occurrence of rhodolith beds is probably not directly constrained by latitudinal variation in temperature, mean solar input, day length, or nutrients.

It has been argued by Halfar et al. (2000a) that high nutrients in the Gulf of California promote the growth of coralline algae, and Halfar et al. (2004b) sampled four sites from Cabo Pulmo to Bahía de Los Ángeles in finding rhodoliths most abundant at what they classified as "mesotrophic" sites, suggesting support for nutrient controls on distribution. Halfar and Mutti (2005) argued that increases in nutrients were at least partly responsible for shifts from coral to coralline algal communities in the Miocene. However, the core of this argument—that coralline algae in general or rhodoliths in particular develop best under moderate to high nutrient conditions—may be an artifact of where Halfar et al. (2004b) sampled at each site. Extensive rhodolith beds are found very near Halfar et al.'s southern, "oligo-mesotrophic" site at Cabo Pulmo (beds 54 and 55 in fig. 3.1), and at their northern, "eutrophic" site in Bahía de Los Ángeles (Dawson 1960b; fig. 3.1). Moreover, it is well known that diverse and abundant invertebrate assemblages occur within, on, and around rhodoliths. Such dense faunal assemblages are capable of producing nutrients that locally contribute to macroalgal growth (Bracken 2004). The nutrients produced by the rhodolith bed fauna, in conjunction with low nutrient requirements suggested by slow growth (discussed later in the chapter), further suggest that nutrient concentrations in the surrounding water have little direct effect on rhodolith distribution.

Current knowledge indicates that the worldwide distribution of rhodolith beds is largely unaffected by the direct influence of wide variation in temperature and nutrients. This is in contrast to reef-building corals that are restricted to the tropics, where temperatures are high and nutrients low (and where grazing on macroalgae is high; reviewed in Miller 1998). It may be that coral formations can dominate rhodoliths for space in such situations (at least in shallow water) as suggested by the rhodolith and coral sequences in cores from some coral reefs (Payri and Cabioch 2004).

It follows from these observations that local, not latitudinal, environmental variation has the greatest effects on the presence and persistence of rhodolith beds. This suggests that—except in high-temperature–low-nutrient tropical environments, where rhodoliths might lose to corals in the competition for space—using the presence (or absence) of living rhodolith beds or rhodoliths in the fossil record to intuit broadscale differences in present or historical environments can be misleading. At local scales, the consensus is that rhodoliths need to be moved to be maintained in a free-living state, and this movement is driven by water motion and bioturbation (reviews in Bosence 1983; Foster 2001). Water motion also reduces the probability of burial by inhibiting sedimentation of fine particles and resuspending such particles after periods of low water motion (Steller and Foster 1995; Marrack 1999; Mitchell and Collins 2004). However, high water motion may break the thalli, especially of delicate species such as *Lithophyllum margaritae* (Steller and Foster 1995), or transport thalli into habitats unfavorable for growth. Distributional data suggest that rhodolith beds generally occur on flat to gently sloping soft bottoms, where rhodoliths and free-living forms of other organisms (such as corals and bryozoans; Reyes-Bonilla et al. 1997; James et al. 2006) can accumulate and where light is sufficient for rhodolith growth.

In the Gulf of California, rhodolith beds have consistently been found at semi-protected (from waves) sites around islands and in bays, and in channels, from north of Puerto Peñasco, Sonora, to Islas Marietas, Jalisco (fig. 3.1). This distribution should be considered tentative, since intensive surveys have only been done in the region around La Paz and Mulege. The shoreline extent and depth of most of these beds are still unknown. Most, however, occur in one of two general types of environments defined by the dominant water conditions. *Wave beds* grow in relatively shallow water (3–12 m) on gently sloping soft bottoms exposed to moderate wave action (e.g., beds 22–30 in fig. 3.1); *current beds* generally grow in deeper water (>12 m) on relatively flat bottoms exposed to tidal currents (e.g., beds 32, 43, and 48 in fig. 3.1). In wave beds, the water motion from waves appears to move the rhodoliths and inhibit the accumulation of fine sediment. In current beds, the large-scale sediment is transported during high currents (D. Steller, pers. obs.); it may be that bioturbation causes most of the movement and that currents inhibit the accumulation of fine sediment. An exception to these categories is a tidally driven "current" bed at the mouth of Santa Rosa estuary (bed 6 in fig. 3.1). At present it is unknown how common these are.

Massive bed relocation as well as transport to shore and deeper water can occur during episodic events such as hurricanes. The former has been observed after hurricanes (M. Foster, M. Johnson, and R. Riosmena, pers. obs.), and the latter is indicated by layers of rhodoliths and rhodolith fragments in sediment cores from depths of 300–700 m offshore from the large bed (bed 48 in fig. 3.1) in Canal de San Lorenzo (Schlanger and Johnson 1969; Foster, pers. obs.). The lower limits of wave beds appear to be determined by sedimentation (Steller and Foster 1995). The deepest current bed found to date is 35–40 m at Punta Perico (bed 53 in fig. 3.1; Vazquez-Elizondo 2005). In areas where suitable substrate and currents occur it is likely that the lower limits of such beds are determined by light. Nutrients may thus have an indirect effect on the depth of rhodolith beds by reducing light via effects on phytoplankton biomass.

These constraints, whose existence is based on local observations, suggest that any larger-scale distribution patterns may result from large differences in nearshore geomorphology, water motion, and sedimentation—a conclusion similar to Nelson's in his review of nontropical shelf carbonates (Nelson 1988). The abundance of rhodolith beds may also be affected by variation in rhodolith growth rates due to temperature and availability of calcium carbonate. The great variation in the occurrence and rhodolith bed composition found within a local site (e.g., Bahía de La Paz; see Foster et al. 1997; Halfar et al. 2000a) argues for caution in attributing general causes after sampling only a few localized modern beds or fossil deposits, no matter how widely separated. Studies of the present have repeatedly verified the fundamental logic that, until small-scale variation is understood, putative large-scale variation may be an illusion (Underwood 1997).

The preceding conclusions are based on limited information concerning spatial distribution and depend largely on qualitative correlation. It is hoped that the increasing use of remote sensing to detect and map rhodolith beds (e.g., De Grave et al. 2000; Forrest et al. 2005), combined with traditional SCUBA sampling and continuous, bed-scale *in situ* recording of appropriate environmental variables (e.g., Mitchell and Collins 2004), will provide the descriptive information necessary to develop more rigorous hypotheses that can be experimentally tested.

Taxonomic Diversity of Rhodoliths

One of the major problems in working with rhodoliths and coralline algae in general is the identification of specimens to the generic or species level. The classification within this order has changed from the recognition of one or two families (Corallinaceae and Solenoporaceae) in the early twentieth century to the recognition of three families (Corallinaceae, Hapalidaceae, and Sporolithaceae) at the beginning of the twenty-first century (Harvey et al. 2003). The changes have been more dramatic at the generic level, and a major evaluation (Woelkerling 1988) was necessary to settle many of the classical problems. However, boundaries of genera and species remain a common problem best solved by regional monographs using consistent diagnostic features (Riosmena-Rodríguez et al. 1999; Harvey et al. 2003). One of the major problems is that many growth-forms can occur in the same species (Woelkerling et al. 1993; Riosmena-Rodríguez et al. 1999).

Riosmena-Rodríguez et al. (1999), Riosmena-Rodríguez (2002), and Vazquez-Elizondo (2005) based their taxonomic surveys on specimens from 55 localities, identifying four main rhodolith species in the Gulf of California: *Lithophyllum margaritae* (fig. 3.2) as well as *Neogoniolithon trichotomum*, *Mesophyllum engelhartii*, and *Lithothamnion muellerii* (= *Lithothamnium crassiusculum*) (resp. fig. 3.3A, B, C). There may be a fifth species, depending on the pending taxonomic resolution of *Lithothamnion australe* (Riosmena-Rodríguez 2002). Additional species in the genera *Archeolithothamnium* and *Lithothamnion* have been recorded from Eocene deposits (Durham 1950;

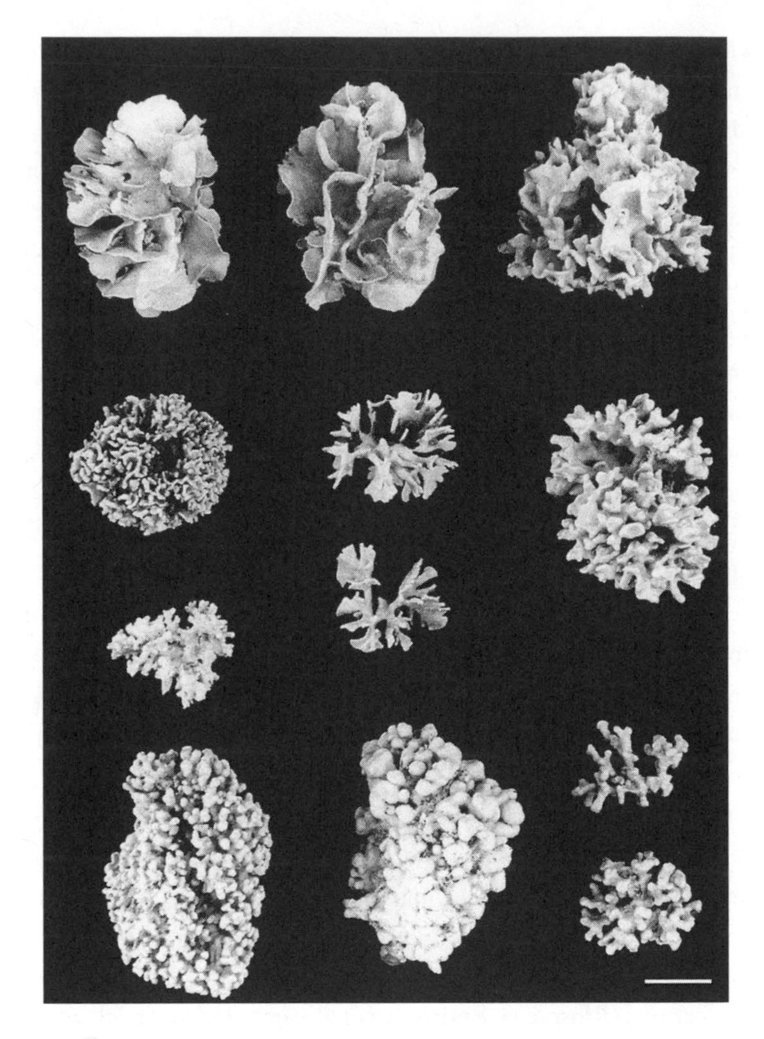

3.2. Growth-form variation in *Lithophyllum margaritae* ranging from foliose (right) to fruticose (left top) to lumpy (left center); transitional forms in the center. From Riosmena-Rodríguez et al. (1999); scale bar = 5 cm.

Squires and Demetrion 1992; Abbott et al. 1993). These need reassessment because *Archeolithothamnium* is a heterotypic synonym of *Sporolithon* (Woelkerling 1988), which means that confirmation of generic circumscription is required before species can be determined. There are no species descriptions for the abundant rhodoliths from Pleistocene strata (Dorsey 1997; Foster et al. 1997; Cintra-Buenrostro et al. 2002). A comprehensive monograph on these is needed to determine the species composition using all spe-

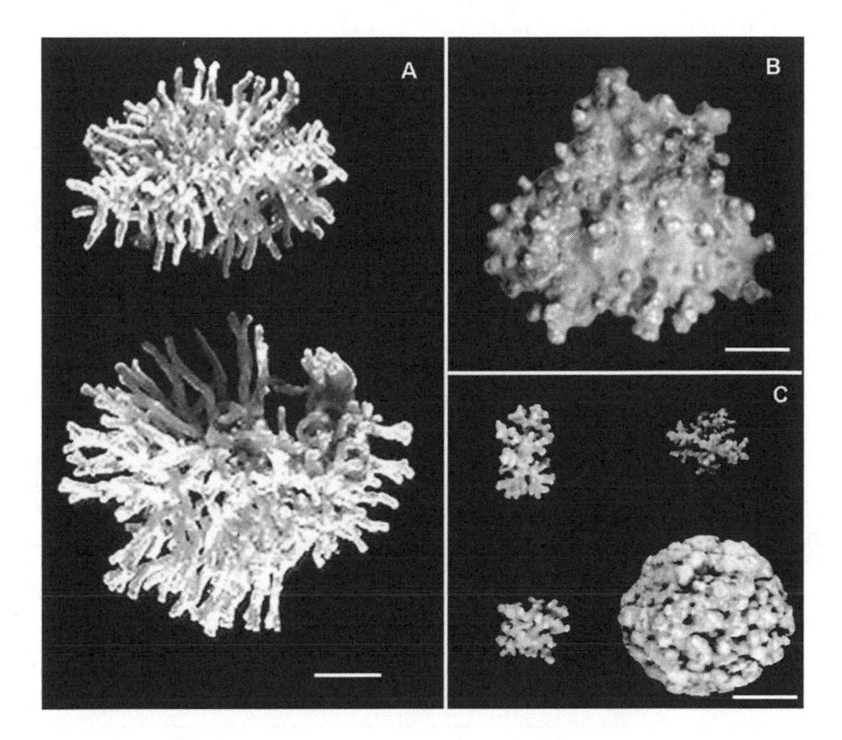

3.3. Other rhodolith-forming species from the Gulf of California with less variable growth-forms than *Lithophyllum margaritae*. (A) *Neogoniolithon trichotomum:* specimen with fruticose growth-form; scale bar = 1 cm. (B) *Mesophyllum engelhartii:* specimen with fruticose growth-form; scale bar = 1 cm. (C) *Lithothamnion muellerii:* specimens with fruticose to lumpy growth-forms; scale bar = 3 cm.

cies in the fossil record. Such a monograph would also facilitate stratigraphic analyses and the interpretation of historic changes in distribution.

Modern taxonomic approaches based on careful evaluation of anatomical structures (Harvey and Woelkerling 2007) combined with molecular analysis as described in Harvey et al. (2003) are crucial to rigorous identification. Assessment of anatomical structures requires thin histological sections, especially of the pore area over conceptacles, in addition to optical and scanning electron microscopy (SEM; Woelkerling 1988; Riosmena-Rodríguez et al. 1999). The main features used to distinguish species in the Gulf are used in the key (table 3.1) and illustrated in figure 3.4. The primary structures that must be assessed are whether the epithelial cells are flared or flat/rounded, the kind of interfilament connections present

TABLE 3.1. Taxonomic key to rhodolith-forming algae from the Gulf of California (terminology based on Woelkerling 1988).

1a.	Foliose to fruticose growth form, thalli pseudoparenchymatous with filaments connected with secondary pit connections (fig. 3.4A, white arrow) *or* cell fusions (fig. 3.4C) present. Epithelial cells flat/rounded (fig. 3.4A, black arrow) but never flared (fig. 3.4B). Uniporate tetrasporangial conceptacles (fig. 3.4D).	2
1b.	Lumpy to fruticose to incrusting growth-form, thalli pseudoparenchymatous with filaments connected and cell fusions present. Epithelial cells flared but never flat or rounded. Multiporate tetrasporangial conceptacles (fig. 3.4E).	3
2a.	Foliose to fruticose growth-form, thalli pseudoparenchymatous with filaments connected but only secondary pit connections (no cell fusions) present and trichocyte absent.	*Lithophyllum margaritae* (Hariot) Heydrich
2b.	Fruticose growth-form and *only* cell fusions present; single trichocyte.	*Neogoniolithon trichotomum* (Heydrich) Setchell & Mason
3a.	Lumpy to fruticose growth-form, flared epithelial cells, no elongated subepithelial cells (fig. 3.4B).	*Lithothamnion muelleri* Lenormand ex Rosanoff
3b.	Fruticose to incrusting growth-form, rounded or flattened epithelial cells, elongated subepithelial cell.	*Mesophyllum engelhartii* (Foslie) Adey

(no evident connections, secondary pit connections or cell fusions or both), and whether the tetrasporangial conceptacles are uniporate or multiporate. Further molecular analyses are being performed to better understand species limits in the Gulf (Riosmena-Rodríguez, unpublished data).

All the recognized species are morphologically variable, but *L. margaritae* is especially so: individuals may have entirely foliose, fruticose, or lumpy growth-forms or may exhibit combinations of these (fig. 3.2). This species is widely distributed in the Gulf and is dominant in the southern Gulf. The morphology of the other three species ranges between lumpy and fruticose (fig. 3.3A–C). *Lithothamnion muelleri* (fig. 3.3C) seems to occur only in the most wave-exposed, shallow water sites, often on patches of sediment among rocks and *Sargassum*. It is also widely distributed in the Gulf. *Neogoniolithon trichotomum* (fig. 3.3A) is more common in beds that are well protected from waves. It is found only in the southern Gulf and has not been found deeper than 12 m. *Mesophyllum engelhartii* (fig. 3.3B) has only been found in bed 53 (fig. 3.1) at depths of ~30 m. The number of rhodolith-forming species is higher in the Gulf of California than in other

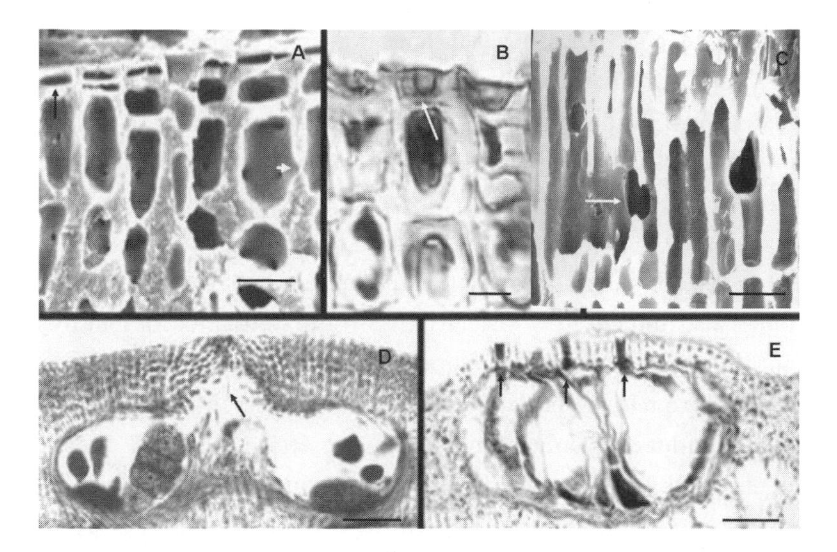

3.4. Main anatomical features used to identify rhodolith-forming species. (A) Longitudinal section showing the flat/rounded epithelial cells (black arrow) and secondary pit connections (white arrow); scale bar = 10 mm. (B) Longitudinal section showing flared epithelial cells (arrow); scale bar = 10 mm. (C) Longitudinal fracture showing cell fusions (arrow); scale bar = 10 mm. (D) Longitudinal section of a uniporate tetrasporangial conceptacle (arrow); scale bar = 100 mm. (E) Longitudinal section of a tetrasporangial conceptacle with a multiporate roof (arrows); scale bar = 100 mm.

geographic regions (with the exception of the northeastern Atlantic and Mediterranean). For example, there is only one species in the temperate northwestern Atlantic (Bird and McLachlan 1992) and three in the tropical northwestern Atlantic (Littler and Littler 2000). There are twelve species in the northeastern Atlantic (Cabioch et al. 1992; Irvine and Chamberlain 1994) and four species in the Mediterranean (Cabioch et al. 1992; Bordehore et al. 2002); the Indian Ocean hosts three species (Verheij 1993), the western Pacific two species (Woelkerling 1996a–e; Littler and Littler 2003), and the temperate northeastern Pacific three species (Athanasiadis et al. 2004).

Rhodolith Morphology and Growth

Rhodolith morphology varies depending on the growth environment (Bossellini and Ginsburg 1971; review in Bosence 1983). In the Gulf, individual *Lithophyllum margaritae* rhodoliths tend to grow larger, be more spherical, and have higher branch densities as water motion increases (Steller and Foster 1995; Foster et al. 1997). Although light often increases with water motion, there is little evidence that light affects morphology. Variation in morphology, including that between wave and current beds, may have influenced genetic exchange among rhodolith populations and growth-forms thereby resulted in geographic isolation (Schaeffer et al. 2002). Such isolation is no doubt also influenced by vegetative reproduction, since most individual rhodoliths appear to grow from fragments. Rhodolith growth-forms thus appear to result from the interplay between the environment and genetics.

Individual rhodoliths may originate from rhodolith fragments, spores, or coralline algal fragments broken from encrusting forms on rocks or shells (reviewed in Foster 2001). Regardless of their origin, however, rhodoliths have persistent growth bands along the branches that can be marked with vital stains. Rivera et al. (2004), using field and laboratory experiments in the southern Gulf with *L. muellerii* (= *L. crassiusculum*) from bed 49 (fig. 3.1), found an average growth rate of 0.6 mm/yr, that 2–3 primary bands are produced annually, and that the number of bands are positively and linearly related to branch length. This relationship was then used to determine the age of the largest *L. muellerii* collected (~15 cm): 120 years. This growth rate was verified for *L. muelleri* from the same bed by Frantz et al. (2000) based on the timing of variation in ^{14}C from atmospheric nuclear testing. This study also found that *L. muellerii* contained a record of ENSO events (fig. 3.5). Halfar et al. (2000b) used isotopic observations to document a persistent pattern of Mg/Ca relative to growth bands in this same species. Old age and the presence of regular growth bands make rhodoliths like *L. muellerii* a potentially excellent biogenetic archive of year-to-year and perhaps seasonal changes in environmental conditions (Halfar 1999; Halfar et al. 2000a,b).

The growth rates of subtropical Gulf of California rhodoliths vary markedly among species with strong seasonal signals. Rivera et al. (2004), using field and laboratory experiments, and Frantz et al. (2000), using variation in ^{14}C, both found that *L. muellerii* (= *L. crassiusculum*) from La Paz (bed 49 in fig. 3.1) had an average apical tip extension growth rate of 0.6 mm/yr. Using methods similar to Rivera et al. (2004), McConnico (pers. obs.) found a similar growth rate for *L. muellerii* at Cabo Los Machos (bed 21 in fig. 3.1) in the central Gulf. Steller (2003) and Steller et al. (2007a), also using vital stains in the field, found higher annual growth rates

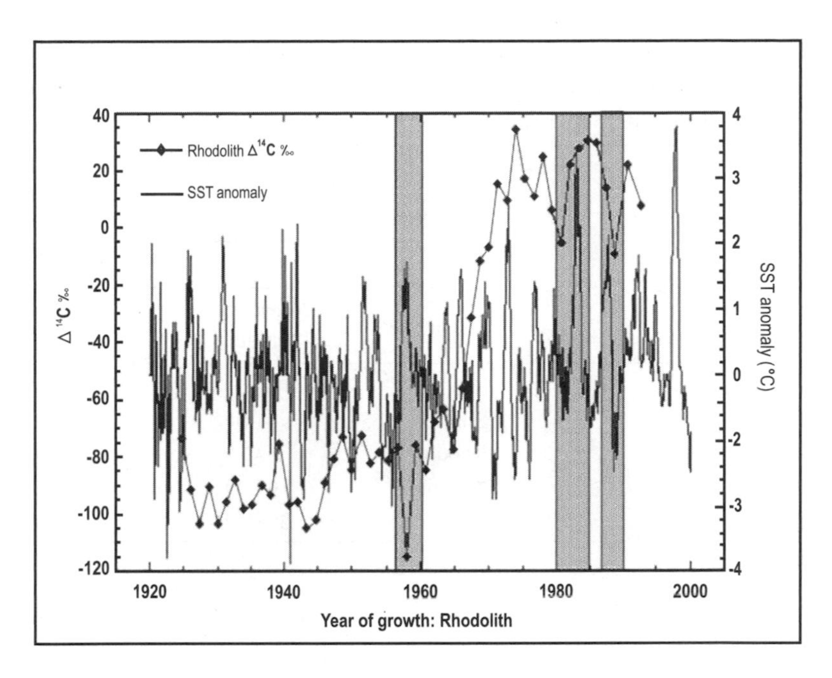

3.5. ^{14}C record from *Lithothamnion muellerii* plotted with the global ocean surface temperature atlas (GOTSA) ENSO 3 sea-surface temperature (SST) anomaly. Three large El Niño events are shown (highlighted in gray) that occurred coincidentally with three large declines in the ^{14}C record of the rhodolith. From Frantz et al. (2000).

in both *L. margaritae* (2.8 mm/yr) and *N. trichotomum* (3.9 mm/yr) at the Isla El Requeson bed (bed 27 in fig 3.1) and in *N. trichotomum* (4.9 mm/yr) at Manto Diguet (bed 42 in fig. 3.1). Strong seasonal growth patterns measured in *L. margaritae* at Isla El Requeson (5.0 mm/yr in summer vs. 0.83 mm/yr in winter; Steller et al. 2007a) and in *L. muelleri* at La Paz (0.2 mm/yr in summer vs. 0.5 mm/yr in winter; Rivera et al. 2004) suggest the need for further studies to understand seasonal growth trends. Overall these slow growth rates strongly suggest that recovery of rhodolith populations from both natural and anthropogenic disturbances will be slow—on the order of decades for all species (Steller 2003).

The Diversity of Rhodolith Communities

It has long been recognized that rhodolith beds harbor a diverse and abundant assemblage of macro organisms and that this high diversity is

TABLE 3.2. Comparison showing that (a) overall rhodolith bed diversity can be much higher than surrounding sand flats, and (b) cryptofauna contribute the most to bed diversity (from Steller et al. 2003).

		Habitat	
	Subhabitat	Sandflat	Rhodolith Bed
A. Richness			
Organism Type		No. species (unique)	No. species (unique)
Epiflora	Surface	2 (0)	4 (1)
Epifauna	Surface	9 (5)	7 (3)
Cryptofauna	Rhodolith	x	33 (26)
Infauna	Sediment	19 (11)	14 (5)
Total		30 (16)	52* (35)
B. Density			
Organism Type		Avg. + S.E.	Avg. + S.E.
Epifauna (#/m^2)	Surface	0.72	1.44
Cryptofauna (#/m^2)	Rhodolith	x	1,402,000
Infauna (#/m^2 to 10 cm depth)	Sediment	1,466	2,473
Estimated total abundance (#/m^2)		1,467	1,404,474

Note: x = subhabitat not present; S.E. = standard error.

* Total number of species in the rhodolith bed is 52 owing to 6 shared species between cryptofauna and infauna.

largely a consequence of the diverse and relatively stable subhabitats provided by the hard, complex structure of rhodoliths and rhodolith-derived sediment (Weber-Van Bosse and Foslie 1904). Subsequent studies have come to similar conclusions based on sampling a number of different sites (reviewed in Steller et al. 2003). Using stratified sampling of a bed in Bahía Concepción (bed 24 in fig. 3.1) with high cover of living rhodoliths, Steller et al. (2003) found that overall rhodolith bed biodiversity was 1.7 times as high as that in the surrounding sand flats and that the greatest contribution to bed diversity was from the cryptofauna: invertebrates that live on, among, and bored into the branches of living rhodoliths (table 3.2). Cryptofaunal diversity also increased as the branch density and volume of the rhodoliths increased, relationships of considerable importance to conservation (discussed in a later section). Steller et al. (2003) estimated that the densities of organisms (>0.5 mm) inside rhodolith beds (table 3.2) are much higher than reported in studies from sites in other regions of the world. This is probably the result of high rhodolith density, the wide range size of classes of highly branched rhodoliths, and especially the high abundance of cryptofauna in rhodoliths at one of the Bahía Concepción sites.

Two additional studies have examined patterns of cryptofauna abundance and distribution in *L. margaritae* and *N. trichotomum* in the southwestern Gulf of California. A comprehensive site (two wave beds and two current beds; beds 33, 42, 43, and 48 in fig. 3.1) and seasonal (winter vs. summer) study of primarily *L. margaritae* cryptofauna by Medina-López (1999) found 118 taxa in the 160 rhodoliths collected. There were about 10 taxa and 32 individuals per rhodolith, with crustaceans being most diverse and abundant. There were surprisingly few differences between bed types or seasons. Hinojosa-Arango and Riosmena-Rodríguez (2004) found a total of 104 cryptofaunal taxa (52 identified to species) in 120 large (4–5 cm diameter) individual rhodoliths, 100 *L. margaritae* and 20 *N. trichotomum* with similar branch densities, collected from two sites in Bahía Concepción and bed 33 (fig. 3.1) at Isla Coronado. As found by Medina-López (1999), crustaceans were the dominant group in all samples. Cryptofaunal diversity and abundance were similar in *L. margaritae* from the two sites in Bahía Concepción, averaging about 10 taxa and 75 individuals per rhodolith. Diversity and abundance were similar between the two rhodolith species from Isla Coronados, with the exception of higher crustacean abundance in *N. trichotomum.*

These studies suggest generally similar diversity and abundance among the cryptofauna in rhodolith beds in the Gulf dominated by *L. margaritae/N. trichotomum.* These rhodoliths have similar morphology. That rhodolith species with substantially different morphologies may harbor different cryptofauna has been shown by surveys at Cabo Los Machos near the mouth of Bahía Concepción (Foster et al. 2007). The wave bed at this site (bed 21 in fig. 3.1) is dominated by high densities (10–50/m^2) of *L. muellerii* (fig. 3.3C) growing in patches of carbonate sediment intermixed with cobbles and boulders. The individual rhodoliths have thick (~1 cm), densely packed branches, so that the cryptofauna consists primarily of boring organisms and other species that occupy spaces produced by borers. Fifteen rhodoliths, five from each of three size classes, were collected at random within the site and then gently broken apart; from these, the cryptofauna were extracted and identified. A total of 117 taxa (107 species) were found, with polychaetes the most diverse group (49 taxa). The average large rhodolith (8.5–12 cm diameter) contained 37.8 (S.D. = 5.6) taxa and often contained mantis (*Neogonodactylus zacae* Manning) and ghost (*Pomatogebia rugosa* Lockington) shrimp. This average is similar to the diversity found in three large *L. margaritae* (table 3.2); and the total diversity from all specimens combined, although from many fewer rhodoliths, is similar to that found by Medina-López (1999) and Hinojosa-Arango and Riosmena-Rodríguez (2004). Although from different sampling methods and times, data from all of these cryptofauna surveys suggest that rhodolith cryptofaunal diversity is high and is similar among species with similar morphologies. Future understanding of rhodolith bed structure and its variation in space and time would be facilitated by the use of similar sampling designs and methods, as described in Steller et al. (2007b).

The species richness of rhodolith beds and associated sediment in the Gulf is further illustrated by the work of Cintra-Buenrostro et al. (2002). They found a total of 142 mollusc taxa larger than 1 cm (85 bivalves, 57 gastropods) in 1 × 0.2 m core samples taken from ten living rhodolith beds in the Bahía Concepción / Punta Chivato region and from the bed in Canal de San Lorenzo (bed 48 in fig. 3.1). Moreover, in addition to rhodoliths, other commonly attached organisms are often found living unattached in rhodolith beds. Reyes-Bonilla et al. (1997) found five species of free-living corals (coralliths) in rhodolith beds in the southwestern Gulf, including several new distributional records. Two bryolith (free-living bryozoans) species were also reported by James et al. (2006) at rhodolith sites in the central and southwestern Gulf.

Considerable taxonomic work remains to be done on invertebrate collections from rhodolith beds in the Gulf, and such work promises to reveal new species and biogeographic insights. For example, Clark (2000) identified eight species of small (2–10 mm) cryptofaunal chitons collected primarily from *L. margaritae* in the southwestern Gulf, four of which were new to science. Clark commented that the chiton fauna of these rhodolith beds is "particularly rich and diverse." It is clear that rhodolith beds and particularly individual rhodoliths are "diversity hot spots" in the Gulf of California.

In addition to the rhodoliths themselves, rhodolith beds often harbor a diversity of other macroalgae, and the seasonal changes in oceanographic conditions in the Gulf of California that are typical of subtropical areas (Alvarez-Borrego 1983) appear to most strongly affect the diversity and cover of these associated organisms. Steller et al. (2003) found as many as thirty associated macroalgal species per bed in four *L. margaritae* beds in the southwestern Gulf in winter but no more than eight species per bed in summer, which is inverse to the pattern observed by Maggs (1983) for maerl beds in Ireland. The winter macroalgal diversity was greater in deeper beds than in shallower beds, perhaps because of lower water temperatures (to 17°C), lower variation in water temperatures, reduced surge, and higher nutrients. Microalgal cover (diatoms and cyanobacteria) was high in summer. The elevated summer water temperatures (to 30°C), potential anoxic conditions at some sites, and thermal stratification likely influence the low summer algal diversity, leading to the disappearance of most foliose red algae and the dominance of more tolerant microalgae (Lechuga-Deveze et al. 2000). Similar seasonal patterns in macroalgal diversity have been described for rocky shores in the southern Gulf of California (Paul-Chavez and Riosmena-Rodríguez 2000) and in Bahía Concepción (Mateo-Cid et al. 1993).

The fishes associated with rhodolith beds are not well known. Our qualitative observations suggest that fish diversity, at least of larger species that spend considerable time in the water column, is low. This is probably because, at spatial scales larger than ~5–10 cm, most rhodolith beds are largely two-dimensional, providing few structures to associate with. Reports of numerous larger fishes in rhodolith beds (see, e.g., Aburto-Oropeza and Balart 2001) may be an artifact of the beds being in close proximity to rocky or other reefs that provide macrostructure. We have noted numerous gobies and blennies on, in, and among rhodoliths, and these fishes may be quite diverse given the complex, small-scale structure of rhodolith beds. This would be an excellent topic for future study. Large, bottom-dwelling or burrowing species can be quite common, including tiger snake eels (*Myrichthys maculosus* Curvier), bullseye electric rays (*Diplobatis ommanata* Jordan and Gilbert) and other rays, and Cortez garden eels (*Taeniconger digueti* Pellegrin).

The distribution of rhodolith beds appears to be largely constrained by abiotic variables (discussed previously). However, within-bed community structure, and perhaps even bed persistence, may be strongly affected by species interactions. For example, bioturbation by invertebrates and fish may, by moving rhodoliths and resuspending fine sediment, facilitate rhodolith growth and bed maintenance (Marrack 1999; James 2000). Nutrients produced by invertebrates may also be important to rhodolith growth. The four new chiton species described by Clark (2000) may require the habitat provided by living rhodolith branches for their existence. Bivalves are abundant in rhodolith beds in the Gulf and elsewhere (Steller et al. 2003; Cintra-Buenrostro et al. 2002; Hall-Spencer 1998, 1999). This may result from larval settlement preferences for coralline algae (Kamenos et al. 2004) or other structured or large-grain substrates, enhanced growth rates of species inside rhodolith beds, and the refuge from predation that beds provide (Steller 2003).

Yet these positive attributes of rhodolith beds create a conservation dilemma because of the degradation that results from commercial fishing (Hall-Spencer 1998, 1999; Hall-Spencer and Moore 2000). Algal species typically found in the diet of the green turtle (López-Mendilaharsu et al. 2003) and its foraging grounds (reviewed by López-Mendilaharsu et al. 2005) in Bahía Magdalena are commonly associated with rhodolith beds or rhodolith sediment (Iglesias-Prieto et al. 2003) and are uncommon in other habitats (Riosmena-Rodríguez, unpublished data). Interaction between biotic and abiotic factors may be especially important. For example, the water motion patterns responsible for rhodolith maintenance may also influence larval delivery of a number of associated species to beds. Such interactions remain to be investigated.

Sediment Production and Fossil Deposits

Foster (2001) described a model for the structure and dynamics of rhodoliths, rhodolith beds, and the geological structures derived from them. Rhodolith fragments can be broken from larger thalli during disturbance by water motion, bioturbation, grazing, and trawling. These processes also produce sand and smaller calcareous particles. Burial in a bed can occur from sedimentation and during/after disturbance. Rhodolith beds are recognized worldwide as "carbonate factories" (James 1979; Nelson 1988) and as important sources of modern carbonate sediments (van Andel 1964; Schlanger and Johnston 1969; Halfar et al. 2000a; Hetzinger et al. 2006) and fossil carbonate sediments (Dorsey 1997; Foster et al. 1997; Libbey and Johnson 1997; Meldahl et al. 1997; Cintra-Buenrostro et al. 2002; de Diego-Forbis et al. 2004) in the Gulf of California.

Rhodoliths, rhodolith fragments, and the hard parts of corals and molluscs are common and often dominant constituents of fossil carbonate deposits in the Gulf. Libbey and Johnson (1997) found that most of the materials in La Palmilla fossil deposits at Punta Chivato are crushed rhodoliths. Sewell et al. (2007) estimate that 86.5 percent of the ~30,000 m^3 of material making up a fossil sand dune on Isla Coronado is rhodolith derived. They estimate that it would take 8,640 rhodoliths 5 cm in diameter to make 1 m^3 of material. Johnson and Hayes (1993) described rhodolith beds as one of the most common elements in Pleistocene deposits. Like living beds in modern nearshore areas, fossil rhodolith deposits are often the dominant elements in Pleistocene deposits in the Loreto area (Dorsey 1997), Bahía Concepción (Meldahl et al. 1997), and Punta Chivato (Simian and Johnson 1997). Quantitative methods used by Russell and Johnson (2000) and de Diego-Forbis et al. (2004) revealed a mixture of molluscs and corals with rhodoliths in fossil records from the Gulf. Studies in the Gulf have used fossil rhodoliths to delimit coastal areas near paleo-islands and other paleoecological settings (Johnson and Ledesma-Vazquez 2001).

Forrest et al. (2005) found that rhodolith beds are involved in determining the geochemistry of hydrothermal gas samples from the El Requeson Fault Zone in Bahía Concepción, suggesting that the thermal decomposition of organic and carbonate-rich sediments is a significant source of the N_2, CO, and CH_4 in the gas. The $d^{13}C$ values of the CH_4 were consistent with derivation of the gas by thermal cracking of marine algal kerogens, or pseudokerogens; and the $d^{13}C$ values of the CO_2 indicate that the dissolution of rhodolith calcium carbonates may contribute to the levels of CO_2 in the gas.

Anthropogenic Disturbance and Conservation

As previously discussed, water motion maintains rhodolith populations in a growing, unattached, and unburied state, and it also influences thallus complexity. However, extreme water motion can result in the destruction of plants. The primary persistent natural disturbances in rhodo-

TABLE 3.3. Species found in rhodolith beds that are protected under "The Mexican Directive for Species under Some Threat" (NOM-ECOL 095).

Taxonomic Group	Species
Cnidaria	*Psammocora stellata*
	Porites panamensis
	Porites sverdrupi
	Fungia curvata
	Fungia distorta
Holothuroidea	*Isostichopus fuscus*
Bivalvia	*Pinctada mazatlanica*
	Spondylus calcifer

lith beds are water motion and sedimentation (Steller and Foster 1995; Foster 2001), bioturbation (Marrack 1999; James 2000), and possibly extreme changes in temperature and nutrient regimes. However, anthropogenic factors including decreased water quality, extraction, and bottom fisheries (De Grave and Whitaker 1999; Hall-Spencer and Moore 2000) can also contribute to the burial and destruction of thalli. Thus, while these beds are dependent upon some degree of disturbance for their maintenance, high levels of disturbance can result in reduced thallus density, size, and structure; loss of living thalli; and ultimately a transition into relatively low-diversity carbonate sand flats. The greatest loss in numbers and unique species is predicted to be in cryptofaunal organisms that associate with complex, hard substrate not found in soft sediments (Steller et al. 2003).

In the Gulf of California, commercial fishers dive from small boats using hookah to harvest scallops and other benthic species. Post-fishing observations in rhodolith beds in Bahía Concepción in May 1991 revealed the almost complete extraction of adult scallops and extensive damage to the rhodolith beds caused by boat anchors, dragging of air hoses across the bottom, and direct diver disturbance while harvesting (Steller and Foster, pers. obs.). Large areas of broken rhodolith fragments and areas cleared of living rhodoliths were common in previously intact beds (100 percent rhodolith cover) where fishing operations had taken place. Trawling for shrimp in Gulf of California rhodolith beds has been documented photographically

TABLE 3.4. Species associated with rhodolith beds in the Gulf of California that are currently exploited for fisheries.

Taxonomic Group	Species
Crustacea	*Litopenaeus stylirostris*
	Litopenaeus vannamei
	Litopenaeus occidentalis
	Farfantepenaeus californiensis
Gastropoda	*Stenoplax mariposa*
	Retusa xystrum
	Hexaplex (= Muricanthus) nigritus
Bivalvia	*Megapitaria squalida*
	Argopecten ventricosus (= A. circularis)
	Lyropecten subnodosus
	Pinna rugosa
	Modiolus capax

(Hall 1992), but its extent and severity is unknown. It is clear that trawling for bivalves in rhodolith beds reduces epifauna and in faunal diversity through large-scale homogenizing of the benthos (Hall-Spencer 1999).

It is not only the direct removal of thalli but also the alteration of their morphology that can affect species diversity and abundance. Thus, even disturbances with relatively low impacts may have severe effects at the community level. Whereas the direct impact of species harvest and the loss of habitat structure on species abundance can be measured, indirect effects are more difficult to estimate. For example, the direct effects of trawling or bottom fishing for scallops and other benthic species include decreased abundance of the target species, but it also indirectly influences community structure by decreasing fragile or omnivorous species, increasing species that feed on soft-sediment suspension (Grall and Glemarec 1997), and increasing the number of scavengers (Hall-Spencer and Moore 2000). Long-term effects are difficult to estimate, but it is clear from extremely low coralline growth rates—both worldwide and in the Gulf—that recovery of the substrate after disturbance is likely to be very slow. Findings from the Gulf of California demonstrate that (1) rhodolith beds are important in maintaining the diversity and abundance of benthic species, including economically important species; and (2) as also found by Grall and Glemarec

TABLE 3.5. Comparative analysis of the conservation status of rhodolith/maerl habitat and rhodolith-forming species.

UK[a]	European Union[a]	Australia[b]	New Zealand[c]
Rhodolith/maerl beds as habitat			
Protected under the Special Areas of Conservation (SACs) in England, Scotland, and Northern Ireland	Legally protected by the European Community Council Directive[a]	Protected as part of Great Australian Bight Marine Park	Protected as part of the Kapiti Marine Reserve
Species protected			
Phymatolithon calcareum (Pallas) Adey & McKibbin, *Lithothamnion corallioides* Crouan & Crouan; *Lithothamnion glaciale* Kjellman has been proposed for listing on EC Habitats Directive	*P. calcareum* & *L. corallioides*	No species are specifically listed for protection	No species are specifically listed for protection

[a] Council Directive 92/43/EEC (1992); protected under Annex V.
[b] Director of National Parks (2005).
[c] Department of Conservation (1998).

(1997), maintenance of high cover of intact, branching rhodolith thalli is imperative to maintaining high diversity.

As a consequence of our long-term experience working with rhodoliths, we have been able to identify at least eight species that occur primarily in rhodolith communities (table 3.3) under the Mexican Official Norm to protect species (*Diario Oficial* 2002a). Rhodolith beds are also habitats of special consideration because at least another twelve exploited species (table 3.4) are usually found in this environment (*Diario Oficial* 2004). Rhodolith beds and rhodolith-forming species are listed as worthy of conservation in several parts of the world (table 3.5), are protected by Special Areas of Conservation in Europe and the United Kingdom, and must be considered in the development of marine parks. In the original general model for designing marine reserves in the Gulf, Sala et al. (2002) used rhodolith beds as one of the relevant habitats for conservation. However, only a few marine protected areas (Loreto, Isla Espíritu Santo, Isla San Pedro Mártir, and

Islas Revillagigedos) have included rhodoliths as habitats for conservation. Rhodoliths are widely distributed in the Gulf of California and elsewhere in the waters of Mexico, and greater consideration of this community should be an essential part of marine conservation strategies.

ACKNOWLEDGMENTS

We thank L. McConnico for sharing some of her unpublished data and R. Brusca for his encouragement, help, and patience. The research upon which this review is based would not have been possible without the support of BCSES, CONABIO, CONACYT, USA/México Fulbright Program, Inter American Institute for Global Change Research, Moss Landing Marine Laboratories, UC Santa Cruz, National Geographic Society, Packard Foundation, and UCMEXUS over the last 15 years. We also thank P. Raimondi, J. Ruprow, C. Sánchez, E. Ochoa, and O. Aburto for their ongoing interest in rhodolith beds and reports of many new locations.

CHAPTER 4

Invertebrate Biodiversity and Conservation in the Gulf of California

RICHARD C. BRUSCA
AND MICHEL E. HENDRICKX

Summary

An analysis of the Macrofauna Golfo invertebrate database indicates that the Gulf is home to over 4,900 species of named and described invertebrates. This is estimated to be about 70 percent of the actual invertebrate fauna of the Gulf of California. The most poorly known regions for invertebrates are the open sea and the deeper (below the continental shelf) benthic environment. In the intertidal region of the Gulf 2,158 species occur, although only 45 of these are strictly intertidal in distribution. Thirty-six hundred species occur at or above the 30 m isobath, and 4,078 species occur on or over the continental shelf (200 m and above). Most invertebrates recorded from the Gulf, 4,350 species, inhabit benthic habitats. There have been 329 species recorded from coastal lagoons in the Gulf, 260 of these from mangrove lagoons.

In general, invertebrate diversity increases from north to south in the Gulf. The unique oceanographic attributes and broad seasonal water temperature range of the Northern Gulf create tropical marine conditions in the summer but warm-temperate conditions in the winter. This provides a refuge for many disjunct warm-temperate (i.e., Californian) species in the upper Gulf that are not found in the Central or Southern Gulf. It perhaps also explains the high invertebrate endemicity seen in the Northern Gulf (128 species). Most of the Gulf's invertebrate fauna, however, is tropical and derived from the Tropical Eastern Pacific, and dozens of these species have a transisthmian distribution with populations in the Caribbean. Relative species diversity of invertebrates can be predicted based on substrate and habitat type, although (aside from a few dozen very common species) the actual presence or absence of a species is difficult to predict and largely stochastic in nature. Overall, the most diverse phyla in the Gulf are Mollusca (2,198 species), Arthropoda (1,062 species), Annelida (722 species), Echinodermata (263 species), and Cnidaria (262 species).

The greatest threats to invertebrates in the Gulf are bottom trawling, hand-collecting by humans during low tides, coastal development, and pollution. In addition, artisanal harvesting of molluscs and swimming crabs (*Callinectes*) is a growing threat. On mainland shores, most large-bodied species at most localities are now gone from the intertidal region. Diversity on islands and on some largely inaccessible stretches of shore (especially on the eastern coast of the Baja California peninsula) are critically important refuges for littoral species now largely extirpated from mainland coasts. Industrial shrimp trawling is probably the most destructive form of fishery in the Gulf to invertebrates and also to the ecological integrity of the seafloor. Estero-based aquaculture is also harmful to invertebrate habitats, and the relocation of shrimp farms inland may ultimately be the only way to protect the fragile coastal lagoons of the Gulf and at the same time find a use for old agricultural land that had been ruined by salinization.

Introduction

Much of the information presented in this chapter was mined from the Macrofauna Golfo Project database, the product of a 10-year effort by many scientists in Mexico and the United States. The database catalogs every macrofaunal species (i.e., animals larger than 5 mm; thus ostracods and copepods are excluded) known to occur in the Gulf of California (*Mar de Cortés,* Sea of Cortez). The project was funded by grants from several organizations, including Conservation International, CONABIO (Comisión Nacional para el Uso y Conocimiento de la Biodiversidad), CIAD (Centro de Investigación en Alimentación y Desarrollo), Pronatura-México, and the Arizona-Sonora Desert Museum. Lead investigators for the *Macrofauna Golfo Project* were: Richard C. Brusca, Lloyd T. Findley, Philip A. Hastings, Michel E. Hendrickx, Jorge Torre, and Albert van der Heiden. The database provides information on taxonomy/classification, geographic distribution (in the Gulf and worldwide), depth, and habitat for about 6,000 species (4,916 invertebrates). An abridged version of the invertebrate data was published by Hendrickx et al. (2005). The complete invertebrate database is available at www.desertmuseum.org/center/seaofcortez/database.php. See Hastings et al. (chapter 5 in this volume) for more information on the Macrofauna Golfo Project, including descriptions of the four recognized biogeographic regions (Northern Gulf, Central Gulf, Southern Gulf, and Southwestern Baja California Sur). For a historical review of invertebrate research in the Gulf of California, see Brusca (2004a,b), Brusca and Bryner (2004), or Brusca et al. (2005). For a review of invertebrate diversity and conservation issues in the Northern Gulf of California, see Brusca (2007c). In this and the following chapter (on fishes), the terms Northern Gulf, Central Gulf, and Southern Gulf refer specifically to those biogeographic regions designated by the Macrofauna Golfo Project and database.

Macroinvertebrate Biodiversity in the Gulf of California

Origins of the Invertebrate Fauna

The invertebrate fauna of the Gulf of California is derived from four sources. Most species are tropical eastern Pacific in origin, whereas others had their origin in the Caribbean Sea or the Transisthmian (Panamanian) fauna (before the uplift of the Panama Isthmus), the temperate shores of California (during the 15–20 glacial periods that pushed cold waters southward and into the Gulf during the past 2 million years), and even across the vast stretch of the Pacific Ocean from the tropical Indo-west Pacific (Walker 1960; Brusca and Wallerstein 1979a; Thomson et al. 1979, 2000; Brusca 1980, 2004a, 2007c; Brusca and Findley 2005). These diverse biotic sources have enriched the diversity of the Gulf of California since its opening 5–6 million years ago.

Among the dozens of contemporary transisthmian species identified to date in the Gulf are: *Nereis riisi, Chloeia viridis, Eurythoe complanata,* and *Spirobranchus giganteus* (Polychaeta); *Phascolosoma perlucens* (Sipuncula); *Conopea galeata* (a gorgonian-commensal barnacle); *Rocinela signata, Cirolana parva,* and *Excirolana mayana* (Isopoda); *Ambidexter symmetricus* (a caridean shrimp); *Sicyonia laevigata* (a penaeoid shrimp); *Acanthonyx petiveri, Cycloes bairdii, Pilumnus reticulatus, Aratus pisonii, Geograpsus lividus,* and *Percnon gibbesi* (brachyuran crabs); *Aplysia parvula* and *Pleurobranchus areolatum* (Opisthobranchia); *Encope grandis* (Echinoidea); and *Molgula occidentalis* (Ascidiacea).

During past glacial events, a number of temperate-water "Californian species" were able to extend their ranges into the Gulf when cold isotherms pushed around the tip of the Baja California peninsula. When water temperatures warmed during subsequent interglacial periods, populations of these coldwater-adapted species became trapped in the Northern Gulf. Most of these cold-water species disappeared from the Gulf during the warmer periods (as today), but a few were adaptable enough to survive. The published literature suggests there are several dozen of these California–Northern Gulf disjunct temperate species. However, some of these records probably represent species whose Southern Gulf and/or southern peninsular distribution has simply not yet been documented (i.e., they probably range throughout coastal waters of the Baja California peninsula). Some others probably represent incorrect identifications (e.g., the crabs *Hemigrapsus oregonensis* and *Hemigrapsus nudus* as well as a number of nudibranchs and chitons). However, probably valid California–Northern Gulf disjunct species include: the long-fingered tidepool shrimp, *Betaeus longidactylus;* the hydroids *Plumularia reversa, Campanularia castellata,* and *Antennularia septata;* the cerianthid anemone *Pachycerianthus aestuari;* the actiniarian anemone *Diadumene leucolena;* the flatworm *Pseudostylochus burchami;* the nemerteans *Baseodiscus punnetti* and *Cerebratulus lineolatus;* the polychaete *Polydora nuchalis;* the echiuran *Urechis caupo;* the amphipod *Corophium uenoi;* the muricid snail, *Pteropurpura macroptera;* the opisthobranch, *Aplysia vaccaria* (the world's largest gastropod); the scaphopods, *Dentalium pretiosum* and *Dentalium vallicolens;* the sea stars, *Henricia aspera* and *Odontaster crassus;* the brittle star, *Ophiopholis longispina;* and the bryozoan, *Microporella cribosa.*

It is tempting to speculate that some California–Northern Gulf species have evolved to the point of separation and now represent California–Gulf sister-species pairs. However, no phylogenetic studies on regional invertebrates have shown this.

Some temperate (Californian) species maintain continuous distributions around the Baja California peninsula and throughout the Gulf, including the following free-living coastal species: *Ophiodromus pugetten-*

sis, Syllis elongata, Gycera tesselata, Diopatra ornata, Diopatra splendidissima (Polychaeta); *Sipunculus nudus* (Sipuncula); *Califanthura squamosissima* and *Paranthura elegans* (Isopoda).

Patterns of Invertebrate Diversity in the Gulf of California

The accumulation of species diversity since the Gulf of California opened has produced one of the most biologically rich marine regions on earth. The diverse benthic habitats and the productive pelagic waters of the Gulf are famous for supporting high numbers of species and large population sizes among many marine taxa: invertebrates, fishes, marine mammals, sea turtles, and marine birds (Brusca et al. 2005; various chapters in this volume). Nearly half of Mexico's fisheries production comes from the Gulf of California (Cisneros-Mata, chapter 6 in this volume).

Invertebrate community composition at any given locality in the Gulf comprises a mix of predictable species combined with a much larger and unpredictable suite of species, where the unpredictability is driven by complex networks of interacting physical and biological factors (e.g., short-term oceanographic and weather events, spawning and recruitment success of local species, variable current regimes). However, *relative* species diversity in the Gulf is quite predictable and largely a function of habitat and substrate type. Benthic invertebrate species diversity (i.e., species richness) is highest in the rocky littoral (intertidal) region, on relatively stable shores, and on intertidal or offshore bottoms composed of softer sedimentary rocks such as sandstone and beachrock ("coquina") or eroded volcanic tuffs and rhyolites. Benthic invertebrate diversity is lowest on beaches composed of smooth hard rocks such as granites and basalts and on unstable beaches of coarse sand or cobble, the latter having perhaps the lowest (benthic) diversity of any coastal habitat. Areas that have a variety of substrate types harbor more species than do more homogeneous ones. Today, diversity on islands in the Gulf far exceeds that on mainland shores, but this is due largely to human impacts and the decline of biodiversity on accessible coastlines during the past few decades. Today, the islands of the Gulf are critically important refugia for species that have been extirpated from the mainland coast.

In the Northern Gulf, species diversity and composition are strongly influenced by seasonal climatic and oceanographic conditions; these change markedly through the year, creating extreme seasonal variations in seawater temperatures (Alvarez-Borrego 1983; Maluf 1983; Bray and Robles 1991; Lavín et al. 1998; Brusca 2007c).[1] As a result, Northern Gulf waters are essentially a warm-temperate marine environment during the winter but a tropical marine environment during the summer. The distinct seasonal species turnover in invertebrates and algae in this region is striking, as tropical species (e.g., *Gnathophyllum panamense, Ocypode occidentalis, Pentaceraster cumingi, Nidorellia armata, Callinectes bellicosus* and other portunid crabs) disappear during the cold winters and temperate species (e.g., *Pachygrapsus crassipes, Aplysia californica, Betaeus longidactylus*) vanish during the warm summers. The Central Gulf shows far less seasonality in water temperatures, and the Southern Gulf shows hardly any seasonality in environmental conditions or species diversity.

Overall, marine macroinvertebrate diversity in the Gulf of California is exceptionally high: 4,916 named species, with ~30–40 new species being described annually (tables 4.1–4.3). Because of the presence of many undescribed invertebrate species, including many members of the planktonic and offshore communities, this total is estimated to be approximately 70 percent of the actual macroinvertebrate diversity of the Gulf (table 4.3). The most poorly known invertebrate faunas are those of the open sea and of the deeper (>300 m) benthic environments. Faunal diversity decreases gradually from the south to the north.

Overall macroinvertebrate endemicity in the Gulf is 16 percent (782 species). At the phylum level, highest endemism occurs in Brachiopoda (80 percent), Ctenophora (50 percent), Platyhelminthes (41 percent),

TABLE 4.1. Macroinvertebrate species known from faunal regions of the Gulf of California.

Faunal Region	Total	Percentage of Total Gulf Species
Northern Gulf	2,275	46.2
Central Gulf	3,324	67.6
Southern Gulf	3,173	64.6
Upper Gulf of California and Colorado River Delta Biosphere Reserve (subregion of Northern Gulf)	1,048	21.3
Southwestern Baja California Sur	676	13.8

TABLE 4.2 Ecological distribution of macroinvertebrate species in the Gulf of California.

Habitat	Number of Species	Percentage of Total Gulf Species
Sandy bottoms (all depths)	2,026	41.2
Sandy intertidal habitats	1,928	19.9
Rocky bottoms (all depths)	1,644	33.4
Rocky intertidal habitats	1,075	21.9
Mud bottoms (all depths)	1,313	26.7
Coastal lagoons (esteros and estuaries)	329	6.7
Mangrove lagoons	260	5.3
Associated directly with mangrove plants	41	0.8
Coral reefs	172	3.5
Benthic species (all depths)	4,350	88.5
Pelagic species (all depths)	262	5.3
Occurring in the intertidal zone	2,258	43.9
Occurring *only* in the intertidal zone	45	0.9
Occurring above the 30 m isobath	3,600	73.2
Occurring *only* above the 30 m isobath	918	18.7
Occurring above the 100 m isobath	4,019	81.8
Occurring *only* above the 100 m isobath	2,402	48.9
Occurring on and over continental shelf (above 200 m isobath)	4,078	83.0
Benthic species on continental shelf (above 200 m isobath)	4,044	82.3
Pelagic species over continental shelf (above 200 m isobath)	32	0.7
Species occurring *only* on continental shelf (above 200 m isobath)	3,016	61.4

Note: Some species have been recorded from more than one habitat.

Echiura (25 percent), and Mollusca (21 percent). At lower taxonomic levels, highest endemism occurs among Anthozoa (34 percent), Polyplacophora (26 percent), Gastropoda (26 percent), Porifera (25 percent), and Cumacea (25 percent). However, these figures should be viewed with caution because many taxa (e.g., Porifera, Brachiopoda, Cnidaria, Ctenophora, Platyhelminthes, Echiura, Cumacea, Tanaidacea, micromolluscs, Urochordata, Hemichordata) are poorly studied in the Gulf and the Tropical Eastern Pacific in general.

In the Northern Gulf, notably high biodiversity occurs on the very limited intertidal beachrock ("coquina") formations that occur at just four sites: Puerto Peñasco and Punta Borrascosa (Sonora), and San Felipe and

TABLE 4.3. Known and predicted species diversity in major macroinvertebrate groups in the (entire) Gulf of California.

Phyla and Major Subgroups	Number of Species Recorded	Number of Species Predicted
Porifera	*115*	575
Cnidaria	*262*	526
Hydrozoa	147	292
Anthozoa	108	204
Scyphozoa	7	50
Ctenophora	*4*	20
Platyhelminthes	*22*	110
Nemertea	*17*	30
Sipuncula	*11*	22
Echiura	*4*	7
Annelida	*722*	820
Oligochaeta	1	3
Polychaeta	720	816
Pogonophora	1	1
Arthropoda	*1,062*	1,522
Pycnogonida	15	45
Cirripedia	48	47
Copepoda	?	25
Ostracoda	?	25
Stomatopoda	28	33
Mysida	3	10
Amphipoda	232	464
Isopoda	82	110
Tanaidacea	2	20
Cumacea	8	20
Euphausiacea	14	20
Dendrobranchiata	26	42
Stenopodidea	2	4
Caridea	130	145
Astacidea	1	1
Thalassinidea	20	24
Palinura	8	9
Anomura	131	192
Brachyura	301	336
Mollusca	*2,198*	*2,590*
Monoplacophora	1	2
Polyplacophora	57	62
Gastropoda	1,534	1,630
Bivalvia	566	848
Scaphopoda	20	25

(*continues*)

TABLE 4.3. *(continued)*

Phyla and Major Subgroups	Number of Species Recorded	Number of Species Predicted
Cephalopoda	20	23
Bryozoa (Ectoprocta)	*169*	*338*
Brachiopoda	*5*	*7*
Echinodermata	*263*	*300*
Chaetognatha	*20*	*25*
Hemichordata	*3*	*5*
Chordata	*39*	*292*
Ascidiacea	17	170
Appendicularia	21	40
Cephalochordata	1	2
TOTALS	4,916	7,189

Note: Phylum-level data in *italics.*

Coloradito (Baja California). These small, rare, eroding beachrock habitats harbor disproportionately high species diversity, giving them a priority need for protection. High diversity is also found at Isla San Jorge and Rocas Consag and also on offshore (subtidal) rock outcroppings of the northern Sonora coastal shelf. Exceptionally high biodiversity, including rich pelagic diversity (and abundance) driven by year-round upwelling, distinguishes the Midriff Islands. All of these high-diversity sites serve as important invertebrate refugia and recruitment sources for mainland shores.

The entire benthic region of the Northern Gulf formerly maintained a high species diversity and biomass. However, in subtidal areas susceptible to bottom trawling for commercial shrimps (i.e., shallower than 100 m), much diversity has been lost over the past 50 years as a result of excessive anthropogenic disturbance (see below). Unfortunately, we have almost no knowledge regarding community composition and food web structure for the Northern Gulf's offshore benthic or pelagic communities. One of the most pressing research needs is to achieve an understanding of benthic community structure in this region and an enhanced sense of how profound the effects of bottom trawling have been on this system.

Forty-six percent of the Gulf's macroinvertebrate species occur in the Northern Gulf (2,275 species), and 1,048 (21 percent of the Gulf species) are known from the Upper Gulf of California/Colorado River Delta Biosphere Reserve. In the Northern Gulf, molluscs (1,008 species), arthropods (514 species), and polychaete annelids (285 species) are the most diverse phyla. Within the Mollusca, gastropods and bivalves stand out with 660 and 287 species, respectively. Among Arthropoda, brachyuran crabs, amphipods, and isopods are notably diverse with 167, 126, and 41 species, respectively. Of the macroinvertebrate species known from the Northern Gulf, 128 (5.7 percent) are unique to that area.

Among the species endemic to only the Northern Gulf are two elegant and giant aphroditid polychaetes (*Aphrodita mexicana, A. sonorae*), sometimes called "sea mice," both of which are now greatly reduced in numbers and threatened as a result of excessive bottom (shrimp) trawling. The beautiful coral *Astrangia sanfelipensis,* today known only from the spatially restricted San Felipe/Coloradito "coquina reefs," is also threatened by habitat degradation at those two upper Gulf sites.

A total of 3,324 macroinvertebrate species has been recorded from the Central Gulf (68 percent of the Gulf species), and 3,173 occur in the Southern Gulf (65 percent of the Gulf species). In addition, 676 species (14 percent of the Gulf's invertebrate species) extend their ranges around Baja California's southern tip and up the Pacific coast, between Cabo San Lucas and the northernmost limit of the Bahía Magdalena lagoon complex—a region that extends the Gulf fauna outside the physical boundaries of the Gulf of California.

Examination of tables 4.1 through 4.3 reveals further interesting patterns of invertebrate biodiversity in the Gulf. Although only a single true coral reef occurs in the Gulf (at Bahía Pulmo, south of La Paz: Brusca and Thomson 1977; Robinson and Thomson 1992), 40 species of corals (order Scleractinia) occur in the Gulf (17 in the Northern Gulf, 30 in the Central Gulf, 26 in the Southern Gulf); this makes the coral diversity richer than that of, say, sea anemones (order Actiniaria; 22 species in the Gulf). Corals are most commonly seen on the Gulf's islands, where they are more protected than on mainland shores. Eighteen hermatypic (zooxanthellate) coral species inhabit the Gulf in six genera (*Fungia, Leptoseris, Pavona, Pocillopora, Porites, Psammocora*). Good, seemingly young coral head development can also be seen in Bahía San Gabriel, on Isla Espíritu Santo, where they could be viewed as "patch reefs." The richest area of coral development is in the southwestern part of the Gulf, especially on the islands along that peninsular coastline.

Notably rich diversity also occurs among gastropods (1,534 species), polychaetes (720 species), bivalves (566 species), true (brachyuran) crabs (301 species), echinoderms (263 species), bryozoans (169 species), hydroids (147 species), tidepool (caridean) shrimps (130 species), sponges (115 species), gammaridean amphipods (111 species), hyperiidean amphipods (109 species), isopods (82 species), chitons (57 species), and porcelain crabs (51 species). Also notable is a single species of intertidal marine earthworm (Annelida: Oligochaeta), *Bacescuella parvithecata,* which occurs with rarity in the Northern and Central Gulf.

The 18 species of sea fans (Anthozoa: Gorgonacea) reported from the Gulf (none of which is endemic) are only a small percentage of the actual gorgonian diversity, and we have observed many undescribed species in the region. Similarly, the 7 species of jellyfish reported from the Gulf clearly represent a small fraction—perhaps only 15 percent—of what is actually there. Similarly, the 38 species of tunicates (subphylum Urochordata) reported from the Gulf probably represent only about 15 percent of the actual diversity in this region. The 115 species of sponges (Porifera) recorded from the Gulf probably represent about 20 percent of the region's actual sponge diversity.

Table 4.2 reveals some interesting ecological relationships. As would be expected, most macroinvertebrate species known from the Gulf of California have been reported from shallow waters. There are 2,158 species in the intertidal zone (44 percent of all Gulf species), but of these only 45 (2 percent) are *strictly* intertidal in their distribution. Thirty-six hundred species (73.2 percent) occur at or above the 30 m isobath, and 918 (18.7 percent) occur *only* above the 30 m isobath. There are 4,078 species (83 percent of all Gulf species) occurring on or over the continental shelf (200 m and above), and 3,016 of these (61.4 percent of all Gulf species) occur *only* on or over the continental shelf (i.e., do not occur below the 200 m isobath).

Most macroinvertebrate species known from the Gulf of California—4,350 species (88.5 percent of the Gulf total macroinvertebrate fauna)—are benthic. Only 262 pelagic species (5.3 percent of the total) have been reported, an artificially low number because many undescribed species occur in this region (and because, at the time of this writing, the Macrofauna Golfo Project database excluded ostracods and copepods).

If all depths are considered, then most invertebrates occur on sandy bottoms—2,026 species (41.2 percent of the Gulf's total macroinvertebrate fauna). Rocky bottoms (all depths) harbor 1,644 species (33.4 percent of the total fauna). If only the intertidal zone is examined then these percentages reverse, and rocky intertidal regions harbor 1,075 species (21.9 percent of the Gulf's total macroinvertebrate fauna, including three dozen or so species that occur strictly as algal epiphytes in rocky habitats) whereas intertidal sandy beaches harbor 928 species (19.9 percent). Mud bottoms (all depths) harbor 1,313 species (26.7 percent of the Gulf's total macroinvertebrate fauna).

Coastal lagoons and *esteros* (moderately hypersaline coastal, or tidal, lagoons) are notably diverse areas, and these habitats provide extremely important nursery and feeding grounds for the young of many coastal fish and shellfish species, including most commercial finfish and shrimp that are traditionally exploited by the Gulf's fisheries. There have been no published, comprehensive (i.e., all-taxa) surveys of any esteros, or other wetlands, in the Gulf of California. These coastal lagoons (estuaries and esteros) are home to at least 329 species (6.7 percent of the Gulf's total macroinvertebrate diversity); of these, 260 are from mangrove lagoons, where 41 of these species are reported as specifically associated with the mangrove plants themselves (e.g., oysters, sponges, tunicates, and other invertebrates that inhabit mangrove roots and stalks). Whitmore et al. (2005) reported 212 species of invertebrates from mangrove lagoons of Baja California Sur. However, because many undescribed species of sponges and tunicates occur in mangrove lagoons, most living on the mangroves themselves, these numbers underestimate the actual level of diversity in that ecosystem.

Some Comparisons to Other Faunal Regions

So far as we are aware, no other comparable marine region in the world has a database of every known macroinvertebrate species. However, the fauna of the Mediterranean Sea is very well known (far better than the Gulf of California) and shares many historical and oceanographic similarities that make it useful for comparison. Unlike the Gulf, the Mediterranean Sea is largely physically isolated from the tropical waters of the Old World. However, numerous Mediterranean-occurring invertebrate species have emigrated from the Red Sea via the Suez Canal since its opening. At least 558 alien species—most entering through the Suez Channel—have been recorded, including 189 molluscs, 99 arthropods, 85 chordates, 85 macroalgae, and 48 polychaetes (Guala et al. 2003; Galil 2008).

The current (post-Messinian) Mediterranean Sea is roughly the same age as the Gulf of California: ~5 million years old (earliest Pliocene). Similarly to the Gulf, which has only one true coral reef, the Mediterranean has no true coral reefs. The Mediterranean also has many aquaculture facilities, although they have been in place much longer than those in the Gulf of California. As a result of this history plus heavy shipping traffic, the Mediterranean Sea has many more exotic and introduced species, from around the world, with which to contend.

Overall, about 6,000 species of benthic invertebrates have been reported from the Mediterranean Sea, which could be viewed as a fairly accurate biodiversity estimate given how well known the region is. About 4,350 benthic invertebrate species have been recorded from the Gulf of California. If the known Gulf benthic species count is assumed to represent about 70 percent of the actual invertebrate diversity, than the actual total is close to 6,165, or about the same as in the Mediterranean.

There have been 649 species of sponges (Porifera) recorded from the Mediterranean Sea (597 Demospongiae, 44 Calcarea, 8 Hexactinellida), 48 percent of which are endemic (Pansisni and Longo 2003). In contrast, 115 species of sponges have so far been recorded from the Gulf of California (113 Demospongiae and 2 Calcarea but no Hexactinellida), 29 (resp. 25) percent of which are endemic. The Caribbean–Central American Atlantic sponge fauna is of about the same diversity as that of the Mediterranean, with 640 species. The Sino-Japanese sponge fauna consists of some 589 species and the Indonesian fauna 965 species (Pronzato 2003). These figures suggest that our estimate of only 20 percent of the Gulf's sponge fauna being described so far is "in the ballpark."

The molluscan fauna of the Mediterranean Sea is often said to be the best known in the world, and 2,042 species are listed from the region: 1,482 gastropods, 410 bivalves, 65 cephalopods, and 16 scaphopods (Bello 2003; Oliverio 2003). Molluscs—also one of the best-known invertebrate phyla in the Gulf of California—share a nearly identical diversity, with 2,198 species (1,534 gastropods, 566 bivalves, 20 cephalopods, 20 scaphopods).

The Gulf's bryozoan (Ectoprocta) fauna is still very incompletely described, with 169 named species compared with 476 species of bryozoans reported from the Mediterranean Sea (Rosso 2003). With just 15 named species, the Gulf's pycnogonid fauna is also far from being fully described; note that 56 species have been reported from the Mediterranean (Chimenz Gusso and Lattanzi 2003). The same can be said for the amphipod fauna of the Gulf, which has yielded 232 species to date compared with 466 species reported from the Mediterranean Sea (Bellan-Santini and Ruffo 2003).

A total of 619 species of decapod crustaceans—one of the best-known groups for the area—have been recorded for the Gulf versus only 340 species in the Mediterranean. However, a hefty 27 percent of these Mediterranean species are introduced exotics (D'Udekem d'Acoz 1999). Consequently, the decapod fauna is 2.5 times more diverse in the Gulf than in the Mediterranean. It is interesting that, because of their high market value, the arrival of alien species of shrimps (e.g., *Marsupenaeus japonicus, Metapenaeus monoceros*) and fishes (e.g., *Upeneus moluccensis*) is considered a boon to Mediterranean fisheries (Galil 2007).

Invertebrate Conservation in the Gulf of California

Prior to the 1960s, anthropogenic pressure on the Gulf's environment was minimal, and anyone visiting the region would have witnessed a seemingly endless bounty of sea life that probably did not differ substantially from the diversity encountered by indigenous peoples during past millennia. In the 1960s, a casual walk in the rocky intertidal zone during low tide would reveal dozens of species of large-bodied invertebrates, especially echinoderms, crustaceans, and molluscs. Common in tidepools and at snorkeling depths were large sea stars (*Oreaster occidentalis, Mithrodia bradleyi, Nidorellia armata, Astropecten armatus, Pharia pyramidata, Linckia columbiae, Heliaster kubiniji, Astrometis sertulifera, Luidia columbia* and *L. phragma*), spectacular huge brittlestars (*Ophioderma teres* and *O. panamense, Ophiocoma aethiops* and *O. alexandri*), and large urchins (*Eucidaris thouarsii, Centrostephanus coronatus, Arbacia incisa, Lytechinus pictus, Echinometra vanbrunti*). The dazzling little "barrel shrimp," *Gnathophyllum panamense,* was commonly seen in association with *Eucidaris thouarsii* or on coral heads. Also common were large sea cucumbers, such as *Brandtothuria arenicola* and *B. impatiens, Fossothuria rigida,* and *Isostichopus fuscus.* Large molluscs were also abundant and included many spectacular murexes, cones, olives, and cowries (e.g., *Haustellum elenesis, Phyllonotus erythrostomus, Hexaplex nigritus, Hexaplex princeps, Luria isabellamexicana, Oliva porphyria,* many species of *Conus*). Large beds of sea fans (gorgonians) lived on offshore rocky outcroppings, which were home to rare invertebrates such as basket stars (e.g., *Astrodictyum panamense*). Shallow sandy bottoms were home to enormous beds of sand dollars and heart urchins (e.g., *Encope grandis, Encope micropora,* and *Lovenia cordiformis*), most of which have been decimated by shrimp trawlers (color plate 3).

Except for a few remote stretches of coastline on the Baja California peninsula, there are no longer any sites on the Gulf mainland coast where these large invertebrates exist in abundance in the intertidal zone. Most of these spectacular large-bodied invertebrates have become rare or largely extirpated from the Gulf's mainland shores. Overfishing (for Asian food markets) reduced *Isostichopus fuscus* to so few sites that it is now federally listed in Mexico as a threatened species. A similar fate has befallen offshore trawling grounds. Prior to the 1970s, sorting through a shrimp-trawl haul was a rewarding and exciting experience, and in those days such by-catch provided a living library of the animal kingdom. This is no longer the case, and in areas that have been heavily trawled for decades, life on the seabed is now dominated by scavengers such as skates, rays, and portunid crabs (color plate 4).

Beginning in the 1950s, three factors began to have synergistic negative impacts on the biodiversity of the Gulf. First was the establishment of Mexico's national fisheries program, which led to overgrowth of fishing efforts and subsidized overexploitation of marine resources. Second was the realization that tourism held the potential to generate enormous revenues, which led to national and regional policies that set coastal Sonora, Sinaloa, Nayarit, Jalisco, and the Baja California peninsula on a path toward wholesale destruction of coastal natural resources. The third factor is the disruption of the rivers that once flowed into the Gulf, including all of the once-perennial rivers of Sonora—the mighty Colorado River among them. Exacerbating these impacts has been an explosive and unchecked population growth in southwestern United States and northwestern Mexico (Brusca and Bryner 2004; Stoleson et al. 2005). These environmental challenges are reviewed in some detail in Brusca (2007c), Brusca and Bryner (2004), and Lluch et al. (2007).

Invertebrate Fisheries

Today, every major fishery in the Gulf is probably overfished (Greenberg and Vélez-Ibáñez 1993; Sala et al. 2003, 2004; Brusca et al. 2005; Cisneros-Mata, chapter 6 in this volume). The American Fisheries Society lists the Gulf, especially its northern part, as one of five geographic "hot spots" in North America where numerous fish species are at risk (Musick et al. 2000). Commercially valuable invertebrates are facing the same fates as finfishes, as population sizes of the giant Mexican limpet (*Patella mexicana*), black murex (*Hexaplex nigritus*), pink-mouth murex (*Phyllonotus erythrostomus*), articulate chiton (*Chiton articulatus*), giant sea cucumbers (*Isostichopus fuscus*), octopus (*Octopus bimaculatus* and others), shrimps (Penaeidae), swimming crabs (*Callinectes* spp.), and others have plummeted over the past decade. Even marine algae are overharvested in northwestern Mexico—mainly on the Pacific Baja peninsula, a region that provides about 10 percent of the world production of agarophytes (the most important commercial species being the red alga *Gelidium robustum*, which has been harvested without regulation since 1945).

Industrial shrimp trawling exacts a harsh toll on the Northern Gulf's benthic environment and also along the coasts of southern Sonora and Sinaloa. The ocean bottom in the Northern Gulf was once estimated to be dragged by shrimp nets as frequently as four times per year (Pérez-Mellado and Findley 1985; García-Caudillo 1999; Brusca et al. 2005), although with the recent partial "collapse" of the trawled shrimp fishery in this region that number has fallen. Shrimp trawl nets are indiscriminant killers, raking the seafloor in a clear-cutting fashion, trapping and killing everything in their path (Engel and Kvitek 1998; Watling and Norse 1998; Dayton et al. 2002). The historically high rate of bottom trawling has seriously damaged the Gulf's fragile, soft-bottom, benthic habitats. In addition, trawl nets in the Northern Gulf capture between 10 and 40 kilograms (depending on the location and time of year) of by-catch for each single kilogram of shrimp (Brusca 2004a, 2007c; Brusca et al. 2005). The number of commercial shrimp trawlers in the Gulf grew from 700 in 1970 to a high of 1,700 in 1989 and then decreased to 1,200 in 1999. Until very recently, hundreds of shrimp boats (and artisanal fishers) were still working *within* the upper Gulf's biosphere reserve. "Catch per unit effort" in the shrimp fishery has been declining for decades (documented at least as early as the 1970s; Snyder-Conn and Brusca 1977) while government subsidies artificially sustained the overcapacity of the industrial fishing fleet. Without government subsidies, commercial shrimp trawling would not be economically feasible. In fact, as a result of catch decreases and the advent of shrimp farming in the Gulf (producing cheaper market shrimp), the economics of commercial shrimping shifted so much just after the turn of this century that the number of bottom trawlers working out of the three main fishing ports in the Northern Gulf fell to just 130 boats (115 in Puerto Peñasco, 15 in San Felipe, and none in El Golfo de Santa Clara). Recently, the Mexican government activated a program aimed at reducing the number of industrial shrimpers in the Pacific, paying compensation for any boat willing to cease fishing activity, but the success of that program remains to be seen.

Limited scientific and anecdotal information suggests that sweeping changes in benthic/demersal community structure have taken place over the past 50 years as a result of disturbance from bottom trawling. These changes include an accelerating decrease in the diversity and biomass of the by-catch, possibly heralding a regional benthic/demersal ecosystem collapse (Pérez-Mellado and Findley 1985; Brusca 2007c; Findley, pers. comm.). In the late 1960s, sorting through the by-catch of a shrimp-trawl haul produced hundreds of species of invertebrates (and fishes) in most known phyla. Today, these same bottom trawl nets (in the Northern Gulf) contain only a few dozen species of invertebrates and are dominated by scavenger species (pers. obs.). Invertebrates whose depth range is the same as that dragged by shrimp trawls have suffered enormous destruction, and many are probably on the verge of extinction (e.g., the beautiful giant polychaetes *Aphrodita mexicana* and *A. sonorae;* the sea pen *Ptilosarcus undulatus*), but no empirical studies have been made in this regard. The destruction of the benthic ecosystem has disrupted the food web of the entire Northern Gulf, which has probably altered the pool of available prey for the critically endangered vaquita porpoise, *Phocoena sinus,* and the totoaba, *Totoaba macdonaldi.*

Tourism and Aquaculture

In areas—such as Puerto Peñasco, San Felipe, and San Carlos/Guaymas—of heavy and increasing tourism in the northern and central

Gulf, littoral biodiversity is but a pale shadow of what it was just 25 years ago. Part of the tourism-driven loss is hand-collecting of animals by visitors (and the trampling underfoot of fragile habitats exposed at low tide). But also important is the collection of large molluscs and echinoderms by residents for sale to tourists as curios and of molluscs sold to local restaurants, where they are served in seafood cocktails (e.g., bivalves, gastropods, and octopuses). In the Northern and Central Gulf today, healthy populations of these large-bodied species are found almost exclusively on island refugia or highly inaccessible stretches of the mainland coast, although some still occur in reduced numbers subtidally.

Increasing loss of coastal habitats due to encroaching housing and resort developments, marinas, and aquaculture installations lacking environmental controls are threatening the rich wetlands (estuary and estero habitats) of the Gulf that serve as critical spawning and nursery grounds for shrimp and other invertebrate and fish species. The complex food webs of coastal bays and wetlands also include species not found anywhere else in the Gulf, such as the rare amphioxus (Cephalochordata), *Branchiostoma californiense*.

Much of the coastline of Nayarit, Sinaloa, and Sonora has now been carved up into aquaculture farms (Glenn et al. 2006; Brusca 2007c). Most of these are shrimp farms, and ~95 percent (64 million pounds in 2000) of this farm-raised shrimp makes its way to the United States. About 90 percent of the world's aquaculture facilities are in developing nations, and they are largely "slash and burn" in their approach: bulldozers tear out mangrove forests and other coastal habitats to be replaced with fish or shrimp ponds, many of which cover many square miles. In concept, these coastal ponds are cheap and easy to construct; a pipe at one end of the pond complex pulls clean ocean water in, and a pipe at its other end spits used water out—laden with shrimp (or fish) wastes, excess food, herbicides (used for algal control), antibiotics and other drugs, disease organisms and parasites, and so forth. In recent years, mangroves have not been directly removed during the construction of Gulf shrimp farm operations, but the extent of damage to this ecosystem by proximity to shrimp farming (e.g., changes in estuarine water circulation, siltation and smothering, pollution, eutrophication) has been little examined. Of course, closed and nonpolluting aquaculture systems are possible inland (and required in the United States), but they are more expensive to build and operate.

Coral reefs and coral communities in the Gulf are significantly threatened by divers and boat anchors. Although the Cabo Pulmo Reef has enjoyed various levels of protection for many years and is now a national park, it has also gradually deteriorated over the past two decades because of divers and fishermen as well as several severe El Niño events (and associated sea-surface warming). Although the coral-predating Eastern Pacific crown-of-thorns sea star (*Acanthaster planci*) occurs throughout the Central and Southern Gulf, it apparently does not pose a threat to corals in the region and preys on numerous other invertebrates (as well as on corals). Other coral predators are primarily fishes (e.g., spotted pufferfish, *Arthron meleagris;* parrotfishes, *Scarus* spp.) and a few gastropods, but these also do not appear to be a threat to corals in the Gulf.

Loss of Rivers

All of the rivers that once reached the Gulf of California have been drastically altered or destroyed by overdraft and diversion, and none of the Sonora rivers that once flowed perennially, or semi-perennially, now reaches the sea (these rivers include the ríos Colorado, Magdalena-Altar-Concepción-Asunción, San Ignacio, Sonora, Yaqui, Mayo, and Fuerte). Historically, the Colorado River carried an estimated annual average of 15–18 million acre-feet (maf) of water to its delta (Carriquiry and Sánchez 1999; Cohen et al. 2001; Brusca and Bryner 2004). During the nineteenth century, especially from 1850 to 1880, riverboats steamed from the Gulf of California up the Lower Colorado/Gila River system into Arizona. Until completion of Hoover (Boulder) Dam in 1935, which created Lake Mead, freshwater from the Colorado River flowed into the Northern Gulf throughout the year, with great seasonal floods resulting from spring snowpack melt in the Rocky Mountains. By the time Glen Canyon Dam was completed in 1963, input of Colorado River water to the delta and upper Gulf had completely ceased. For 20 years after completion of that dam, as Lake Powell filled, virtually no water from the river reached the sea. In 1968, flow readings at the southernmost measuring station on the river were discontinued, since there was nothing left to measure.

Today, 20 dams (58 if the Colorado River's tributaries are included) and thousands of kilometers of canals, levies, and dikes have converted the Colorado River into a highly controlled plumbing system in which every drop of water is carefully counted, managed, and litigated. The original water allocation estimates were made in the 1920s and, based on data from an unusually wet time period, assumed an average river flow of about 22 maf per year. However, the river's average annual flow during the last 500 years has actually been about 14 maf/yr. Hence there are now more legal claims to the water than are possible to meet, so it is no wonder that today almost no water reaches the delta. Additionally, most of the delta's wetlands have been converted into farmland or urban sprawl. What was once 2 million acres of wetlands has been reduced to about 150,000 acres (Glenn et al. 1992, 1996, 1999, 2001). As a result of the greatly reduced freshwater flow, the powerful tides of this region now overwhelm the Lower Colorado River channel. During high tides, seawater creates an estuarine basin (estero), for 50–60 km upriver, that averages 2–8 km in width and 16 km wide at its mouth. This marine intrusion has killed most of the freshwater flora and fauna that once lived along the lowermost river corridor.

Prior to construction of Hoover Dam, the annual sediment discharge from the Colorado River into the Gulf was enormous: estimates range from 45 to 455 million metric tons. Accumulated river sediments on the delta are thousands of feet thick. The entire Northern Gulf is considered the "Colorado River Sedimentary Province." However, the reduction of freshwater input and sediment discharge since 1935 has modified the hydrography and oceanography of the Colorado River delta–upper Gulf system, initiating a regime of deltaic erosion. New deltaic deposition no longer takes place, and the entire delta is now exposed to the dynamic forces of extreme tides, currents, and storms, which promote re-suspension and erosion of ancient river sediments as well as the gradual export of sediments out of the delta region. These changes are altering the littoral wetlands and biological equilibrium of the region. They are also destroying habitat for an estimated 340 species of marine macroinvertebrates that inhabit the sand/mud benthic environment of the delta region.

It is likely that the reduction of freshwater input into the upper Gulf, in combination with other anthropogenic factors, has driven some species to (or nearly to) extinction. However, we have so few historical or baseline data for marine organisms of this region that extinctions (or local extirpations) would go unnoticed for commercially unimportant or otherwise little-known species. There has never been a comprehensive dedicated survey of the marine fauna of the upper Gulf and Colorado River delta ecosystem.

The delta clam, *Mulinia coloradoensis,* was probably once one of the most abundant animals of the uppermost Gulf. Windrows of its shells line the beaches of the delta and western shores of the upper Gulf. This species was thought to be extinct until its recent rediscovery in small numbers near the mouth of the river (Kowalewski et al. 2000; Rodríguez et al. 2001a; Cintra-Buenrostro et al. 2004). It has been suggested that the near demise of this species is the result of decreased benthic productivity resulting from upstream diversion of the Colorado River's flow. However, there is no evidence that nutrient levels (and hence productivity) have decreased significantly in the Northern Gulf, and nutrients that have been lost by depletion of riverine input may have been regained in the form of agricultural runoff and deltaic erosion (release of ancient trapped nutrients). Therefore, the near extinction of this clam may be linked to another factor, still unknown, that is related to reduction of freshwater input to the delta.

Freshwater input from the Colorado River is also important to the life history of commercial shrimps of the region. Commercial shrimp catches have been falling since the 1960s, which is due to a combination of overfishing and loss of habitat for young. It has been estimated that an annual influx of just 250,000 acre-feet of Colorado River water could double shrimp production in the Northern Gulf (Galindo-Bect et al. 2000). The young of these shrimp utilize the shallow wetlands and esteros of the region (including the tidelands of the delta) as a nursery, migrating into these areas after their offshore planktonic larval phase. When the shrimp reach a juvenile or subadult stage, they migrate offshore once again.

Rescuing Invertebrate Biodiversity

Since the mid-1980s, a growing conservation movement has emerged in northwestern Mexico led by such nongovernmental organizations as Agrupación Sierra Madre, ALCOSTA, the Arizona-Sonora Desert Museum, CEDO (Centro Intercultural de Estudios de Desiertos y Oceanos), COBI (Comunidad y Biodiversidad), Conservation International-

Mexico, ENDESU (Espacios Naturales y Desarrollo Sustentable), ISLA (Conservación del Territorio Insular Mexicano), Marisla, Naturalia, The Nature Conservancy, Sociedad de Historia Natural Niparajá, Noroeste Sustentable (NOS), The David and Lucile Packard Foundation, Proesteros, Pronatura, ProPeninsula, Wildcoast, World Wildlife Fund–Mexico, and other organizations often associated with local communities. Such organizations have had a powerful influence on natural resource conservation in the Gulf. In addition, the government sector has increasingly stepped up its conservation efforts, especially SEMARNAT (Mexico's ministry of environment and natural resources) and its national commission for protected natural areas (CONANP) and "Islas del Golfo" program. The active participation of these organizations was critical to establishing the Upper Gulf of California and Colorado River Delta Biosphere Reserve, developing conservation priorities for the Gulf and its islands, working with artisanal fishers and indigenous peoples to develop sustainable fisheries, and working with state and federal governmental agencies to push for more protected areas and better protection of the marine and coastal environment. As a result of the efforts of these groups over the past two decades, fisheries laws have tightened up, gillnetting is on the verge of becoming illegal, bottom trawling is becoming better regulated (and, it is hoped, will soon be banned), and high-visibility species such as totoaba and vaquita are attracting the attention of conservationists all over North America (summarized in Brusca and Bryner 2004; Brusca et al. 2005; Brusca 2007c; Lluch-Cota et al. 2007; Carvajal et al., chapter 11 in this volume). Recently, new laws were passed that prohibit use of gill nets with mesh sizes greater than 6 inches and that protect against "destruction of the marine floor" (e.g., shrimp trawling) in all protected areas in the Gulf, including the Upper Gulf of California and Colorado River Delta Biosphere Reserve. These new environmental laws could go a long way toward reducing the incidental take of vaquita and sea turtles and toward protecting the seafloor; however, it will be up to the federal government (PROFEPA, the enforcement arm of SEMARNAT) to enforce them, and many fishers are still protesting or ignoring them.

There remain many fundamental but unanswered questions about the Gulf's ecosystems. What is the nature of the benthic sediment–water column food web in shallow Gulf waters, and how does energy flow through that system? How has that system been affected by bottom (shrimp) trawlers during the past few decades? How are commercial species such as shrimp affected by freshwater input (e.g., from the Colorado River), and how important are annual freshwater pulses from the Colorado River to the marine ecosystems? What are the biological relationships between the Gulf's estuaries/esteros and its open-water (pelagic) ecosystem? How effective are the fully protected marine reserve areas with "no-take zones" for the recovery of marine species (e.g., San Pedro Mártir and Bahía de los Ángeles Biosphere Reserves)?

Despite the considerable damage that has been inflicted by humans on Gulf environments and despite the many lingering threats, there is cause for optimism. If the conservation movement in the Gulf of California continues with its present momentum, then critical new areas will receive protection and better enforcement of currently protected regions should follow. Most urgent is to: (1) ban *all* bottom trawling in the Gulf so that the benthic/demersal ecosystem can, if possible, recover; (2) implement a sustainable management program for fisheries; (3) protect the four "coquina reefs" in the upper Gulf; (4) improve enforcement of existing laws for protected areas; (5) increase public education; (6) ban the take of all marine life from the intertidal zone, except that done through a regulated fishery basis; and (7) better understand the marine ecosystems of the Gulf. Fortunately, one still can find island and isolated coastal refugia, areas not easily accessible by road or large fishing boats, that serve as important shelters for species extirpated elsewhere in the Gulf.

The jointly developed Monterey Bay Aquarium/Arizona–Sonora Desert Museum "Southwest Seafood Watch Cards" have taught and inspired seafood consumers (since 2004) on both sides of the border to restrict their purchases to sustainably harvested seafoods from the Gulf of California and elsewhere. Despite resistance from fishers in Mexico and from U.S. and Mexican seafood purveyors and restaurateurs, evidence suggests that the sustainable seafood programs are having an impact on both sides of the border, though much remains to be done.

NOTE

1. As defined by the Macrofauna Golfo Project, the *Northern Gulf* (GCN) faunal region extends from the marine-influenced Colorado River delta southward to (and in-

cluding) the Midriff Islands (*las Islas del Cinturón*), the largest of which are islas Tiburón and Ángel de la Guarda, and to Bahía San Francisquito (Baja California) and Bahía Kino (Sonora). Within the Northern Gulf is the subregion of the Upper Gulf of California and Colorado River Delta Biosphere Reserve, extending from the delta to a line running from Punta Pelícano (= Roca del Toro; the southern margin of Bahía Cholla and the larger Bahía Adair), Sonora, across the Gulf to Punta Machorro (= Punta San Felipe) at San Felipe, Baja California. The *Central Gulf* (GCC) faunal region ranges from Bahía San Francisquito (Baja California) and Bahía Kino (Sonora) to Punta Coyote (Baja California Sur) and Guaymas (Sonora). The *Southern Gulf* (GCS) faunal region extends southward to Cabo Corrientes, Jalisco, on the mainland and to Cabo San Lucas on the Baja California peninsula. (See fig. 5.4 in the next chapter.)

CHAPTER 5

Fishes of the Gulf of California

PHILIP A. HASTINGS, LLOYD T. FINDLEY,
AND ALBERT M. VAN DER HEIDEN

Summary

The Gulf of California has a rich history of exploration and scientific discovery, and its fishes have been an important part of that history. The first major oceanographic expeditions collecting fishes in the Gulf were conducted by the U.S. Fish Commission steamer *Albatross* between 1889 and 1911. The ship's naturalist, C. H. Gilbert, described 168 Gulf fish species as new to science over his career. Numerous other reseachers from around the world contributed to our current knowledge of the systematics of Gulf fishes. To date, over 900 species of fishes have been recorded from Gulf waters, and trends in species descriptions indicate that other new species remain to be identified, described, and named. The Gulf owes its faunal diversity, in part, to its location at the juncture of the tropical and temperate regions of the eastern Pacific Ocean. The Gulf fish fauna is dominated by tropical species (87 percent of Gulf fishes also occur south of the Gulf), but it also includes a significant number of species known from temperate areas (31 percent of Gulf fishes also occur north of the Gulf, in the temperate waters of coastal North America). Nearly 10 percent of Gulf fish species are endemic: found nowhere else in the world. This includes a diverse array of species with varied phylogenetic and geographic affinities. The northern Gulf includes several endemic species historically associated with the Colorado River estuary, and these are of obvious special concern to conservation biologists. The diversity of fishes increases as one moves southward in the Gulf, with nearly twice as many species known from the southern Gulf as from the northern Gulf.

The clear and often calm waters of the Gulf of California, together with its striking biodiversity, have made it an ideal setting for the study of the biology, ecology, and behavior of fishes. This microcosm of the larger eastern Pacific and world oceans provides unlimited opportunities for studying and understanding the processes of evolution and ecology in the marine realm. Conservation of this incomparable region and its biotic resources will ensure the continued heritage of this remarkable region for future generations.

Introduction

Fishes play a prominent role in the ecology, evolution, and economics of the Gulf of California (Sea of Cortez). Fishing activities led to early establishment of several towns surrounding the Gulf, especially in its northern region (Munro Palacio 1994; Bahre et al. 2000), and fishes and fisheries products continue to contribute significantly to the economy of northwestern Mexico (see Cisneros-Mata, chapter 6 in this volume). The abundance and diversity of Gulf fishes, together with the Gulf's often calm waters, support a thriving tourist industry and continue to lure numerous biologists from around the world. Herein we review some of the history of ichthyology in the Gulf of California, provide an overview of the diversity of Gulf fishes, and briefly review past studies on the ecology and evolution of Gulf fishes.

A Brief History of Gulf of California Systematic Ichthyology

By the early 1800s, nearly 50 widespread fish species that would eventually be recorded from Gulf waters had been named—based mostly on specimens collected in other parts of the world. These included several well-known species such as dorado or dolphinfish, *Coryphaena hippurus* Linnaeus, and the white shark, *Carcharodon carcharias* (Linnaeus). By the mid-1800s, this number had risen to nearly 140 species, but it was not until the latter half of that century that the bulk of species now known from Gulf waters were described (fig. 5.1). This upsurge resulted from extensive collections of fishes from other areas of the Pacific, especially the coast of Central America, that were studied by zoologists from around the Western world. European scientists studying Gulf fishes included Albert Günther of the

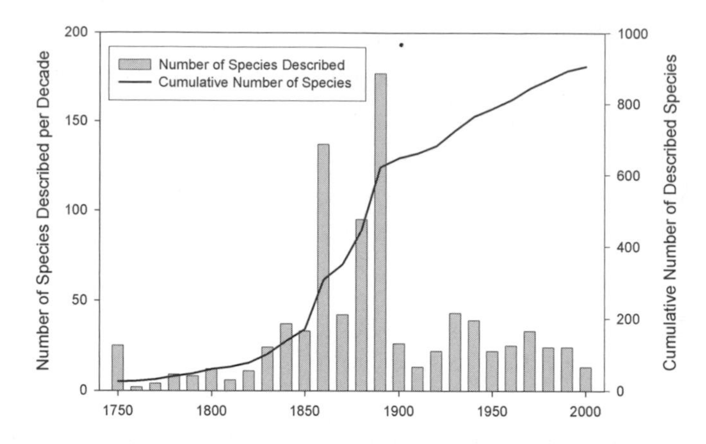

5.1. Number of currently valid species of Gulf of California fishes by decade of scientific description.

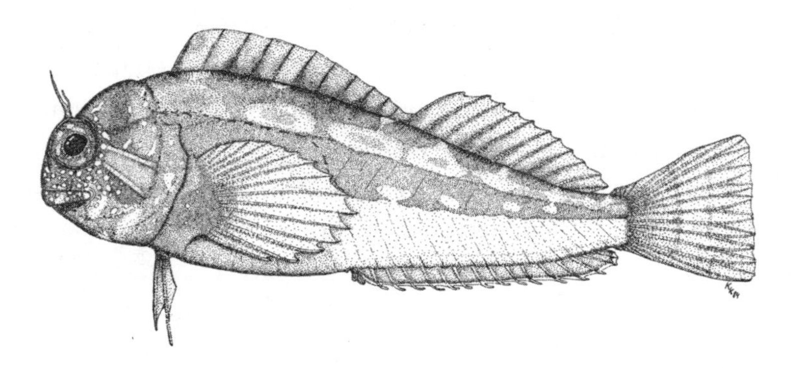

5.2. The Barnacle Bill blenny or borracho vacilón, *Hypsoblennius brevipinnis* (Blenniidae), a species found throughout the Gulf and tropical eastern Pacific, was described by Albert Günther in 1861. Drawing by K. Kotrschal.

British Museum, who described 60 species of fishes from the region (e.g., Günther 1864; see fig. 5.2), and Franz Steindachner of the Natural History Museum, Vienna, who described 31 species of Gulf-occurring fishes (e.g., Steindachner 1877). However, many of these newly collected specimens fell into the hands of North American scientists: Samuel Garman of Harvard's Museum of Comparative Zoology, who described 36 species of Gulf fishes (e.g., Garman 1899); David Starr Jordan of Stanford University, who, alone or with collaborators, described 114 species of Gulf-occurring fishes

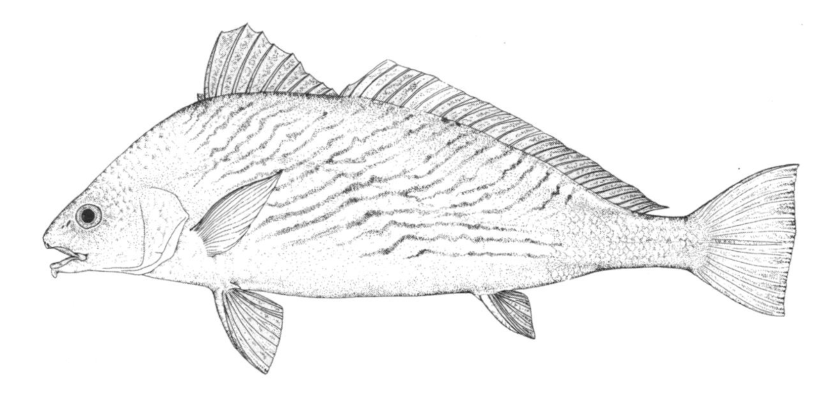

5.3. The yellowfin croaker or berrugata aleta amarilla, *Umbrina roncador* (Sciaenidae), a northern disjunct species found in the northern Gulf and along the coast of California and western Baja California peninsula, was described by David Starr Jordan and Charles Henry Gilbert in 1882. Drawing by T. Hansen.

(e.g., Jordan and Gilbert 1882a,b); and Theodore Gill of the Columbian University (now George Washington University) and the Smithsonian Institution, who described 58 species of Gulf fishes—many of which had been collected by John Xántus, a tidal observer for the U.S. Coast Survey stationed in Cabo San Lucas (e.g., Gill 1862–1863).

Yet extensive collections of fishes within the of Gulf of California were not made in earnest until the Gulf cruises of the U.S. Fish Commission steamer *Albatross.* This large research ship was reassigned from the Atlantic to the west coast of the United States in 1888 to study the aquatic life and "hydrology" of the eastern Pacific, and it made five cruises to Gulf waters between 1889 and 1911 (Allard 1999). During much of this time, Charles Henry Gilbert was chief naturalist on board this legendary vessel, a post that led to his extraordinary influence on ichthyology in the eastern Pacific (Dunn 1997). From *Albatross*-collected specimens alone, Gilbert described 176 currently recognized species of fishes from the western coast of North America (e.g., Gilbert 1890, 1892). He named 168 fish species that are known from the Gulf and another 67 as second author with his mentor David Starr Jordan, widely recognized as the "father of American ichthyology" (e.g., Jordan and Gilbert 1892a,b; see fig. 5.3). Summaries of the diversity of Gulf and other tropical eastern Pacific fishes were reported by a few early authors, most notably Jordan (1895), who compiled an inventory of the fishes of Sinaloa (updated by van der Heiden and Findley 1990), and Jordan and Evermann (1896–1900), whose landmark three-volume series on the fishes of North and Middle America included the Gulf of California region (e.g., Brittan 1997).

Following these and other early efforts (e.g., Streets 1877; Jenkins and Evermann 1889; Evermann and Jenkins 1891) to name and describe the fishes of the Gulf and adjacent waters of the eastern Pacific, much of our knowledge of their taxonomy and distributions has come from detailed systematic studies of groups that included species inhabiting Gulf waters. This was especially true for small benthic fishes, which are not collected using normal fishing gear, are usually overlooked by casual scuba divers, and are often difficult to identify. Notable in this regard are the studies on clingfishes by Briggs (1951, 1955), on blennies by Hubbs (1952, 1953), on labrisomid blennies by Springer (1959) and by Rosenblatt and his students and colleagues (Rosenblatt and Parr 1969; Rosenblatt and Taylor 1971), on triplefin blennies by Rosenblatt (1959), on chaenopsid blennies by Stephens (1963), and on sand stargazers by C. Dawson (1974, 1975, 1976). From the framework provided by these and other systematic ichthyologists, knowledge of Gulf fishes grew and eventually permitted the publication of more recent overviews of the Gulf's fish fauna, including reef fishes (Thomson et al. 1979, 2000) and selected other fishes (Thomson and McKibbin 1976, 1978), commercially important fishes (e.g., Fischer et al. 1995a,b; Cudney Bueno and Turk Boyer 1998), and all coastal fishes of the tropical eastern Pacific (Allen and Robertson 1994, 1998; Robertson and Allen 2002; Wilson 2002).

Knowledge of the diversity and distribution of Gulf fishes was first summarized by B. Walker (1960) in a symposium volume focusing on the biogeography of the region. This important work was based on years of collecting in Gulf waters with students and colleagues from UCLA and Scripps Institution of Oceanography, and it resulted in his extensive, unpublished running checklist of fishes recorded from the Gulf—maintained and updated over many years by himself, his students, and Wayne Baldwin (Montgomery 2002). More recently, Castro-Aguirre et al. (1995) summarized much of the knowledge on the entire Gulf ichthyofauna. In addition, numerous studies have documented fishes occupying specific habitats, including soft bottoms (e.g., Berdegué 1956b; Castro-Aguirre et al. 1970; Pérez-Mellado and Findley 1985; van der Heiden 1985, 1986; Plascencia-González 1993; Nava-Romo 1994; Amezcua-Linares 1996; Balart et al. 1997), reefs and rocky shores (e.g., Thomson et al. 1979, 2000; Brusca and Thomson 1977; Molles 1978; Villarreal, 1988; Pérez-España et al. 1996; Aburto-Oropeza and Balart 2001; González-Cabello 2003; Viesca-Lobatón 2003), coastal lagoons/estuaries (e.g., Findley 1976; Warburton 1978; Yépiz-Velázquez 1990; Varela-Romero 1990; Balart et al. 1993; Grijalva-Chon et al. 1996; Castro-Aguirre et al. 1999; Galván-Magaña et al. 2000; Minckley 2002), and the deep sea (e.g., Lavenberg and Fitch 1966; Robison 1972; Brewer 1973; De la Cruz-Agüero and Galván-Magaña 1992). Still other studies have summarized the fish fauna from particular regions within the Gulf, including Baja California Sur (De la Cruz-Agüero et al. 1997), Bahía de La Paz (Chávez 1985, 1986; Abitia-Cárdenas et al. 1994; Balart et al. 1995, 1997; González-Acosta et al. 1999; Galván-Piña et al. 2003), Isla Cerralvo (Galván-Magaña et al. 1996), Isla Espiritu Santo (Rodríguez-R. et al. 2005), Bahía Concepción (Rodríguez-R. et al. 1992, 1994, 1998), and Bahía de los Ángeles (Viesca-Lobatón et al. 2008).

Our knowledge of the systematics of Gulf fishes continues to accumulate. There has been a steady increase in the number of Gulf-occuring fish species described, even during the most recent decade (fig. 5.1). It is likely that this trend will continue for the foreseable future as new species are discovered and as previously known (but yet unnamed) species are formally described (Hastings and Robertson 2001).

Biogeographic Context of the Gulf and Its Fishes

The Gulf of California lies at the intersection of two vast coastal regions of the eastern North Pacific that differ significantly in their thermal regimes and biogeographies. To the north is the temperate Californian Province, while to the south lies the tropical Panamic region (Briggs 1974; Brusca and Wallerstein 1979a). Although the continental shelves of both regions are relatively narrow, they support diverse ichthyofaunas that, overall, are quite different from one another owing to the striking differences in thermal regimes (see below). A recent accounting of fishes recorded in the region north of the Gulf (from the Alaska–Yukon border to the southern tip of Baja California Sur at Cabo San Lucas) reported 1,450 species of principally coldwater-adapted species (Love et al. 2005). This number includes coastal, pelagic, and deep-sea fishes recorded within 300 miles of the coastline as well as several warmwater-adapted species with tropical affinities reported from Cabo San Lucas that are rarely found north of the southwestern region of the Baja California peninsula (e.g., at Bahía Magdalena or at most—during years of normal water temperature—north to Punta Eugenia; see fig. 5.4). South of the Gulf, a similar number of fish species occurs in the tropical eastern Pacific. Robertson and Allen (2002) reported nearly 1,100 species of coastal and pelagic fishes from inshore waters of the tropical eastern Pacific from Bahía Magdalena and from the north-central Gulf of California southward to Cabo Blanco in northern Peru. This number excludes most species of deep-sea fishes, which (if included) would increase the number of species known from the tropical regions of the eastern Pacific considerably, probably to a number comparable to or exceeding the number of fish species reported from north of the Gulf.

Significant habitat gaps separate the eastern Pacific, including the Gulf of California, from other biogeographic regions of the world. These include the so-called east Pacific barrier (Briggs 1961; Leis 1984), a vast expanse of open water separating the eastern Pacific and its offshore oceanic island groups (e.g., Islas Galápagos, Islas Revillagigedos) from the central/western Pacific and Indian Oceans (the Indo-west Pacific). This expanse of open ocean between the eastern Pacific and the distant islands at or near the western margin of the massive Pacific plate (e.g., Easter Island, Marquesas Islands) is inhabited by many "high seas" pelagic and deep-sea fishes. Thus several of these fishes that can be found in the eastern Pacific and Gulf of California have broad distributions that extend well beyond the limits of its waters—some to all oceans (Briggs 1960).

The situation is quite different, however, for coastal fishes (those occurring on or over the continental shelf in less than 200 m depths). For most of these, the east Pacific barrier is formidable (Leis 1984; Robertson et al. 2004), and numerous lineages of neotropical fishes have failed to cross it (Grigg and Hey 1992). This includes a number of genera and even some families of fishes restricted to the neotropics, such as sand stargazers (Dactyloscopidae) and labrisomid blennies (Rosenblatt 1967).

Although neotropical organisms rarely cross the east Pacific barrier westward (Grigg and Hey 1992), a number of tropical coastal Indo-west Pacific fishes have managed to cross it eastward, sporadically or even regularly (Rosenblatt et al. 1972; Rosenblatt and Waples 1986; Robertson et al. 2004). Among reef fishes, these include mostly species with relatively long-lived larval stages (e.g., surgeonfishes and moray eels) that can survive in the water column for long periods of time. A number of these long-distance travelers are widely distributed in the tropical eastern Pacific and thus have also been recorded in the waters of the Gulf of California. Well-known examples include the moorish idol, *Zanclus cornutus;* convict surgeonfish, *Acanthurus triostegus;* longnose hawkfish, *Oxycirrhites typus;* and zebra moray, *Gymnomuraena zebra.*

A small number of fish lineages with representatives in the Gulf appear to have dispersed into the western Pacific, not westward across the east Pacific barrier but instead northward along the continental margins of the temperate regions. These include several coldwater-adapted groups such as the rockfishes (Sebastiinae), whose center of diversity is in the temperate northeastern Pacific but who also have representatives in the northwest Pacific and in the Gulf of California (Chen 1975; Love et al. 2002).

To the east, the eastern Pacific is isolated from the Caribbean Sea and the rest of the Atlantic Ocean by the continental land mass of the Americas (Woodring 1966). This impenetrable barrier (for marine fishes) is relatively young compared to the east Pacific barrier. The trans-Panamic seaway, which once connected the Atlantic and Pacific, finally closed completely about 3.5 million years ago (Coates and Obando 1996). This event had a strong effect on the evolution of marine habitats in the Caribbean (Jackson and D'Cruz 1998) and significantly affected the evolution of marine fishes in the neotropics (Topp 1969; Robins 1972; Collins 1996; Bermingham et al. 1997; Hastings 2000). Consequently, and notwithstanding the artificial semi-connection provided by the Panama Canal, very few fish species are now shared between the tropical eastern Pacific and the Caribbean (Rosenblatt 1967; McCosker and Dawson 1975; Castro-Aguirre 1980).

Within the large expanse of the eastern Pacific there exist strong temperature gradients both northward and southward (Briggs 1974; Brusca and Wallerstein 1979a). The Gulf of California lies in the northern transition zone, with warm tropical waters to its south and cool temperate waters to its north. This transition is most evident along the outer coast of the Baja California peninsula, a consequence in part of the westward deflection of the southward-flowing cold California Current near the peninsula's central region. This offshore movement of the coastal water mass has the effect of pulling warmer, southerly waters northward along the coast of central and southern Mexico and into the southern region of the Gulf and then northward along the southern, outer part of the peninsula (Alvarez-Borrego 1983; Bray and Robles 1991; Alvarez-Borrego, chapter 2 in this volume). This general northward flow of warm, tropical water, combined with solar heating within the Gulf, makes the Gulf waters typically warmer than those at a comparable latitude along the outer coast of the Baja California peninsula. The resulting steep thermal gradient along that outer coast, which is present year-round, represents a formidable barrier to northward movement of the tropical component of the Gulf's biota (Brusca and Wallerstein 1979a; Brusca et al. 2005). However, many Gulf-occurring fishes may temporarily expand their distribution northward during El Niño events, when warm tropical waters move farther northward along that outer coast (Hubbs 1948; Lea and Rosenblatt 2000). In fact, some such species now seem to be establishing longer-term residence in those northern waters, a result perhaps of increasing water temperatures in the region.

A seasonal thermal gradient is also present within the Gulf itself. Because the Gulf extends northward into warm-temperate latitudes (to 31.7° N), the surface waters of its northern part are exposed to low air temperatures during the winter, resulting in a fairly steep thermal gradient along the main axis of the Gulf. This gradient breaks down during warm months owing to solar heating of Gulf waters and the surrounding deserts, at which time the surface temperature of the northern Gulf may even exceed that of the southern Gulf (Alvarez-Borrego 1983; Brusca and Findley 2005; Brusca et al. 2005; Alvarez-Borrego, chapter 2 in this volulme). As a consequence, the northern Gulf is a thermally more variable environment than either coastal California, which is consistently cooler, or the southern Gulf (e.g., at Mazatlán and San Blas), which is usually warmer (Thomson and Lehner 1976). In addition, there are seasonally consistent upwelling events within the Gulf that bring cool, nutrient-rich waters to the surface, and these events vary in location and intensity (Alvarez-Borrego et al. 1975a; Alvarez-Borrego 1983, chapter 2 in this volume).

Fish Diversity in the Gulf of California

In 1994, the Macrofauna Golfo research group was formed to update and summarize our knowledge of Gulf-occurring vertebrate and macroscopic invertebrate animals via the construction of a computerized searchable database, printed checklists, and other publications (Brusca et al. 2005; Hendrickx et al. 2005; Findley et al., in press; Findley, in prep.). The fishes subgroup (L. Findley, A. van der Heiden, P. Hastings, H. Plascencia, J. Manuel Nava, and J. Torre) compiled data from their own fieldwork, the published literature, and records of species in selected museum holdings of all fishes recorded from Gulf of California waters. A special effort was made to compile records from the major museums containing Gulf of California specimens, including the California Academy of Sciences, Scripps Institution of Oceanography, University of California at Los Angeles, University of Arizona, Natural History Museum of Los Angeles County, and the smaller collections at the Instituto Tecnológico y de Estudios Superiores de Monterrey–Campus Guaymas, the Instituto de Ciencias del Mar y Limnología–Unidad Académica Mazatlán (of the Universidad Nacional Autónoma de México), and the Centro de Investigación en Alimentación y Desarrollo, A.C.–División Mazatlán.

For this fish survey (as well as for the macroinvertebrates), we considered the boundaries of the Gulf to include all waters between the Baja California peninsula and mainland Mexico north of a line extending from Cabo San Lucas, Baja California Sur, to Cabo Corrientes, Jalisco (Thomson et al. 1979, 2000; Findley and Brusca 2005; see fig. 5.4). This definition largely corresponds to a "geological/geophysical" definition of the Gulf, representing the waters northward of the tip of the Baja California peninsula and the point on the mainland from which much of the southern peninsula broke free and rifted to the northwest during its geological evolution (Lyle and Ness 1990). It also corresponds to a commonly used "oceanographic" definition of the Gulf (Roden 1964; Brusca et al. 2005; Brusca and Findley 2005; Alvarez-Borrego, chapter 2 in this volume). We also delineated three major subregions or faunal regions within the Gulf (Findley and Brusca 2005; see fig. 5.4) based primarily on previous biogeographic studies of fishes (Walker 1960; Thomson et al. 1979, 2000). For each species recorded from Gulf waters and incorporated into the database, the

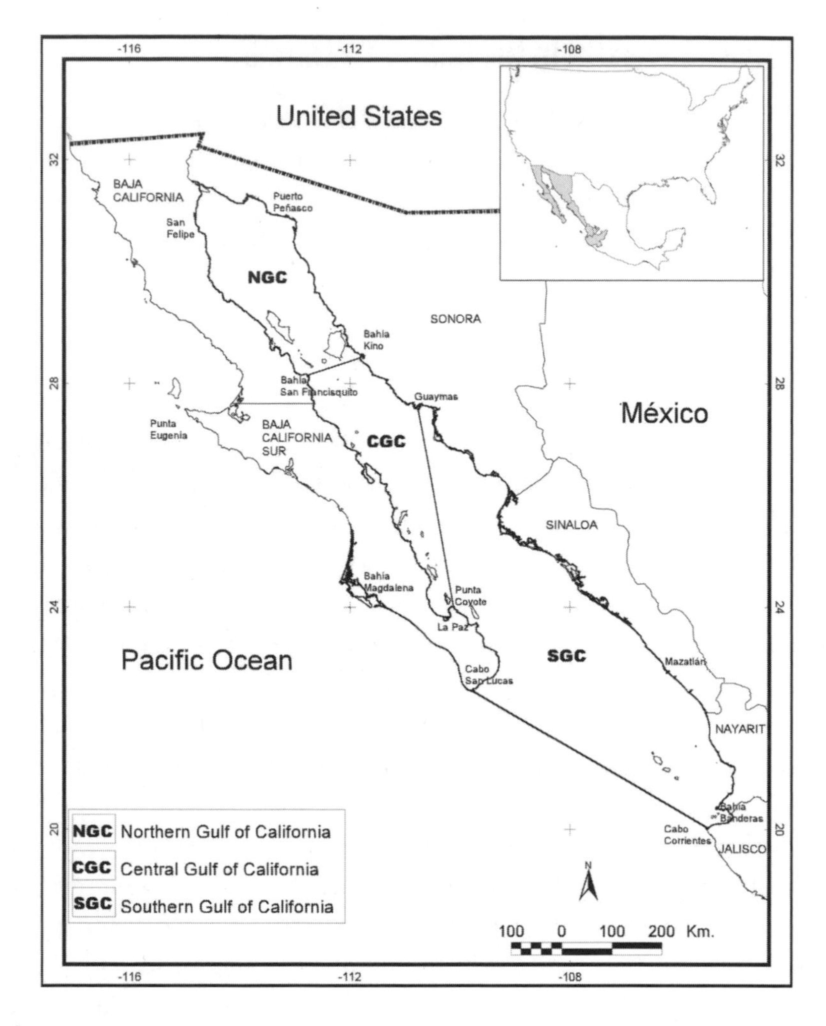

5.4. Map of the Gulf of California showing three faunal regions (as defined by the Macrofauna Golfo Project).

Macrofauna Golfo survey included details of its taxonomy and nomenclature, including important synonyms, the northernmost and southernmost documented occurrence of the species on both the mainland and peninsular sides of the Gulf, its occurrence in the subregions of the Gulf, its known geographical distribution outside the Gulf, its general ecological community affiliation and habitat preferences, and pertinent literature references (Findley et al., in press). We here present a summary of these findings for

TABLE 5.1. Fish diversity in the Gulf of California by faunal regions.

Group	SGC	CGC	NGC	Total Species Known from Gulf
Myxini	1	2	1	3
Chondrichthyes	81	74	61	87
Actinopterygii	741	531	356	821
Total species recorded in region	823	607	418	911

Notes: SGC = Southern Gulf of California; CGC = Central Gulf of California; NGC = Northern Gulf of California.

the fishes and briefly discuss their implications for our understanding of the origins and evolution of the Gulf ichthyofauna.

The Macrofauna Golfo survey has revealed confirmed records of 911 species of fishes within the Gulf of California. This diverse ichthyofauna includes 3 species of primitive, jawless hagfishes (Class Myxini) and 87 species of cartilaginous fishes (Class Chondrichthyes: sharks, rays, and chimaeras) but is dominated by ray-finned fishes (Class Actinopterygii), with 821 species known from Gulf waters (table 5.1). This number includes several species known from only a single or a few sporadic records. More of these "rare" species certainly remain to be documented, as do fishes from poorly sampled habitats such as deep reefs.

In his summary of the Gulf ichthyofauna, Walker (1960) reported 526 species of fishes. The significant increase in the number of Gulf fish species reported here is the consequence of several factors. First, Walker's southeastern limit of the Gulf was placed at Topolobampo, in northern Sinaloa, whereas ours extends southward to Cabo Corrientes, in northern Jalisco. Thus our survey includes approximately 159 species known from the extreme southeastern Gulf that are not known from the Baja California peninsula or from north of Topolobampo along the mainland. Second, Walker essentially excluded deep-sea fishes whereas our tabulation captures all known species, including midwater and deep-sea fishes reported from the Gulf. Third, the taxonomy and systematics of many groups of fishes have since been clarified. This includes 117 new species described in the years since 1960, a process that continues today (e.g., Castro-Aguirre et al. 2002; Bussing and Lavenberg 2003; Castro-Aguirre et al. 2005; Møller et al.

2005; Pérez-Jiménez et al. 2005; van der Heiden and Plascencia-González 2005). Finally, our survey is based on more than thirty years of additional collecting efforts in the Gulf, during which time many "extra-limital" species have been newly recorded from Gulf waters (e.g., Tavera et al. 2005; Trujillo-Millán et al. 2006).

The diversity of Gulf fishes varies considerably from north to south, with nearly twice the number of species having been recorded in the southern Gulf (SGC, fig. 5.4) as in the northern Gulf (NGC, table 5.1). In the Gulf, this pattern is exaggerated for at least three reasons. First, cold winter water temperatures of the northern Gulf limit the survivorship of some tropical fishes (Thomson and Lehner 1976; Lehner 1979). Second, the northern Gulf generally lacks some habitats found in the central and southern parts of the Gulf, especially the deeper ocean basins (Alvarez-Borrego 1983; Brusca et al. 2005; Alvarez-Borrego, chapter 2 in this volume). Third, the northern Gulf lacks a direct connection with marine waters to the north and west, which precludes the easy movement of temperate fishes into the region.

Biogeographic Origins of Gulf Fishes

Because the Gulf of California lies at the broad intersection of the temperate and tropical faunal regions of the northeastern Pacific Ocean, its ichthyofauna represents a mixture from those two regions in addition to a number of both widespread and endemic species. Despite earlier arguments to the contrary (e.g., Briggs 1974), it is clear that the Gulf's ichthyofauna is primarily tropical. Eighty-seven percent of Gulf fish species (like the one shown in fig. 5.2) also occur south of the Gulf in the tropical eastern Pacific, whereas only 31 percent of Gulf fish species also occur in the temperate northeastern Pacific along the outer coast of the Baja California peninsula and northward. Many of the latter species in the Gulf are found in these more northerly areas mainly as vagrants—especially during El Niño events, when tropical species recruit to the temporarily warmer waters to the north (e.g., Lea and Rosenblatt 2000). Only 2 percent of continental shelf fishes in the Gulf have primarily temperate distributions. These are the *northern disjuncts* (Walker 1960; see fig. 5.3): species that occur in the northern Gulf and along the outer coast of the Baja California peninsula and/or off California but are generally absent from the southern Gulf.

Northern Disjunct Fishes. The prominent winter thermal gradient in the Gulf has led to the persistence of a distinctive temperate component of the fish fauna in its northern part. This includes a few endemic species of temperate origin plus the so-called northern disjunct species, which are also present along the outer Baja California peninsula and coastal California but are absent from the southern Gulf (Walker 1960). During glacial and interglacial cycles, the transition zone between cooler and warmer winter waters along the outer peninsula and within the Gulf shifted to the south and to the north, respectively. This had the effect of periodically isolating coolwater fish species of the northern Gulf from those on the outer peninsula and coastal California (as they are today) and variously recombining them during colder intervals. The geologically/oceanographically recent isolation of these fishes has led to genetic divergences between several Gulf and outer peninsular populations, but the degree of divergence varies considerably (Stepien et al. 2001; Bernardi et al. 2003; Jacobs et al. 2004; Sandoval-Castillo et al. 2004). Some species—such as the mussel blenny, *Hypsoblennius jenkinsi* (see Present 1987), opaleye, *Girella nigricans* (Terry et al. 2000), Mexican rockfish, *Sebastes macdonaldi* (Rocha-Olivares et al. 2003), and sargo, *Anisotremus davidsonii* (Bernardi and Lape 2005)—present forms that are morphologically and genetically similar in the two regions, so the populations are considered conspecific. Other species, such as the grunions (Moffatt and Thomson 1975), have significant morphological differences and are thus considered distinct species, one of which is endemic to the Gulf.

Gulf Endemic Fishes. A prominent component of the Gulf ichthyofauna is its endemics (Walker 1960), or species restricted to its waters (e.g., *Axoclinus nigricaudus;* fig. 5.5). Eighty-two species, or nearly 10 percent of Gulf fishes, are found nowhere else. This number includes several species with more northerly affinities, but it is dominated by fishes from primarily tropical groups. Fishes endemic to the Gulf are found in all coastal habitats and include benthic and demersal soft-bottom species such as croakers (three species), some families of eels (four species), anchovies (three species), New World silversides (three species) and gobies (four species) as well as reef-associated species such as damselfishes (two species), clingfishes (six

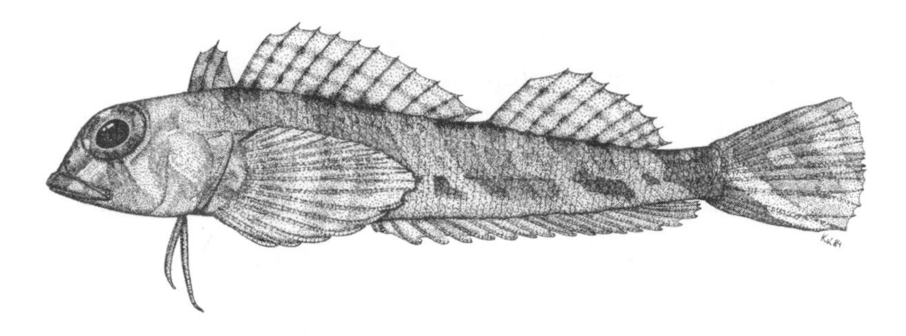

5.5. The Cortez triplefin or tres aletas colinegra, *Axoclinus nigricaudus* (Tripterygiidae), a Gulf endemic, was described by Gerald R. Allen and D. Ross Robertson in 1991. Drawing by K. Kotrschal.

species), labrisomid blennies (seven species), tube blennies (eight species), and gobies (seven species). Pelagic offshore and deep-sea fishes tend to have broad distributions and, accordingly, none is endemic to the Gulf.

The phylogenetic relationships of most Gulf endemics are not well studied, but a preliminary analysis shows them to be quite varied, reflecting the complex nature of the Gulf ichthyofauna. Among the few species whose relationships have been hypothesized or can be inferred from their systematics, the majority are most closely related to more southerly tropical eastern Pacific species. This includes, for example, the northern fraildisc clingfish, *Pherallodiscus funebris;* it is closely related to the southern fraildisc clingfish, *P. varius,* a similar species occuring along the Pacific coasts of southern Mexico and Central America (Briggs 1955). This pattern is seen in at least 14 species of Gulf endemic fishes and implies allopatric speciation of these forms within the eastern Pacific (Hastings 2000). Another common pattern is for Gulf endemics to be most closely related to warm-temperate species occurring off the coasts of California and/or the outer side of the Baja California peninsula. This includes the Gulf grunion, *Leuresthes sardina,* whose only known congener (the California grunion, *L. tenuis*) occurs along the west coast of North America from San Francisco to Baja California (Moffat and Thomson 1975). This pattern is seen in at least six Gulf endemics and implies allopatric speciation via isolation of these relatively coldwater-loving forms by the southern extension of the Baja peninsula into tropical waters (Walker 1960). Similarly, the northern-disjunct component of the Gulf ichthyofauna shows varying degrees of genetic isolation consistent with this mechanism of isolation (Huang and Bernardi 2001; Stepien et al. 2001; Bernardi et al. 2003; Jacobs et al. 2004; Sandoval-Castillo et al. 2004). A third pattern is seen in at least four Gulf endemics (two pairs) that are most closely related to other Gulf-occurring species: the North American silverside genus *Colpichthys* (Crabtree 1989) and the chaenopsid genus *Emblemaria* (Stephens 1963). This pattern implies sympatric or microallopatric speciation within the Gulf. At least five Gulf endemic species are most closely related to fishes from much more distant areas. This includes the slow goby, *Aruma histrio,* whose closest relative is found along the Pacific coast of South America (Hoese 1976); the saddlebanded goby, *Barbulifer mexicanus,* whose closest relative occurs in the Caribbean (Hoese and Larson 1985); and the Cortez pipefish, *Syngnathus carinatus,* which is most closely related to an Indo-Pacific species (Fritzsche 1980; Dawson 1985). The phylogenetic relationships of most Gulf endemic fishes remain unknown, and the complex evolutionary history of this endemic ichthyofauna is clearly deserving of further study.

Within the Gulf, there is relatively little regional endemism. Of the 418 fish species known from the northern Gulf (NGC, table 5.1), only 13 are restricted to this region. These include species such as the delta silverside, *Colpichthys hubbsi* (Crabtree 1989), which are likely associated with the historical estuarine conditions of the northern Gulf, and several species of endemic rockfishes of the genus *Sebastes* that are restricted to the Midriff Islands region (Chen 1975; Love et al. 2002). Even fewer narrow-range endemics are known from the central (CGC) and southern (SGC) portions of the Gulf. Only 5 of the 607 species recorded from the central Gulf have not been reported elsewhere, whereas only 9 of the 823 species known from the southern Gulf are restricted to this region. Most of these are poorly known species that may well have wider distributions within the Gulf and possibly outside it.

One particularly interesting pattern, documented by Thomson and Gilligan (1983) for endemic reef fishes of the Gulf, is that endemic species are often more abundant than more widespread related species. In a survey of the distribution and abundance of primary reef fishes (i.e., those living in direct association with reefs), the authors found that eight of the thirteen (61 percent) most abundant species are Gulf endemics. These are all small species and include seven blennioids and one clingfish. This pattern of narrow-range endemic species being the most abundant in their region of occurrence has been observed in fishes from other regions (e.g., Randall 1976), but its underlying causes have scarcely been studied. It may result from selection for a particular set of traits ideally suited to a particular microhabitat or region that is not shared by more widely distributed species (i.e., specialist versus generalist strategies). Alternatively (or in addition), this pattern may be related to larval characteristics of endemic species that promote their retention on local reefs compared to more widespread species (Brogan 1994; Victor and Wellington 2000; Robertson 2001). The underlying causes of this pattern remain one of the many mysteries surrounding fishes in the Gulf of California.

Variation of Fishes within the Gulf

Most populations of fish species show little if any morphological variation within the Gulf. This is mirrored by a general lack of genetic variation in the very few species whose genetics have been studied, including the Panamic fanged blenny, *Ophioblennius steindachneri* (see Riginos and Victor 2001), the Pacific red snapper, *Lutjanus peru* (Rocha-Olivares and Sandoval-Castillo 2003), and the flag cabrilla, *Epinephelus labriformis* (Craig et al. 2006). Such apparently highly dispersive species exhibit minimal genetic variation within the Gulf and, indeed, little genetic variation throughout the remainder of their broad ranges in the tropical eastern Pacific. In contrast, significant morphological variation within the Gulf has been demonstrated for a few small, reef-associated, endemic blennies, including the browncheek blenny, *Acanthemblemaria crockeri* (see Lindquist 1980), the lizard triplefin, *Crocodilichthys gracilis* (Rosenblatt 1959), and the redrump blenny, *Xenomedea rhodopyga* (Rosenblatt and Taylor 1971). Although the genetic basis for variation in these species has not been studied, significant genetic variation is evident in northern and central Gulf populations of the Gulf triplefin, *Axoclinus nigricaudus* (Riginos and Nachman 2001; Riginos and Victor 2001; fig. 5.5). This genetic break corresponds to similarly located genetic breaks in four other Gulf fishes (Riginos 2006) and roughly coincides with the location of a probable ancient seaway across the mid-part of the Baja California peninsula (Upton and Murphy 1997; Grismer 2000; Riddle et al. 2000a,b; Bernardi et al. 2003; Riginos 2006; but see Jacobs et al.

2004). Other fine-scaled genetic studies of fishes within the Gulf are needed to establish the generality of these patterns and their underlying bases.

Another aspect of variation within Gulf fishes that remains to be fully explored is the presence of so-called Bergman size clines. Several species of small, reef-dwelling triplefin blennies (Gilligan 1991) and gobies (Findley, pers. obs.) are smaller in the southern Gulf than in the northern Gulf. The underlying or causal mechanism for this trend in species being smaller in lower latitudes, which is common for many organisms, has not been resolved—especially for ectothermic species like most fishes (but see Van Voorhies 1996).

The Gulf of California as an Ecological and Biological Observatory

A considerable body of knowledge exists on the ecology and behavior of Gulf of California fishes, especially those associated with reefs. For decades, the often clear and protected waters have lured ecologists and ethologists interested in fishes to the Gulf. A landmark study in this arena was Hobson's (1968) survey of predator–prey interactions among a wide array of reef fishes in southeastern Baja California Sur. Since that time, numerous researchers from a variety of institutions have contributed to our knowledge of the ecology and behavior of Gulf reef fishes. Hobson's study on predatory fishes was followed by a number of studies on predatory sharks (e.g., Klimley and Nelson 1981, 1984; Galván-Magaña et al. 1989; Klimley et al. 1993; Villavicencio-Garayzar 1996), herbivorous fishes (e.g., Montgomery 1980, 1981; Montgomery et al. 1980), and interspecific feeding associations (e.g., Montgomery 1975; Strand 1977, 1988). The mating systems and reproductive behavior of a variety of Gulf reef fishes have been studied, including groups as diverse as angelfishes (Moyer et al. 1983), wrasses (Hoffman 1983, 1985), damselfishes (Petersen and Marchetti 1989; Hoelzer 1990, 1995), sea basses (Hastings and Petersen 1986; Petersen 1987), snappers and groupers (Sala et al. 2003), labrisomid blennies (Petersen 1988), triplefin blennies (Petersen 1989; Neat 2001), tube blennies (Hastings 1988a,b), and sharpnose puffers (Kobayashi 1986). In addition to reef fishes, the reproductive behavior of some of the Gulf's New World silversides has also been studied in some detail, especially the unusual Gulf grunion, *Leuresthes sardina* (Thomson and Muench 1976), and the intertidally spawning false grunion, *Colpichthys regis* (Russell et al. 1987).

The study of Gulf fishes has also played a role in debates on the nature of reef fish communities (Sale 1980). A notable early study by Thomson and Lehner (1976) documented the seasonality of rocky intertidal fishes in the far northern Gulf and provided evidence supporting a deterministic or competitively based regulation of reef fish communities. Several studies have explored various aspects of the ecological biogeography of Gulf fishes. For example, Lehner (1979) analyzed geographic distribution and community structure of several species of rocky-shore fishes from numerous localities along a latitudinal gradient stretching from the northern Gulf of California to northern Peru, and Molles (1978) used Gulf of California reef fishes to explore the theory of island biogeography. Thomson and Gilligan (1983) used Gulf reef fishes to analyze differences between island and mainland fish communities, concluding that islands support a greater diversity than adjacent mainlands for fishes directly associated with reefs. A few studies have explored the use space by Gulf reef fishes: Lindquist (1985) studied competition of blennies for space; and Strand (1977) explored space use and foraging associations in Gulf reef fishes. There have also been several investigations on the diversity and ecology of reef fish communities at different places in the Gulf (e.g., Pérez-España et al. 1996; Sanchéz-Ortiz et al. 1997; Jiménez-Gutiérrez 1999; Aburto-Oropeza and Balart 2001; Arreola-Robles and Elorduy-Garay 2002; González-Cabello 2003; Campos-Dávila et al. 2005). Finally, fishes of the Gulf have been included in recent analyses of the broad distributional patterns of eastern Pacific shore fishes (e.g., Hastings 2000; Robertson et al. 2004; Mora and Robertson 2005a,b).

A number of studies have been conducted on aspects of the biology and ecology of selected species or groups of species in the Gulf. Notable in this arena are the studies of Kotrschal and Lindquist (1986), Kotrschal and Thomson (1986), and Kotrschal (1988) on feeding ecology and morphology of the Gulf's diverse blennioid fishes. Similarly, Gulf fishes have been the focus of numerous insightful physiological studies, notably in the areas of temperature and salinity tolerances (Heath 1967; Reynolds and Thomson 1974; Rowell et al. 2005), tolerance to low oxygen levels (Barlow 1961; Todd and Ebeling 1966), and biochemistry of larval metamorphosis (Pfeiler and Luna 1984; Pfeiler 1986), among others.

Several studies have been conducted on larval fishes in the Gulf. These include early inventories (e.g., Moser et al. 1974), which continue today (e.g., Aceves-Medina et al. 2003a,b), as well as a variety of studies describing the development of certain species (e.g., Brogan 1996). Brogan (1994) explored the fine-scale distribution of reef fish larvae in the central Gulf, reporting that larvae of most groups of fishes occur far from reefs but that a few—such as tube blennies (Chaenopsidae) and triplefin blennies (Tripterygiidae)—remain within a few meters of reefs throughout development. This has important implications for reducing dispersal (and thus gene flow) and for increasing the probability of speciation in such fishes (Riginos and Victor 2001).

Conservation of Gulf Fishes

A wide review of the conservation status of Gulf of California fishes is beyond the scope this chapter. Nevertheless, we take this opportunity to add our voices to those of others calling for conservation measures in the Gulf. Fishes of the Gulf are deserving of our care and protection in their own right. In addition, they are—in one way or another—of great economic importance to Mexico, and their sustainability requires increased and continued conservation efforts. Not only are they a significant source of revenue via fisheries catches and the burgeoning aquarium trade, but Gulf fishes also provide an irresistible attraction for tourists eager to experience them firsthand either by angling or by observing them underwater in the often clear, warm waters of the Gulf.

Fishes in different habitats and different regions of the Gulf face numerous and varied challenges to their well-being. During the past several decades, benthic and demersal soft-bottom fishes of the continental shelf of the northern, east-central, and southeastern Gulf have been subjected to nearly unrelenting seabed trawling by industrial shrimp fishing (Magallón-Barajas 1987; Galindo-Bect et al. 2000; García-Caudillo and Gómez-Palafox 2005). Not only are adults and juveniles of many fish species captured as by-catch and discarded dead (Ortíz de Montellano 1987; Nava and Findley 1994), but trawling greatly reduces habitat complexity: it changes these once-biodiverse areas to essentially two-dimensional barren wastelands, reminiscent of clear-cut forests (Watling and Norse 1998; Dayton

2004). Although few comparative quantitative data are available (Guevara-Escamilla 1974; Rosales-Juárez 1976; Romero-C. 1978; Young and Romero 1979; Pérez-Mellado and Findley 1985; van der Heiden 1985; Nava-Romo 1994; García-Caudillo et al. 2000), it is clear that both the abundance and diversity of fishes in the heavily trawled areas of the Gulf have declined dramatically.

Estuarine fishes, a historically dominant component of the northern Gulf and Colorado River delta (Hastings and Findley 2006), have also experienced major habitat alterations due to the damming and diversion of the Colorado River starting in the early part of the last century (Galindo-Bect et al. 2000; Brusca and Bryner 2004; Brusca et al. 2005; Rowell et al. 2005). The lack of consistent freshwater flow into the northern Gulf has changed the system from typical estuarine conditions with a prominent salt wedge to a hypersaline system (Lavín et al. 1998; Lavín and Sánchez 1999). The impact of these changes may never be fully known because we generally lack quantitative knowledge of the fish communities prior to these fundamental changes in freshwater flow regimes (but see Rowell et al. 2005). Although comparative data are sparse, these same threats face estuarine (*estero*) fishes along the southeastern coast of the Gulf, where rivers are being increasingly diverted for agriculture (Ruíz-Luna and Berlanga-Robles 1999) and where coastal lagoons are losing important nursery and refuge habitats to an ever-accelerating pace of aquaculture development (e.g., shrimp farms) and small-boat marina development. Similarly, many fishes in the Gulf depend upon mangroves for early development, but these extremely important habitats continue their precipitous decline in response to increased coastal development (Balart et al. 1993; Castro-Aguirre et al. 1999; Whitmore et al. 2005).

Large pelagic fishes in Gulf waters face threats from ever-increasing gillnet and long-line fisheries. These nets and hooks are particularly harmful to shark populations that, even without fishing pressure, can recover only slowly because of their low fecundity and slow growth to maturity (Applegate et al. 1993; Musick et al. 2000), and to other large fishes such as the totoaba (Flanagan and Hendrickson 1976; Barrera-Guevara 1990; Cisneros-Mata et al. 1995a) and Gulf corvina (Rowell et al. 2005).

Similarly, threats to Gulf reef fishes are increasing. Many areas near fishing communities and urban centers are heavily overfished, forcing fishers to venture farther afield and thus reducing fish populations over increasingly larger areas (Sala et al. 2004; Sáenz-Arroyo et al. 2005a,b). The nascent large-scale harvesting of aquarium fishes (Aburto-Oropeza and Sánchez-Ortiz 2000), if unregulated, is a potential threat to otherwise unexploited species such as damselfishes, butterflyfishes, angelfishes, and wrasses.

These and other threats are best dealt with by implementation of well-designed marine protected areas together with strict enforcement of their rules and regulations. Overwhelming evidence indicates that marine protected areas work to protect adult populations and provide sources of recruits to other areas (Agardy 1997; Sobel and Dahlgren 2004). Marine conservation plans are available for the Gulf (e.g., Sala 2000; Sala et al. 2002), some of which have begun to be implemented. These include most notably the Upper Gulf of California and Colorado River Delta Biosphere Reserve in the northernmost Gulf (*Diario Oficial* 1993; INE 1995) and the marine park in the Loreto area of Baja California Sur (Campos-Dávila et al. 2005), among others (Carvajal et al. 2004). The islands and protected areas of the Gulf of California were recently declared a World Heritage site (IUCN 2005). We applaud these efforts.

Reporting on the 1911 *Albatross* cruise to the Gulf, Charles H. Townsend stated: "In cruising about the Gulf the many large-sized fishes to be seen leaping indicate their abundance" (Townsend 1916, p. 452). It is hoped that conservation efforts and their enforcement will reverse the sad decline in the abundance and diversity of Gulf fishes, eventually restoring the abundance of fishes that characterized these waters nearly a century ago.

ACKNOWLEDGMENTS

We thank the several researchers in the Macrofauna Golfo project who assisted in compiling data on fishes and in other technical aspects, especially Jorge Torre, J. Manuel Nava-Romo, Héctor Plascencia, Mauricia Pérez-Tello, Rocio Güereca, and J. A. Barragán. For review of some data sets we thank the following specialists: Bruce Collette, Héctor Espinosa-Pérez, Phil Heemstra, Cynthia Klepadlo, Carlos Navarro-Serment, Richard Rosenblatt, Bill Smith-Vaniz, and H. J. Walker. We thank the curators and collection managers who provided access to fish specimens and data records in their charge, including David Catania, Tomio Iwamoto, and Bill Eschmeyer at CAS; R. Rosenblatt, C. Klepadlo, and H. J. Walker at SIO; Jeff

Seigel and Bob Lavenberg at LACM; Don Buth at UCLA; and Peter Reinthal at UAZ. The initial phase of the Macrofauna Golfo project was funded by grants from CONABIO (Comisión Nacional para el Conocimiento y Uso de la Biodiversidad), Instituto Nacional de la Pesca-SEPESCA (via Pronatura), and USAID and the Homeland Foundation (via Conservation International Mexico–Región Golfo de California). Continuing support came from CEMEX and Ford Motor Company (via Conservation International), Conservation International (CI), CIAD, CIDESON (now IMADES), SIO, and Arizona-Sonora Desert Museum. Special thanks are due Alejandro Robles of CI, who conceived the project along with LTF and Juan Carlos Barrera (then of CIDESON), and very special thanks go to Machangeles Carvajal, the director of CI's Región Golfo de California Program, who has been the project's staunchest supporter over the years, along with A. Robles and (more recently) Inocencio Higuera and Alfonso Gardea of CIAD. Special thanks are also due Rick Brusca for the invitation to participate in the "Gulf of California Symposium 2004" and for his continued patience with the "fish boys."

CHAPTER 6

The Importance of Fisheries in the Gulf of California and Ecosystem-Based Sustainable Co-Management for Conservation

MIGUEL Á. CISNEROS-MATA

Summary

The waters of the Gulf of California are well known for their high productivity, which results from a complex array of physiographic and oceanographic attributes. This high year-round productivity supports large populations of sea birds, marine turtles and marine mammals; it also supports important fisheries (industrial and small-scale) of small pelagic fishes, predatory fishes, and jumbo squid. The combination of beautiful coastlines and fishery products of high market value has been a strong attractant for human settlement: the Gulf coastal region is inhabited by 8 million people, and the main economic activities are shrimp cultivation, fisheries, and tourism. The Gulf of California is commonly regarded as Mexico's foremost fishing region because 40–50 percent of the country's commercial fisheries are located here. This chapter provides insight into the socioeconomic and ecological importance of fisheries in the Gulf. A historical account identifies five stages in the history of fishing in the Gulf, from exploration (1930s to late 1960s) to the present, critical phase. Four commercial fishing sectors in the Gulf are identified: artisanal or small-scale, industrial, sport or recreational, and subsistence fishing; only the first two are discussed here. Artisanal fishers, typically using hand-operated fishing gear in *pangas* (small boats) with outboard gasoline motors, catch about 80 species, whereas industrial boats with diesel engines generally fish for about 6 species guilds. An estimated 50,000 artisanal fishers operate 25,000 pangas, and some 10,000 fishers work on approximately 1,300 boats in the industrial sector. Fisheries in the Gulf are highly socioeconomically relevant in terms of jobs and regional income generated per capita. Artisanal fisheries are the region's only primary sector in which two direct jobs are generated for each investment of about $10,000 (U.S. dollars throughout), and, in addition, they immediately obtain food and goods for trade. A rough analysis is presented here on the importance of Gulf fisheries with respect to job generation, and shrimp is identified as a key driver in this respect. Nearly 90 percent of all pangas operate during the shrimp season (September to March); as the shrimp fishing comes to an end, fishers shift to other fishing resources or stop fishing. Most seasonal artisanal fisheries thus actually generate labor days rather than permanent jobs. In a given year there are nearly 18,000 pangas that work in the Gulf with an average of nearly 36,000 labor days. Industrial fisheries in the Gulf include 1,280 active boats with crews that vary from five to eleven members. If we assume a mean ratio of 1:6 for direct to indirect jobs, then total labor days generated by fishing activities in the Gulf amount to an average of about 271,000 labor days per year. The commercial catch in the Gulf of California during 2002 was 597,153 metric tons (mt); artisanal fisheries yielded 26.8 percent or an average 8.9 mt/boat, while industrial fisheries averaged 336.2 mt/vessel. In that year, the firsthand or "beach" value of Gulf fisheries amounted to $332 million or 27.7 percent of the country's fisheries' total value, with an estimated $8,881/year per panga and $132,796/year per industrial fishing vessel (costs not subtracted from firsthand value). In terms of landed weight, 90 percent of the total catch includes ten species groups, of which the most important for artisanal fisheries are shrimp, jumbo squid, and clams; most important for industrial fisheries are sardines, shrimps, and tunas. Sardines and squid constitute almost 60 percent of the total Gulf catch (cultivated shrimp, which is highly valued, is not included in the analysis because it is not caught in the wild). Excluding squid and sardines, fisheries landings in the Gulf represent 10 percent (130,000 mt) of Mexico's total landings by weight, yet with this reduction (from 60 to 10 percent) the total value of landings decreases by only 12 percent. The squid and sardine fisheries generate 9,849 jobs per year in the Gulf, which represents 3.7 percent of the Gulf's total fisheries labor days. At specific ports where the bulk of squid and sardine landings take place, these two fisheries are clearly of high economic importance.

Management and conservation efforts for Gulf fisheries should consider both ecological and socioeconomic impacts. This paper offers an analysis, based on a World Wildlife Fund publication, of current challenges and opportunities in that regard. Artisanal fisheries and industrial fisheries are considered priority threats to the ecological stability of the Gulf, along with urban and tourist development. Currently, 85 percent of the Gulf's fisheries are either at their maximum sustainable yield or overexploited. Sardines and jumbo squid may have room for careful and regulated increased catch rates. Remedial actions are needed for development and conservation in the Gulf's artisanal and industrial fisheries. New paradigms should be adopted to address de facto open access and economic and social inefficiency due to market problems. Three key elements should guide sustainability: maximizing the value of fish products (instead of the short-term gross catch); establishing exclusive user rights in artisanal fisheries; and considering the ecosystem considerations, including management reference points, in fisheries co-management. Bottom-up grassroots interventions, enforcement by communities and authorities, and structural changes are mandated in the new National Fishing Law and should be implemented, along with FAO's Code of Conduct for Responsible Fisheries. Regionalizing fishing and promoting a sense of ownership should reduce mortality of exploited resources, minimize illegal fishing, and reduce ecosystem impacts. Regionalization has the potential to articulate marine protected areas and traditional fisheries management tools such as closed seasons, no-take zones, and size limits. Finally, the frequent claim that the Gulf of California is the most important fishing region in Mexico can be contested by pointing out that, when compared with the relative *weight* of the catch (mostly jumbo squid and sardines), the *economic* contribution is relatively small (even though these species remain highly relevant to the ecosystem). Biodiversity conservation and improved livelihoods of fishers can be achieved by combining fisheries management and conservation tools. Both goals must be explicitly considered in management plans, which need to be implemented with the active participation of fishers.

Introduction

With a surface of about 540,000 km^2, the Gulf of California (herein defined as the waters bounded by the states of Baja California, Baja California Sur, Sonora, Sinaloa, and Nayarit) is one of the most significant re-

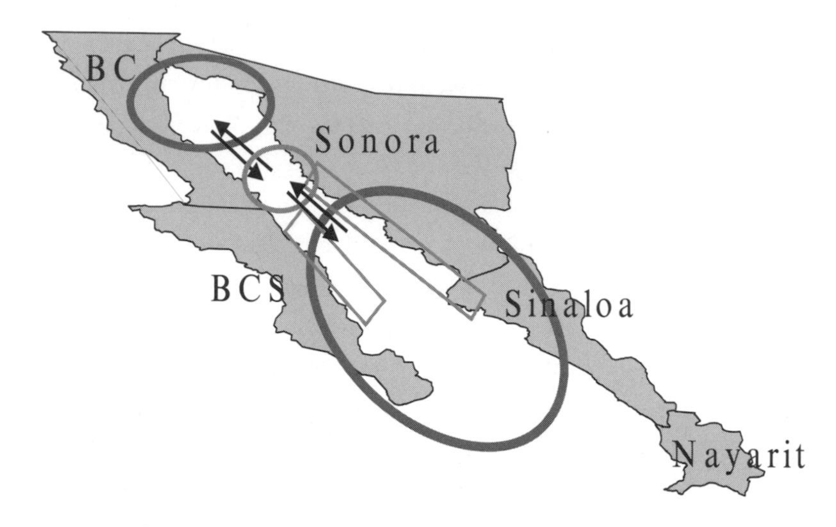

6.1. Map of three main physiographic regions: Northern, Midriff Islands, and Central-Southern.

gions on earth in terms of natural productivity (Alvarez-Borrego 1983). In this semi-enclosed sea, high primary productivity is driven by a complex topography, year-round high solar input, advection of deep waters, thermohaline circulation, and wind-driven upwelling in coastal shallows (fig. 6.1). The Gulf supports one of the world's most important concentrations of small pelagic fishes (anchovies and sardines) (Jacobson et al. 2001), which are a key food source for jumbo squid, predatory fishes, sea birds, and marine mammals (Alvarez-Borrego and Lara-Lara 1991; Velarde et al. 2004).

The Gulf coastal region is inhabited by 8 million people, and the principal human activities along the coastal zone are fisheries and tourism. An important and growing aquaculture industry produces about 95 percent of the cultivated shrimp along the east coast of the Gulf (SAGARPA 2004). Other relevant industrial activities are mining, livestock, and farming. Tourism is an emergent industry at specific sites in northwest Mexico; Fondo Nacional de Fomento al Turismo (http://www.fonatur.gob.mx/mar_de_cortes/Escalas_Nauticas/es/pdfs/sierra_madre.pdf) reported 2.1 million tourists in 2002.

Fisheries in the Gulf

It is commonly stated that the Gulf of California is Mexico's most important fishing region (SEMARNAP 2000; SAGARPA 2004) because half of the country's total annual commercial fisheries landings are caught here. However, there has been no thorough analysis contrasting the socio-economic and ecological importance of fisheries in the Gulf. Proper fisheries management and biodiversity conservation in the Gulf requires understanding the complex dynamics and evolution of regional fisheries in order to learn from that experience and adapt to new challenges (see Alcalá 2003 for a historical account of fisheries in Mexico; see Cisneros-Mata 2004 for reviews of fisheries growth and dynamics in the Gulf of California). Five general stages can be identified in the history of fishing in the Gulf as follows.

1. *Exploration and establishment* (1930 to late 1960s). Artisanal sector: sharks, totoaba (*Totoaba macdonaldi*), and turtles. Industrial sector: shrimp and, more recently, sardines.
2. *Massive investment* (1970s). Growing yet unnoticed competition between artisanal and industrial fishers for fish resources in industrial shrimp by-catch; growth in jumbo squid fishery (Ehrhardt et al. 1983; Ehrhardt 1991; Hernández-Herrera et al. 1998; Markaida et al. 2005); shrimp season set at 8 months, October to May.
3. *Stabilization* (1980s). Growth of artisanal shrimp fishing; shrimp farming established; beginnings of social unrest regarding growth of artisanal fishing.
4. *Saturation* (mid-1990s). Sardine overfishing and El Niño–related fishery collapse. Major growth of jumbo squid fishery. Growth of artisanal fishing of "trash" or alternative species: triggerfish, crabs, rays, jellyfishes. Some industrial shrimp boats turn to trawl-fishing rockfishes. Explosive development of shrimp farming.
5. *Critical phase* (2000s). Growth of illegal artisanal fishers, many practicing bare subsistence fishing. Shrimp season effectively lasts just 4–7 months (legally, September to March); industrial shrimp vessels in crisis; blue and white shrimp stocks near coastal lagoons overexploited; fuel (gasoline and diesel) subsidies fully implemented for commercial vessels.

The two main commercial fishing sectors in the Gulf have historically been industrial fisheries and artisanal or small-scale fisheries. Sport fishing and subsistence fishing are not addressed in this chapter, although a small but increasing number of artisanal fishers provide sport fishing and other tourist services in San Blas, Mazatlán, Puerto Peñasco, San Felipe, La Paz, and Loreto (Cisneros-Mata, pers. obs.). Artisanal fishers typically use 21-foot pangas with outboard gasoline motors and hand-operated fishing gear (long lines, gill nets, cast nets, hook and line, pots, traps) to catch about 80 species of fishes, crustaceans, and molluscs. Industrial fishers use larger (≥70 ft) boats powered by diesel engines and with a holding capacity of 15 metric tons or more as well as mechanized gear to catch small pelagic fishes with purse seine nets, shrimp with paired bottom trawl nets, rockfish with bottom trawl nets, hake with midwater trawl nets, and sharks with long lines and gill nets.

The dynamics of Gulf fisheries are driven by seasonality of available resources and de facto open access (WWF 2005a), so the actual extent of the fishing effort and its impact on natural resources is not well known. An estimated 50,000 artisanal fishers operate 25,000 pangas and some 10,000 fishers work on about 1,300 boats in the industrial sector. A significant number of small-scale shrimp fishers in Sinaloa are allowed to use a single trawl net per panga to fish in shallow coastal waters. Although these boats do not use engines to retrieve the net, an ingenious system is employed (see INP 2002 for details) that allows us to consider this a semi-industrial fishery.

Fisheries in the Gulf are definitely socioeconomically relevant in terms of jobs and regional income generated per capita. Artisanal fisheries are probably the region's only primary sector area where two direct jobs can be immediately generated with an investment of about $10,000 (Cisneros-Mata, unpublished data). Moreover, artisanal fishers, unlike farmers or livestock growers, can readily obtain their harvest for sustenance and trade.

Because of the Gulf fisheries' complex dynamics, it is difficult to estimate the number of jobs generated by artisanal fishing activities. Two strong attractors for fishers to enter and remain in the activity are the prospect of a job and the high economic value of particular seasonal resources. Nearly 90 percent of all pangas in the Gulf are utilized during the open shrimp season (September to March), targeting the highest-valued fishing resource (see below). As the shrimp season progresses and comes to an end, fishers either abandon the activity or shift to other fishing resources (e.g., Spanish mackerel, sharks, crabs). Therefore, regarding some seasonal artisanal fisheries one should speak of seasonal labor days, or day's wages,

TABLE 6.1. Monthly estimated number of laborers in artisanal fisheries of the Gulf of California.

Month	% Shrimp	% Other	Pangas	Laborers/day
Sep	90	10	25,000	50,000
Oct	85	10	23,750	47,500
Nov	80	10	22,500	45,000
Dec	70	10	20,000	40,000
Jan	50	20	17,500	35,000
Feb	30	40	17,500	35,000
Mar	10	50	15,000	30,000
Apr	1	60	15,250	30,500
May	1	70	17,750	35,500
Jun	1	60	15,250	30,500
Jul	1	50	12,750	25,500
Aug	1	50	12,750	25,500

Notes: Two fishers per panga and a total of 25,000 pangas were used as the base for calculating numbers of laborers working each month. Last five rows correspond to the season during which shrimp fishing is banned; the 1% figure is a rough estimate.

rather than permanent jobs created. Table 6.1 shows a rough calculation of monthly fishing effort in terms of pangas and labor days for artisanal fishers in the Gulf. The (weighted average) number of pangas in the Gulf is 17,917, with an average of 35,833 labor days annually. In addition, given an estimated ratio of 6:1 for direct to indirect jobs (Alcalá 2003), one can estimate that artisanal fisheries generate a total of about 215,000 indirect labor days.

Industrial fisheries in the Gulf include a total of 1,280 active boats that mainly fish for sardines (Guaymas, Yavaros, and Mazatlán), shrimp (Guaymas, Yavaros, and Mazatlán), jumbo squid (Mazatlán), and hake and rockfish (Puerto Peñasco and Mazatlán). The number of crew members ranges from eleven on sardine and squid vessels to seven on shrimp trawlers. The weighted average is 7.1 crew members per vessel, for a total of 9,334 fishers. This means that, given the same ratio (6:1) of direct to indirect jobs, industrial fisheries in the Gulf generate nearly 56,000 labor days altogether. Thus, total labor days generated by fishing activities in the Gulf amounts to almost 271,000 (215,000 + 56,000) annually.

An in-depth analysis of commercial catch contributes to our understanding of the socioeconomic importance of fishing activities in the Gulf. Total commercial catch in the Gulf of California during 2002 was 597,054 mt,

TABLE 6.2. Commercial fisheries landings and "beach" value in the Gulf of California during 2002.

Fleet	BC	BCS	Son	Sin	Nay	Total
A. Landings (in metric tons)						
Artisanal						
Direct	4,759	52,519	66,594	25,412	10,755	160,040
Industrial						
Direct	2,693	1,154	92,813	23,336	1,010	121,007
Indirect	125	10	250,620	65,352		316,107
Total industrial	2,818	1,164	343,433	88,689	1,010	437,114
State total	*7,439*	*62,134*	*433,565*	*138,583*	*16,595*	
Gulf Total						**597,153**
B. Value (in thousands of U.S. dollars)						
Artisanal						
Direct	6,827	22,504	49,427	57,455	22,907	159,120
Industrial						
Direct	3,703	1,299	52,671	84,971	8,273	150,917
Indirect			16,464	5,253		21,717
Total industrial	3,703	1,299	69,135	90,225	8,273	172,635
State total	*10,529*	*23,803*	*118,563*	*147,680*	*31,180*	
Gulf Total						**331,755**

Notes: BC = Baja California, BCS = Baja California Sur, Son = Sonora, Sin = Sinaloa, Nay = Nayarit. Table developed from data at http://www.conapesca.sagarpa.gob.mx/wb/cona/cona_anuario_estadistico_de_pesca.

which amounted to 44 percent of Mexico's total in that year (table 6.2A). The artisanal fisheries catch was 26.8 percent of the total Gulf catch, or an average of 8.9 mt/boat; industrial catches contributed 73.2 percent of total Gulf catch, or an average of 336.2 mt/vessel. Sonora had the highest reported catch by both fishing sectors. In terms of economic value, in 2002, fisheries in the Gulf amounted to almost $332 million in firsthand or "beach" value (table 6.2B), which corresponds to 27.7 percent of the country's fisheries' economic total. Landings by artisanal and industrial fishers made up (respectively) 48 percent and 52 percent of the total value of catches in the Gulf. These figures yield an estimated $8,881/year per panga and $132,796/year per industrial fishing vessel. In terms of value of landings by state, Sinaloan fisheries are the most important for both fishing sectors.

These calculations are not expressed as revenues because costs were not subtracted from firsthand value. One study (WWF 2005b) found an average return of 30 percent in artisanal fisheries in the upper Gulf of California, although this is probably higher than normal because fish resources in that region are highly valued and have a marked seasonality (Cudney-Bueno and Turk-Boyer 1998). A study on industrial shrimp trawling (García-Caudillo and Gómez-Palafox 2005) found that about half the Gulf's vessels do not make a profit in an average fishing season.

In terms of catch volume, the most important groups of species for artisanal fisheries are shrimp, jumbo squid, and clams; 90 percent of the total catch consists of just ten species groups (table 6.3A). In terms of tonnage landed, sardines are by far the most important group of species for industrial fisheries, followed by shrimps and tunas (table 6.3B). Although Gulf fisheries supply 44 percent of Mexico's overall catch by weight, in terms of economic value the situation is quite different. As mentioned previously, my calculations—based on official government data—show that firsthand value of landings in the Gulf represent less than a third of the total of Mexico's fisheries. The reason for this disparity is twofold. First, sardines and squid constitute almost 60 percent of the total Gulf catch; second, cultivated shrimp—though highly valued and rapidly increasing—is not included in this analysis because it is not regarded as a wild-caught species.

If squid and sardines are not included in the calculations, then marine-caught landings in the Gulf represent only 10 percent (130,000 mt) of Mexico's total landings by weight. Yet if those species are excluded then the total value of landings in the Gulf decreases by only 12 percent, or $40 million. In other words, a 60 percent relative importance in volume landed signifies only 12 percent in economic value. The sardine fishery is based on a total of 37 sardine purse seiners, each with a crew of 11. The squid fishery is based on about 1,000 pangas with two fishers on board each (Nevárez-Martínez et al. 2000). For the squid fishery this amounts to 2,000 seasonal laborers. Given that the squid fishery runs for 6 months, on an annual basis 2,000 laborers would correspond to 1,000 direct jobs (2,000/2) or 7,000 total jobs per year (i.e., 1,000 × 6 = 6,000 indirect plus 1,000 direct jobs). On the other hand, the sardine fishery generates 407 direct jobs or 2,849 total jobs per year. Therefore, the squid and sardine fisheries can be

TABLE 6.3. Commercial fisheries landings in the Gulf of California by species group during 2002 (metric tons).

Species group	BC	BCS	Son	Sin	Nay	Total	USD/ mt	Cumm%
A. Artisanal fisheries								
Jumbo squid		32,142	46,775	1,546	5	80,468	213	50
Shrimp	240	50	4,700	7,800	2,700	15,490	5,510	59
Clams	311	9,601	26	290	18	10,246	590	66
Mojaras	58	180	799	5,027	3,192	9,256	935	71
Crabs	154	241	3,713	3,117	24	7,249	857	80
Sharks	506	1,004	1,457	1,842	447	5,256	934	76
Croakers	691	445	2,915	240	185	4,476	957	83
Spanish mackerel	80	334	1,983	923	574	3,894	889	86
Bass	286	3,231	68	6		3,591	477	88
Mullets	64	192	338	2,179	640	3,413	579	90
B. Industrial fisheries								
Sardine, indirect use			245,955	64,544		310,499	47	71
Sardine, human consumption			81,259			81,259	97	89
Shrimp	250	70	8,000	14,000	1,500	23,820	5,510	95
Tuna	2,800	1,100	750	9,000	10	13,660	830	98
Anchovy, indirect use			2,923			2,923	44	99
Mackerel			2,742			2,742	83	99
By-catch	125	10	1,742	808		2,685	2,754	100
Jumbo squid				773	5	778	213	100
Skipjack			121	35	8	164	578	100

Notes: BC = Baja California, BCS = Baja California Sur, Son = Sonora, Sin = Sinaloa, Nay = Nayarit. USD/mt = firsthand (beach) value per metric ton in U.S. dollars; Cumm% = cumulative landings. Table developed from data at http://ganaderia.sagarpa.gob.mx:8083/compass?scope=conapesca.

viewed as generating 9,849 jobs per year in the Gulf, which represents just 3.7 percent of the Gulf's total fisheries' labor days.

These conservative calculations indicate that the squid and sardine fisheries of the Gulf represent 60 percent volume landed, 12 percent economic value, and only 3.7 percent job or labor days creation. Thus, we can see that these fisheries should probably be valued more in terms of their impact on the Gulf's ecosystem functions (i.e., in terms of biomass removed from the ecosystem) than in terms of their socioeconomic importance. This might not be the case at specific sites such as Santa Rosalía, Guaymas, Mazatlán, and Yavaros, where the bulk of squid and sardine landings take place, respectively. At those sites, these two fisheries are of high local economic importance, although a reduced market for jumbo squid has limited the economic benefits for fishers (De la Cruz-González et al. 2005). This analysis suggests that sustainability of Gulf fisheries should take into consideration a balance between ecological and socioeconomic impacts, notably in the squid and sardine fisheries. Maintenance of healthy populations and trophic dynamics of sardines and jumbo squid are important interrelated components of the pelagic ecosystem of the Gulf, and their population sizes have direct effects on many other species, as well as on fisheries' take (Ehrhardt 1991; Jacobson et al. 2001; Ruíz-Cooley et al. 2004; Velarde et al. 2004). Because both the sardine and jumbo squid stocks are highly variable (Nevárez-Martínez et al. 2000; Sánchez-Velasco et al. 2002), their fisheries should be carefully monitored and regulated to avoid a collapse of the pelagic ecosystem and a socioeconomic crisis in the Gulf region.

Current Status and Challenges

An analysis conducted by World Wildlife Fund (2005a,b) in the Gulf of California identified main threats to biodiversity and sustainable development in the region. Threats were prioritized based on the following three criteria: surface area of the Gulf where the threat exists, severity of the threat per surface unit, and urgency of the threat (table 6.4). That analysis indicated artisanal fisheries and industrial fisheries as priority threats, along with urban and tourist development in the Gulf. A second group of threats includes shrimp farms, excessive freshwater use, and marine turtle consumption. In what follows I discuss potential interventions to address threats by unsustainable fishing practices.

Poor regulation of fishing activities has led to a condition where 85 percent of the Gulf's fisheries are either at their maximum sustainable yield or overexploited (Díaz de León-Corral and Cisneros-Mata 2000). None of the important artisanal fisheries allows for increased catch, and most require a reduction of fishing effort (table 6.5). The only important fisheries currently identified as having room for careful and regulated in-

TABLE 6.4. Threat analysis of human economic activities in the Gulf of California with regard to biodiversity and economic development.

Threat	Area	Severity	Urgency	Total	Priority
Artisanal fisheries	10	7	10	27	1
Industrial fisheries	8	8	6	22	3
Urban and tourism development	5	10	9	24	2
Aquaculture	4	9	8	21	4
Excessive water use	6	6	7	19	5
Turtle consumption	9	4	4	17	6
Pollution	7	5	5	17	7
Sport fishing	2	2	2	6	9
Logging	3	3	3	9	8
Substrate destruction due to sport diving	1	1	1	3	10

Note: "Threat" refers to specific unsustainable practices in the indicated activities (WWF 2005a,b).

creased catch rates in the Gulf are sardines and jumbo squid. In the jumbo squid fishery, economic considerations to promote development of the fishery (added value) need to be considered (De la Cruz González et al. 2005; SAGARPA 2004). For the sardine fishery, the growing trend of shrimp farming and, more recently, tuna ranching will most likely increase demand for fish meal and fresh-frozen sardines (WWF 2000; ATRT 2005). Therefore, sardine fisheries management in the Gulf will need to adopt strict control of fishing mortality to avoid overexploitation of this economically and ecologically critical fish stock.

Strict remedial actions are needed to orient future economic development and conservation in the Gulf's artisanal and industrial fisheries. The current state of fisheries in the Gulf is a consequence of insufficient fishing policies in Mexico (Alcalá 2003). New paradigms need to be adopted within a sound planning process that deals with the root issues of fisheries in the Gulf. One such analysis (WWF 2005a,b) identified the following as root causes for fisheries failures: (a) de facto open access due to insufficient enforcement and regulations; (b) illegal fishing, especially artisanal; (c) dated and inappropriate technologies in shrimp trawling; (d) economic and social inefficiency due to market problems; (e) lack of skills among the majority of fishers; and (f) poor knowledge resulting in insufficient regulations and saturated fishing effort.

TABLE 6.5. Status of the main artisanal fisheries of the Gulf of California.

Resource/fishery	Status	Recommendation
Clams	Deteriorated in BCS and Sinaloa; at its maximum in Sonora and BC	Do not increase fishing effort
Jumbo squid	With potential for development	Consider economic yield, in addition to biological yield
Shrimp	Sonora and upper Gulf of California: Blue and brown shrimps at their maximum sustainable yields Sinaloa and Nayarit: Brown shrimp at its maximum, with symptoms of deterioration; white and blue shrimps overexploited	Do not increase fishing effort
Snails	Panocha snail in BCS, and chino negro snail in BCS, at their maxima	Do not increase fishing effort
Crabs	In all states within the Gulf of California, fisheries at their maxima	Do not increase fishing effort
Gulf coney, basses	Fisheries at their maxima	Do not increase fishing effort
Croakers	Fisheries at their maxima	Do not increase fishing effort
Snappers	Fisheries at their maxima	Do not increase fishing effort
Jacks	In BCS, the fishery has potential for further development; in all other states, fisheries at their maxima	Do not increase fishing effort
Snooks	Deteriorated	Do not increase fishing effort
Spanish mackerels	Fisheries at their maxima	Do not increase fishing effort
Sharks	Fisheries at their maxima	Do not increase fishing effort

Note: Analysis based on information in SAGARPA (2004).

A policy for guiding future fisheries management in the Gulf will need to embrace three key elements.

1. Promote long-term fisheries goals and maximum value of fish products instead of short-term maximization of catch (Wilen 2005).
2. Implement exclusive user rights in artisanal fisheries to generate a sense of ownership that would foster co-management of resources (e.g., Pomeroy et al. 2004).

3. Explicitly implement ecosystem considerations, including reference points, in fisheries co-management (cf. Link 2002).

Before modern fisheries management in the Gulf can function as a tool for biodiversity conservation, profound changes and broad compliance with clear rules and regulations will be necessary (fig. 6.2). Institutional support should be available to promote sustainable practices. Maximizing added value requires training of artisanal fishers in particular.

Sound management and conservation in the Gulf require integrated, bottom-up grassroots interventions (Pomeroy et al. 2004); enforcement by communities and authorities; and structural changes in the form of modern regulations, including fisheries management plans as mandated in the new National Fishing Law. Elements for sustainable fishing and conservation in the Gulf could be implemented following guidelines contained in the Code of Conduct for Responsible Fisheries (FAO 1995). Unfortunately, Mexican legislation has yet to explicitly incorporate the FAO Code. The

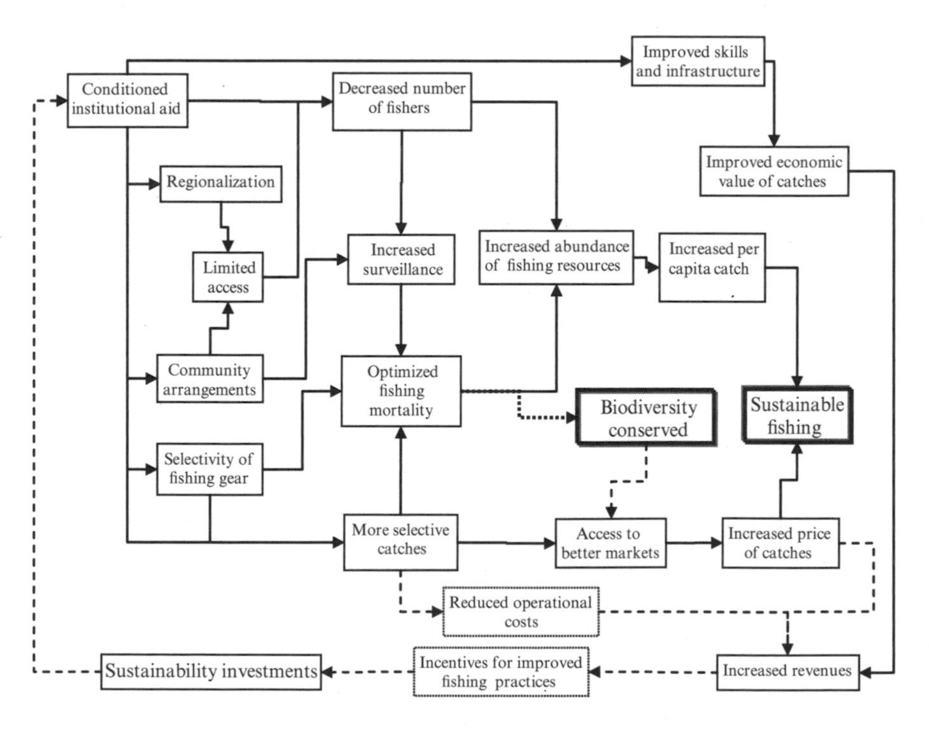

6.2. Conceptual model for sustainable fishing and biodiversity conservation.

National Fisheries Chart, a Mexican public fisheries management document, recently recognized the need for a precautionary approach and sustainable fishing according to FAO standards: "Excessive fishing and world overcapitalization of the fishing industry have led to implementation of control measures to regulate activities. This resulted in adoption, in 1995, of the FAO Code of Conduct for Responsible Fisheries, of which Mexico is a key proponent" (SAGARPA 2004).

A management/conservation plan for the Gulf should recognize the region's heterogeneity. Regionalizing fishing will promote a sense of ownership through exclusive access to fishing resources. Explicit goals of this strategy should be to reduce fishing mortality of exploited resources, to minimize illegal fishing, and to reduce ecosystem impacts. The latter is particularly true in the case of shrimp trawling in order to (a) minimize damage to the seafloor and (b) reduce the economic impacts to artisanal fishers, since numerous by-catch species in the shrimp trawl would otherwise be captured by artisanal fishers (García-Caudillo et al. 2000; García-Caudillo and Gómez-Palafox 2005). Regionalization could become a key component in the process of optimizing fishing effort because it has the advantage of articulating marine protected areas, which is currently the most relevant conservation tool for marine resources (Hastings and Botsford 1999; Gell and Roberts 2003), as well as traditional fisheries management tools such as closed seasons, no-take zones, and size limits.

Concluding Remarks

For many decades, the lack of a clear fisheries policy in Mexico led to the current state: little room for increasing capture fisheries and the majority of exploited resources at or beyond their maximum sustainable level of annual catch rates. Therefore, future exploitation of most resources should be done carefully.

The Gulf of California is actually not the most important fishing region in Mexico, given that the socioeconomic contribution in the region by two massive fisheries (jumbo squid and sardine) is relatively small. Even so, those two stocks—whose availabilities in the Gulf are highly variable—are crucial to the ecosystem and hence should be properly managed to avoid both ecological and socioeconomic crises. These factors must be explicitly considered in the management plans, which are now mandated by law, that need to be implemented and enforced.

Management and conservation strategies for Gulf of California fisheries should be implemented in an integrated manner. Elements should include: (1) maximizing value, not volume; (2) promoting rights-based access to reduce fishing mortality; and (3) implementing ecosystem-based co-management. Institutional aid should be tailored to best fishing practices that are oriented to increasing added value and reducing impacts on the environment. Fishers should be the stewards of their target resources and should be allowed to participate actively in management decisions and in research and monitoring activities. A set of structural changes should also be implemented in the Gulf's fisheries; these include the FAO Code of Conduct, regionalized fisheries, fisheries management plans, and enforcing the new National Fishing Law to promote recovery of depleted stocks. Biodiversity conservation should and can be achieved with a combination of fisheries management and conservation tools and with an explicit goal of improving the livelihoods of legal fishers.

CHAPTER 7

Sea Turtles of the Gulf of California

Biology, Culture, and Conservation

JEFFREY A. SEMINOFF

Summary

The productive waters of the Gulf of California provide important feeding and developmental habitats for five of the world's seven sea turtle species. The most abundant species in coastal waters is the green turtle, known locally as the black turtle (*Chelonia mydas*). The hawksbill turtle (*Eretmochelys imbricata*) also frequents nearshore waters, although it is more prevalent in the southern Gulf. The olive Ridley (*Lepidochelys olivacea*), leatherback (*Dermochelys coriacea*), and loggerhead (*Caretta caretta*) may also be present along the coast, but these species are more common in the Gulf's offshore waters. Exploitation of eggs and turtles for food, degradation of marine and nesting habitats, and incidental mortality relating to marine fisheries have reduced the local populations of all species throughout the region. Industrial-scale exploitation of sea turtles in the region began in the early twentieth century, and hunting was focused primarily on the green turtle. By the 1960s the harvests reached a production peak, primarily for domestic consumption. At the same time, large-scale harvest of nesting females and eggs at nesting beaches, as well as individuals of both sexes from nearshore waters, occurred farther south on the Pacific coast. As it did earlier in other parts of the world, the demand for green turtles, as well as other species, outstripped the ability of these slow-growing animals to regenerate. A total ban on all sea turtle exploitation was declared by a Mexican presidential decree in 1990. Although this legislation set the legal framework for the protection of sea turtles, only recently has there been a perceived decrease in the illegal hunting of these marine species. This has been due largely to a strong network of sea turtle conservation programs throughout the region, centered largely on the Baja California peninsula. This locally based conservation movement augments a regional nesting beach conservation program that has been in place throughout Pacific Mexico since the 1960s. Together, these efforts have contributed to increases in olive Ridley and green turtle populations throughout the region. Unfortunately, the outlook for loggerheads, hawksbills, and leatherbacks is grim, which is mostly due to unabated by-catch in fisheries gear. Recovery for these species may still be possible but the opportunities are rapidly running out. The leatherback in particular teeters at the edge of extirpation in the Pacific Ocean, and immediate and drastic conservation measures are required at remaining nesting sites, feeding areas, and migratory corridors. Rather than rely solely on laws that are difficult to enforce, conservationists in Mexico are embracing education, ecotourism, stronger fisheries management, and other alternative conservation strategies that together could restore all sea turtle species in the Gulf as well as the entire Pacific Ocean.

Introduction

> When the "Albatross" visited San Bartolome on April 11, 1889, a very remarkable catch of green turtle was made. The U.S.S. "Ranger" was there at the same time and a seining party was made up consisting of members of the crew and that of the "Albatross." In a single haul of a seine 600 feet long we brought ashore 162 green turtles, many of them of large size. Probably half as many more escaped from the seine before it could be beached; there being a continual loss by turtles crawling over the cork lines during the entire time we were hauling it. (Charles Haskins Townsend in "Voyage of the 'Albatross' to the Gulf of California in 1911")

When early explorers first set eyes on the Gulf of California, the coastal waters hosted a remarkable abundance of fish, invertebrates, marine mammals, and sea turtles. The populations of many of the most conspicuous species were so healthy that they seemed to be an inexhaustible resource to those who hunted and fished for subsistence or sale to outside markets. Today, however, this former abundance lives only in our collective memory, as decades of exploitation have depleted most of the target species—perhaps none more so than the green turtle (*Chelonia mydas*). Without a historical

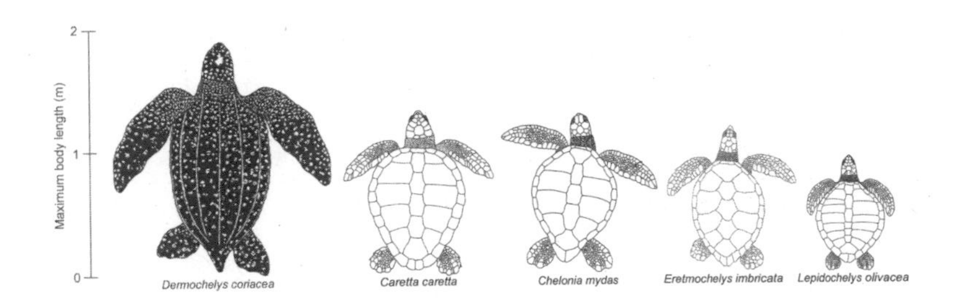

7.1. The five sea turtle species found in the Gulf of California. Maximum body size is for females at their respective nesting beaches. Illustrations courtesy of Thomas MacFarland.

baseline, it is difficult to believe that during pre-Columbian times sea turtles were the most abundant large vertebrate group in the Gulf of California.

Humans have indeed changed the face of the Gulf and its desert shoreline, but the importance of this area for sea turtles has remained. Although some nesting occurs in the Gulf, its greatest value for sea turtles is as a feeding area. Diverse marine habitats—including marine algae and seagrass pastures, rocky reefs, deep-water benthic invertebrate forests, and open water—provide important food resources. Of the seven species found worldwide, five occur in this semi-enclosed sea (fig. 7.1; color plate 5): the green turtle, the loggerhead turtle (*Caretta caretta*), the olive Ridley turtle (*Lepidochelys olivacea*), the hawksbill turtle (*Eretmochelys imbricata*), and the leatherback turtle (*Dermochelys coriacea*). Sadly, regional populations of all five species have declined in response to overexploitation of eggs and turtles at distant nesting areas, illegal hunting of turtles, incidental mortality relating to marine fisheries, and degradation of marine and nesting habitats. As a result, all are considered vulnerable, endangered, or critically endangered by the International Union for the Conservation of Nature (IUCN 2008; see table 7.1).

As ubiquitous as sea turtles once were to undersea habitats, their role in coastal ecosystems has eroded to the point that travelers to the Gulf today are unlikely to see evidence of their presence aside from discarded turtle shells cast upon a trash heap or strewn along a beach. However, if we could step back in time, the presence of sea turtles and their ecological value within coastal ecosystems would be highly apparent. Prior to human exploitation, sea turtle population abundance was controlled not by harpoon

TABLE 7.1. Status of sea turtles occurring in the Gulf of California.

Species	ESA Status	IUCN Red List	CITES	Historical Level in Gulf	Current Level in Gulf
Leatherback, *Dermochelys coriacea*	Endangered	Critically endangered	Category 1	Uncommon	Extremely rare
Green turtle, *Chelonia mydas*	Endangered*	Endangered	Category 1	Highly abundant	Uncommon
Loggerhead turtle, *Caretta caretta*	Threatened**	Endangered	Category 1	Common	Rare
Olive Ridley turtle, *Lepidochelys olivacea*	Threatened	Vulnerable	Category 1	Common	Uncommon
Hawksbill turtle, *Eretmochelys imbricate*	Endangered	Critically endangered	Category 1	Uncommon in north, common in south	Rare

* The U.S. Endangered Species Act (ESA) lists all but the Pacific Mexico population of green turtles as threatened.

** At time of publication, the North Pacific loggerhead population was under petition to be listed by the ESA as endangered.

points and fishing nets but rather by ecological carrying capacity. Their incredible abundance, and their consumption of vast quantities of benthic prey, contributed to their important role in ecosystem function. Much as in other portions of the world's oceans, sea turtles—particularly the green turtle—acted as biological regulators (Jackson 1997; Bjorndal and Jackson 2003). Through their constant consumption and egestion of marine algae and seagrass, green turtles made the nitrogenous components and other nutrients of these marine plants available to other consumers more quickly than if the plants died and degraded naturally. This "short-circuiting" of the detritus cycle accelerated nutrient turnover, promoting high productivity in coastal ecosystems (Thayer et al. 1982). Through their nesting activity and deposition of eggs, sea turtles were integral for energy transfer from marine to terrestrial systems (e.g., Bouchard and Bjorndal 2000). And perhaps most importantly they served as keystone predators, shaping the structure of benthic communities through their constant culling of selected prey. However, the steady influx of human activity and the increased frequency of sea turtle harvests eventually diminished this ecological role. Over the last century, for example, sea turtles have been cherished not as ecological cornerstones but instead as a human food and economic commodity.

Clearly, humans have exacted a heavy toll on sea turtles in this region. Yet no matter how desperate their conservation status, the resilience and biology of these marvelous creatures suggests that there is still time to prevent local extinctions. In this chapter I describe the diverse life histories of sea turtles in the Gulf of California, discuss their cultural importance to coastal communities, identify threats to their survival, and suggest a number of conservation alternatives that will promote the recovery of local sea turtle stocks.

Sea Turtles and Human History in the Gulf

Sea turtles are unique in that they are an important component of coastal marine ecosystems and also vital to the culture and traditions of past and present human settlements of the Gulf of California. In earlier times sea turtles provided a source of nutrition, medicine, and material and were often held in high esteem during ceremonial events. As European-derived cultures expanded along the Gulf's shores, marking the emergence of a cash economy, the importance of sea turtles gradually shifted to that of an economic commodity waiting to be captured, bought, sold, and shipped to settlements many miles away. Today, although protected by strong laws in Mexico (*Diario Oficial* 1990), the surviving sea turtle populations continue to be exploited in Mexico and throughout the world (Nichols et al. 2002a,b; Seminoff et al. 2003a; J. Seminoff, pers. obs.). Fortunately, sea turtles have withstood this onslaught; although now much fewer in number, their survival speaks to their resilience.

Indigenous Peoples

Since prehistory, native cultures have been closely tied to the Gulf's marine, brackish, and freshwater ecosystems, benefiting from the abundance of natural resources for both food and nonfood uses. These included the Pericu and Cochimí along the Gulf coast of the Baja California peninsula; the Cucapá, who inhabited the delta region and lower reaches of the Colorado River; the Hia C-ed O'odham, who occasionally made use of marine resources along the Alto Golfo's Sonoran coast; the seafaring Comcáac farther south along the Sonora coast; and the Yaqui, who lived along the shores of southern Sonora and Sinaloa. Sea turtles naturally occurred within the marine and/or aquatic range of each of these cultures, and in some cases they played a key role in their daily lives and ceremonial traditions (e.g., Aschmann 1966; Felger and Moser 1985).

Along the Baja peninsula's Gulf coast it appears that sea turtles had cultural importance to the Pericu, a group that once inhabited southern Baja, and the Cochimí, who were present throughout much of the central and northern peninsular deserts. Yet because these groups—along with a third peninsular group, the Huaycuras—vanished long ago, the prehistoric human–sea turtle connections in Baja are poorly understood. We do know, however, that turtles were valued for food and material resources. For example, at Valle de los Muertos just south of La Paz, turtle bones with burn marks have been found at Pericu refuse sites (C. Mandujano, pers. comm.). That these bones were burned suggests that sea turtle meat was part of the local Pericu diet, although without direct accounts it is difficult to know just how commonly sea turtles were consumed. Evidence of material worth comes from historical writings about an area much farther north, near San Luis Gonzaga, where it appears that the sea turtle shells were used. During his voyage to the Colorado delta in 1539, Francisco de Ulloa, leading three ships sent by Cortez to explore the region, encountered people (probably Cochimís) at this northern Gulf shoreline area. He later wrote: "They had a little enclosure of woven grass without any cover over the top, where they lodged, ten or twelve paces from the sea. We found inside no sort of bread or anything resembling it, nor another food except fish, of which they had some which they killed with well-twisted cords that had thick hooks made of tortoise shell bent in fire" (Wagner 1929, p. 36).

Most likely this passage referred to the hawksbill turtle (known locally as *la carey*), a species whose shell—often called tortoiseshell—is renowned for its value in jewelry and figurines. This is the only mention of tortoiseshell hooks in the Gulf of California, although such use has been

7.2. Sea turtle images drawn by Gulf peoples. *Left:* Drawing of *moase-hihp-polihuh,* the giant leatherback turtle and creator of the Seri world (original by Jose Torres; published image is sketch of photograph taken by William Smith, courtesy University of Arizona Library Special Collections). *Left center:* Sketch of sea turtle pictograph by Cochimí or Huaycura at Canipole near Loreto, Baja California Sur, Mexico (drawing measures 36 cm from tip to tip of front flippers; pictograph in iron oxide (brick-red color). *Right center and far right:* Sketches of petroglyphs by Cochimí or Huaycura at Canipole near Loreto, Baja California Sur, Mexico (etchings are 26 cm and 21 cm, respectively, from tip to tip of front flippers; lower left portions of both etchings were eroded from rock face).

documented within other indigenous cultures around the world (Frazier 2003). Perhaps the most intriguing and also the most unequivocal evidence that Baja's local inhabitants were aware of sea turtles in their local seascape are the sea turtle pictographs and petroglyphs on desert rockscapes in places like San José de Comundú and Canipole near Loreto (fig. 7.2). Although the culture responsible for turtle paintings at these and other sites remains unknown, these artistic renditions suggest a deeper, perhaps spiritual, connection.

The Cucapás also relied on aquatic ecosystems of the Gulf for food resources, but in their case it was the fresh, brackish, and seawater environments of the Colorado River. During historic times, when the Colorado was the dynamic freshwater ecosystem of legends past, the vastness of its delta created a giant aquatic corridor into the interior of the Sonoran Desert. Sea turtles were known to occur near the Cucapá village of El Mayor, some 80 km from the Gulf proper (Richard Felger and Amadeo Rea, unpublished data). During a field trip in 2002 to the Colorado delta, Richard Felger was told by local fisher Miguel Ángel Romo Buston that sea turtles were once fairly common in the river. Señor Buston commented that nowadays he sees only two or three per year in the delta (Richard Felger, unpublished notes). Although we don't know what role sea turtles played in Cucapá culture, it is likely that this indigenous group was aware of their presence.

To the east of the Cucapá territory, in the land of the Hia C-ed O'odham (the "Sand Papago"), sea turtles were apparently consumed on occasion. For example, during a nineteenth-century survey for potential railroad routes, P. R. Brady (in Michler 1857) described indigenous campsites in the Pinacate region as having a variety of food discards including seashells and turtle shells (Felger 2007). And while recounting his experiences with the Hia C-ed Oodham sometime around 1900 near Bahía Adair, Ramon Anita commented that "when the tide came in, they would catch fish and big turtles (caguamas)" (Thomas 1991).

Perhaps no group has been more famous for their cultural connection with sea turtles than the Comcáac, or Seri, a nomadic people who resided along the Sonoran coast (Felger et al. 1976; Felger and Moser 1985; Nabhan 2003). Today they live primarily in the villages of Punta Chueca and El Desemboque along the shores of the Infiernillo Channel north of Bahía Kino, but their territory once stretched from Guaymas northward to at least Puerto Lobos; the Comcáac may have ventured as far north as the Colorado River (Felger and Moser 1985; Moser 1963) and probably crossed the Gulf to the Baja peninsula (Bowen 2000). Sea turtles feature prominently in traditional Comcáac culture as both a food resource and as the cultural foundation of their communities (Felger and Moser 1985). Sixteen kinds or variants of sea turtles were recognized: ten green turtles, two loggerheads, two hawksbills, the olive Ridley, and the leatherback. It was primarily the green turtle (generally referred to as *moosni*) that was the focus of organized turtle hunts and the primary food resource, but it was the leatherback (known as *mosnípol*) that was cherished in traditional beliefs and oral history as an integral part of creation stories and Comcáac legend (Felger and Moser 1985; Nabhan 2003; Seminoff and Dutton 2007). It was both the hunting and the worship that strengthened the Comcáac communities and provided opportunities for the teaching of traditional knowledge to the younger generation. However, accelerated exploitation by local European-derived cultures has so drastically affected sea turtle populations in northwestern Mexico that the Comcáac have gradually shifted away from their reliance on sea turtles. Though ceremonies and tribal activities that embrace sea turtles continue on occasion, there has been a breakdown in the information exchange between elders and youths and thus a gradual loss in the wealth of Seri knowledge of sea turtles. Fortunately, there is a new generation of Comcáac youths who have formed the Equipo de Conservación de Tortuga–Nación Comcáac and have become actively involved in sea turtle monitoring and research within the Infiernillo Channel. The Equipo's active involvement in the region's sea turtle conservation network is a prime example of how human groups can temper traditional wildlife use practices and change social norms with contemporary conservation activities.

The Yaqui of southern Sonora and Sinaloa also had, and continue to have, a connection with sea turtles. Yaqui elders possess knowledge regarding the ecology of sea turtles in their territory, and in decades past they relied on these marine creatures as a food resource. For example, during a field trip to Yaqui territory in 1988, Richard Felger (unpublished notes) learned from Guillermo "Checho" Amarillas that the green turtle was recognized by the Yaqui as a creature that lived in their territory and that it was consumed locally. It is interesting that the Yaqui name for green turtle (*mosenim*) is remarkably similar to that of the Seri (*moosni*), whereas the second name used by the Yaqui (*cahuama*) reflects the local Mexican name for sea turtles, often specifically the green turtle. Checho also stated that the *mosenim* ate *Zostera,* which we know to be true in the Infiernillo Channel as well (Felger et al. 1980).

Early Explorers and the Opening of the Gulf

In addition to indigenous cultures, there was a knowledge of sea turtles by early explorers, missionaries, and prospectors. The mention of sea turtles in historical documents suggests that this knowledge goes back nearly 500 years, as evidenced by the previously quoted passage regarding tortoiseshell hooks and the Cochimís from Francisco de Ulloa in 1539 during his voyage to the Colorado River delta. This would be the first of several mentions of sea turtles in historical documents. A few of the more notable quotes follow.

In 1765, the Spanish missionary Padre Norberto Ducrue (O'Crouley 1774; translated in Aschmann 1966, p. 45) noted after a voyage of several months throughout the Gulf that the hawksbill turtle, although present in southern waters, was not found in the Upper Gulf region: "There are turtles in great abundance in both seas but the one with the transparent shell known as carey exists only in the southern extremity on the Pacific side, and in the Gulf is found up to latitude 27 1/2° N. In areas farther to the north it is not found, either in the Pacific or in the Gulf."

In 1851, Lt. George Horatio Derby (1851; in Faulk 1969, p. 37) described his journey to Bahía de los Ángeles along the Gulf's west coast, mentioning the great abundance of sea turtles in the area: "The bay is probably well known to the people of the interior, as we found many traces of old encampments, piles of oyster shells, heaps of ashes, and many mule-tracks leading to the southward. There are plenty of turtle in the harbor, but we did not succeed in taking any."

In 1889, Charles Townsend (1916, p. 445) also remarked on sea turtle abundance, writing that green turtles "are plentiful in the Gulf of California, and the Albatross obtained specimens in the vicinity of Willard Bay, on the Peninsula near the head of the Gulf in 1889."

Writing about a second voyage to the Gulf by the *Albatross,* in 1911, Townsend (1916, p. 445) gave the first mention of nesting in the area: "Turtles are said to abound near the mouth of the Río Colorado where their eggs are deposited in the sand. The inhabitants of the Peninsula seem to have no difficulty in obtaining a supply of them."

Economics and the Era of Commercial Sea Turtle Harvest

Sea turtles have been a part of the Gulf's human history through the centuries, and their value for food and nonfood uses continues today in local Mexican cultures. However, whereas the pre-modern cultures were relatively small groups that, even in times of maximal catch, had little impact on sea turtle populations, contemporary cultures—because of their burgeoning populations and high demand for sea turtle products—have extracted sea turtles at an unparalleled rate. This take of turtles was, and continues to be, primarily for food, although sea turtles have also been sought for medicine, leather, and oil. As a result of their utility through the ages, sea turtle populations have paid a heavy price.

Of the five sea turtle species present in the Gulf, green turtles have been the most important for the human population. By the early 1900s, the commercialized green turtle fishery of Baja California had begun (Agler 1913; Averett 1920; Craig 1926). Hunting was widespread, and virtually all size classes were targeted. Yet despite the abundant catches, the large distance and difficulty of trucking turtles to the north made it difficult to earn a profit. By the 1950s, however, the construction of roads and the greater availability of overland motorized vehicles made life much easier for turtle fishermen of the Gulf. Green turtles landed along the Sonoran coast were commonly shipped from coastal villages to places like Hermosillo, Nogales, Tucson, Phoenix, and Yuma; those landed in Baja were trucked and boated up to Ensenada, Tijuana, San Diego, Los Angeles, and San Francisco (Craig 1926; Carr 1961; Caldwell 1963; O'Donnell 1974; Cliffton et al. 1982).

Green turtles were abundant throughout the Gulf, and the population supported a lucrative fishery for decades. The turtle fishery was particularly active in areas such as Guaymas, Bahía Kino, Puerto Libertád, Puerto Peñasco, San Felipe, Bahía Gonzaga, and Loreto. In addition, turtle hunters would frequently depart from these areas to fish for extended periods in the Midriff Islands. The principal turtle fishing base in the Gulf, especially during the 1950s to early 1970s, was Bahía de los Angeles (Caldwell 1962, 1963). Extraction was so heavy and economically important that, during their research in the area, Caldwell and Caldwell (1962) called green turtles the "black steer" of the Gulf of California. Referring to Bahía de los Ángeles turtle hunting, Caldwell (1963, p. 147) wrote: "I saw over 500 landed in a 3-week summer period in 1962 at Los Angeles Bay alone, and a comparable number, considering fishing effort, per week in winter." In total, 186 tons of sea turtle were reported landed that year from Bahía de los Ángeles (Márquez 1984). An even more exasperating figure is provided by Olguin-Mena (1990), who reported that, during a 6-week period in the early 1960s, 2,600 turtles were captured by the ten-panga 'Sociedad' turtle cooperative operating throughout the Bahía de los Ángeles–Isla Ángel de la Guarda region. Not surprisingly, this population crashed within a short period: in 1982, only 11 tons of sea turtle were fished from Bahía de los Ángeles—a 96 percent drop in annual landings in only two decades (Márquez 1984; Olguin-Mena 1990). As a result of the overhunting and declining populations, the sea turtle fishery entered a "bust" cycle during which the local cooperatives were much less active. During the period 1981 to 1985, for example, fewer than 200 turtles were reported landed each year by the Canal de Ballenas cooperative in Bahía de los Ángeles (A. Resendiz, unpublished data). Despite the population crash, many fishermen continued to hunt what few turtles they could for personal consumption and for sale to independent buyers until the 1990 nationwide ban on turtle fishing (*Diario Oficial* 1990; A. Resendiz, pers. comm.). The impacts apparently took their toll; as at least one turtle fisherman recalled, in the early 1990s there were "no more turtles" (F. Savín de Smith, pers. comm.).

Sea Turtle Exploitation in the New Millennium

Subsequent to the close of commercial sea turtle fishing operations, local populations have dwindled even further as the demand for green turtles (and other species) has continued to outstrip the ability of these slow-growing animals to regenerate. Today, most populations remain a small fraction of their historical abundance. With national conservation initiatives ongoing since 1966 and a presidential decree outlawing the use of sea turtles and their products in 1990, one might expect that all sea turtle populations would be on their way to recovery. Unfortunately, this has not been the case, which is due in part to the vastness of the country and difficulty of enforcing protective laws but due primarily to the rich tradition of sea turtle use in virtually all segments of society, from fishermen in small coastal villages to politicians and lawmakers in cosmopolitan centers.

In addition to illegal hunting, sea turtle populations in the Gulf, as elsewhere in the world, are threatened by incidental mortality in a variety of marine fisheries. Shrimp trawling in the Gulf was the cornerstone of the local fisheries-based economy and, although it continues at diminished levels, it remains a tremendous problem for sea turtles that live and forage in the same areas where shrimp occur. In addition, drift nets set in offshore waters to catch sharks and rays—as well as set nets anchored in shallow nearshore habitats to capture a variety of finfish such as halibut and guitarfish—are detrimental to the survival of sea turtles. In many parts of the world, bycatch mortality is the primary culprit preventing sea turtle population recovery, and if it were not for the ongoing directed sea turtle harvest then the same would be true in the Gulf of California.

With respect to habitat degradation, coastal terrestrial and marine habitats are relatively free from contamination—one of the Gulf ecosystem's fortunate traits. But even though habitats may not yet be contaminated, there are several areas where coastal development is a problem. This development is particularly detrimental to sea turtle nesting, as many of the pristine nesting beaches in areas like Mazatlán in Sinaloa and the Cape region of Baja have been transformed into tourist centers. Numerous development projects may also affect turtles in the marine environment; the most notorious of these was the Escalera Náutica project, which had the potential to destroy coastal habitats throughout the Gulf (Nichols 2003b) but has been substantially downsized in recent years, although several splinter projects continue to threaten foraging areas in a variety of once-pristine areas. Fortunately, environmental awareness and economic challenges are slowing the progress of many development projects, but with Mexico in dire need of economic revitalization, it is increasingly possible that coastal development will someday triumph over environmental stewardship.

Sea Turtle Life History

Sea turtles have a remarkable life history, during which individuals inhabit broadly separated localities over the course of their lives (fig. 7.3). From hatchling to adult, sea turtles undergo ontogenetic shifts in habitat

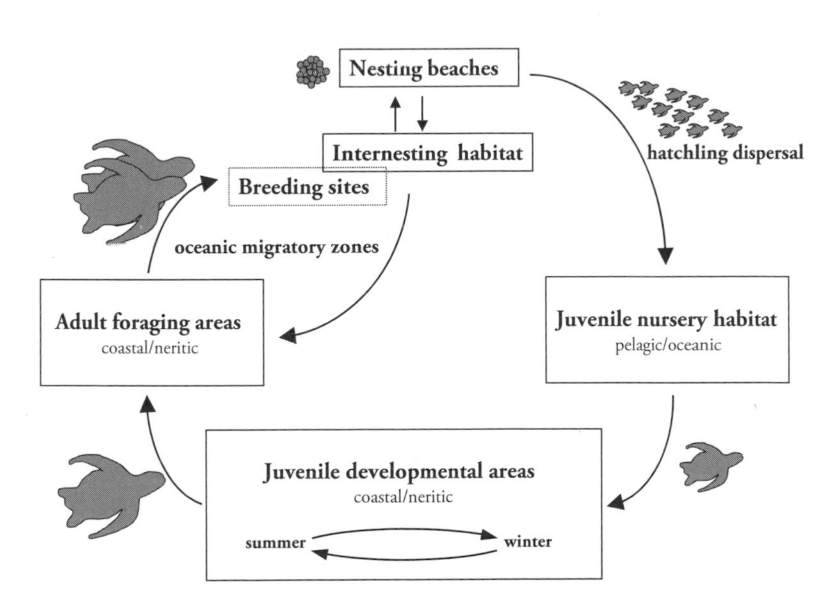

7.3. Generalized life history of the hard-shelled sea turtles. Not included in this group is the leatherback, a species with juvenile developmental areas and adult foraging areas in the pelagic zone.

TABLE 7.2. Nesting beach origins of sea turtles found in the Gulf of California.

Species	Nesting Beach Origins	Minimum Travel Distance	Evidence	Citations
Green turtle (*Chelonia mydas*)	(1) Michoacán, Mexico (2) Revillagigedos Islands, Mexico	(1) 800 km (2) 400 km	(1) Satellite telemetry; flipper-tag recovery (1, 2) Genetic analysis	Nichols 2003a; Dutton, unpublished data; Resendiz, unpublished data
Loggerhead turtle (*Caretta caretta*)	Japan	10,000 km	Satellite telemetry; flipper tagging	Resendiz et al. 1998; Nichols et al. 2000
Hawksbill turtle (*Eretmochelys imbricate*)	Unknown	Unknown	No data	
Olive Ridley (*Lepidochelys olivacea*)	Assorted sites on Mexico Pacific and in Gulf of California	0 km	Direct observation	Seminoff, unpublished data; López-Castro, Carmona, and Nichols 2004
Leatherback (*Dermochelys coriacea*)	(1) Western Pacific (2) Pacific Mexico	(1) 10,000 km (2) Unknown	(1) Genetic analysis (2) Direct observation	(1) Dutton, unpublished data

use that encompass nesting beaches, migratory corridors, juvenile developmental habitats, and adult foraging areas. Upon emergence, hatchlings depart the nesting beach to begin the oceanic phase of their life cycle, floating passively for a year or more in major current systems (gyres) that serve as open-ocean developmental grounds (Carr and Meylan 1980; Carr 1987). While offshore, small juveniles commonly associate with rafts of flotsam that provide important food resources and offer protection from predators (Nichols et al. 2001). Most species eventually move into coastal developmental habitats rich in food resources, where they forage and grow until maturity (Musick and Limpus 1997; Seminoff et al. 2002b–d), although the pelagic leatherback remains in offshore waters. Depending on the species, age to maturity may be between 10 and 50 years, with species such as leatherbacks and olive Ridleys maturing the fastest and green turtles maturing the slowest. Upon maturity, sea turtles carry out breeding migrations that take them from foraging grounds to their natal nesting beaches. These migrations are undertaken every few years by both males and females and may traverse oceanic zones, often spanning thousands of kilometers (Nichols et al. 1999, 2000; Nichols 2003a; Seminoff et al. 2002c). During nonbreeding periods, adults reside at coastal neritic feeding areas that sometimes coincide with juvenile developmental habitats. Through these ontogenetic migrations, sea turtles inhabit a variety of habitats and, as with many migratory species, their movements often traverse international boundaries (table 7.2). This transboundary nature of their life history has created many conservation challenges for sea turtles throughout the world.

The Gulf and Sea Turtle Foraging

In the context of sea turtle life history, the Gulf of California is most important as a foraging and developmental area. Places like the Infiernillo Channel, the Midriff Islands, Bahía de los Ángeles, Bahía Loreto, Bahía San Rafael, and Canal de San José are all critical areas for sea turtle feeding. Coastal habitats of these and many other sites host marine algae, seagrasses, benthic invertebrates, and gelatinous megaplankton, all of which are important foods.

Marine habitats of the Gulf are highly dynamic. Given the tremendous seasonal upwelling and annual fluctuation in seawater and air temperatures, the distribution and abundance of potential sea turtle food items can vary substantially between seasons and years (e.g., Pacheco-Ruíz and Zertuche-González 1996b). Many marine algae species that are abundant during winter and spring periods become very patchy during warm summer periods. Some species may vanish altogether during extreme temperatures, such as those that occur during El Niño Southern Oscillation (ENSO) events (Carballo et al. 2002).

During summer months, the most important habitats for sea turtles are the nearshore marine algae pastures and seagrass habitats lining the Gulf. Among the most important foods for green turtles are the red algae *Gracilariopsis lemaneiformis, Gracilaria* spp., and *Gigartina* spp. as well as the green algae *Ulva lactuca* and *Codium* spp. (Seminoff et al. 2002c). Although limited in distribution, seagrasses, particularly eelgrass (*Zostera marina*), are important diet items (Felger et al. 1980). In addition, the richness and diversity of invertebrate species in the Gulf (Brusca 1980) enhances the value of coastal habitats for foraging sea turtles—especially the hawksbill, which is largely a sponge-feeder, and the loggerhead and olive Ridley turtles, which commonly ingest crustaceans and large molluscs. Invertebrates such as sea hares (*Aplysia* spp.) and sea pens (*Ptilosarcus undulatus*) are consumed by these species as well as by green turtles (Seminoff et al. 2002c).

As winter approaches, sea turtles undergo substantial changes in their behavior and food intake. As ectotherms, these marine reptiles require heat energy from their surrounding environment to sustain normal metabolic activity and digestion. Hence they become much less active during the cold winter months of December to March. During this period, green turtles and, to a lesser extent, loggerheads enter a torpid state during which they may lie motionless on the sea floor. This behavior remains poorly understood, although it has facilitated large-scale harvesting. Overwintering turtles have long been targeted by Comcáac turtle hunters, but it wasn't until the early 1970s that Mexican lobster divers from Bahía Kino came upon this easily collected resource, a discovery that spelled certain doom for overwintering turtles in the Midriff area (Felger et al. 1976). Using hookah-diving (i.e., compressor) equipment, this harvest lasted only a few years before torpid turtles became economically extinct (Felger et al. 1976). In the decades since then, overwintering *Chelonia* have been discovered on rocky ledges and caves throughout the central and northern Gulf of California, presumably in a similar state of torpor (Seminoff and Nichols, unpublished data). Even today, divers harvesting sea cucumbers and octopus opportunistically take torpid sea turtles when they find them, although nowhere approaching the levels of the Midriff Island harvests of the 1970s.

The Gulf and Sea Turtle Nesting

When speaking about sea turtle nesting, the primary points of interest are located at the southern extremities of the Gulf: along the shores of Sinaloa to the southeast and in the Cape region of the Baja California peninsula to the southwest. In both areas the primary nester is the olive Ridley, although leatherbacks and (to a much lesser extent) green turtles may also deposit eggs (Olguin-Mena 1990; Seminoff 1994; Sarti et al. 2000; López-Castro et al. 2004; G. Tiburcio-Pintos, unpublished data). Nesting along the Cape region occurs from Cabo Pulmo around to just north of Todos Santos (Agua Blanca). The northernmost nesting beach in Sinaloa is near Cruz de Elota, roughly 100 km north of Mazatlán (Seminoff 1994).

Nesting has been reported in the central and northern Gulf on occasion, but the region has probably never been a particularly important area for such activity. The olive Ridley is by far the most common nester in these regions, and it has been observed nesting in areas such as Punta Chivato (Baja California Sur) and near the towns of Bahía Kino and Bahía de los Ángeles (in Sonora and Baja California, respectively) (Caldwell 1962; W. J. Nichols, unpublished data; Resendiz and Seminoff, unpublished data). It is interesting that the olive Ridley's distinction as the most common nesting sea turtle in the Gulf is consistent with indigenous knowledge; the Comcáac once described this species as the only sea turtle that might have shelled eggs in the oviduct upon capture (Felger et al. 2005).

As for other species nesting in the Gulf, the only species for which we have apparent confirmed nesting activity is the leatherback. Caldwell (1962) tells of a hotel proprietor in San Felipe who had seen leatherback hatchlings near San Felipe in July of 1962. Likewise, on separate occasions both Richard Felger and I have been told by Comcáac informants that leatherback hatchlings have been seen along the beaches of the Infiernillo Channel (unpublished data). In addition to these accounts, nesting by green turtles has been suggested but remains unconfirmed. For example, during the aforementioned voyage of the *Albatross* into the Gulf of California during the late nineteenth century, Charles Townsend (1916, p. 445) spoke of sea turtle nesting near the mouth of the Colorado River. Although this mention was in the context of the green turtle, it is not certain that this is the species Townsend was discussing. More recently, Juan de la Cruz (pers. comm.) from the peninsular community of Juncalito claimed to have seen green turtle nesting on the island of Monserrate, but this too is unconfirmed. Large green turtles with shelled eggs have reportedly been captured by fishermen in Bahía de los Ángeles (Caldwell 1962; Nichols and Seminoff, unpublished data), indicating that nesting may occur in the area. Green turtles reportedly occasionally nest also in the Cape region (G. Tiburcio-Pintos, unpublished data), and there are anecdotal reports of green turtle nesting along the sandy beaches of Isla Margarita near Bahía Magdalena (W. J. Nichols, pers. comm.).

Species Accounts of the Gulf's Sea Turtles

The waters of the Gulf of California host five of the world's seven sea turtle species: the leatherback, green turtle, hawksbill, olive Ridley, and loggerhead (figs. 7.1 and 7.2). The only species not found here are the flatback turtle (*Natator depressus*), an Australian endemic, and the Kemp's Ridley (*Lepidochelys kempii*), a quasi-endemic to the Gulf of Mexico. These seven species are distributed among two families: Dermochelyidae, whose sole extant member, the leatherback, has been virtually unchanged for 100 million years; and Chelonidae, comprising six hard-shelled species, all of which have remained morphologically consistent for tens of thousands of years (Ernst and Barbour 1989). The Chelonidae has two tribes, Carettini (represented by *Caretta*, *Eretomchelys*, and *Lepidochelys*) and Chelonini (represented by *Chelonia* and *Natator*). Separation between these two groups may have occurred as many as 40 million years ago (Bowen et al. 1993). Amazingly, however, numerous examples of hybridization between species in these tribes have been documented (e.g., Seminoff et al. 2003c). How could crossbreeding occur between the Carretini and Chelonini after tens of millions of years of divergent evolution? The answer undoubtedly involves the slow rate of genomic and morphological evolution of sea turtles (Bowen et al. 1993).

Not only have sea turtles remained relatively unchanged through time, they also provide a living record of how the morphology and physiology of organisms are perfectly suited to their ecology. An array of physiological adaptations such as high aerobic capacity and the ability to dive to extreme depths—coupled with the benefits of large body size and unique beak, esophageal, and digestive tract morphology for diet specialization—have allowed sea turtles to succeed in the marine environment. I next describe these adaptations as well as the ecology and conservation status of each of the five species occurring in the Gulf of California.

Leatherback (*Dermochelys coriacea*)

The leatherback is the largest of the extant sea turtles. Locally known as *siete filos* or *laúd,* the species attains a shell length of nearly 2 m and a weight of up to 800 kg (Pritchard and Trebbau 1984). This massive turtle has a flexible, leathery shell and a teardrop-shaped body; it possesses dorsal and ventral keels (on the dorsum it possesses seven keels, hence the name *siete filos,* "seven ridges") that enable it to move through the water with maximal efficiency. In addition to the advantage of efficient swimming that is afforded by their body shape, their large size provides for considerable thermal inertia, which prevents excessive cooling or overheating. This endotherm-like thermoregulation, termed *gigantothermy* (Paladino et al. 1990), enables leatherbacks to access abundant food resources in pelagic waters that are too cold for other sea turtle species. Together, these attributes create a species that is perfectly adapted for life at sea. Leatherbacks have a more pelagic existence than any other species, living off the abundant sea jellies and other gelatinous megaplankton that associate with the deep scattering layer in offshore waters (Eckert et al. 1989).

Leatherback nesting occurs along the Pacific coast of the Americas from Mexico to Costa Rica. However, genetic analysis and satellite telemetry have established that at least some leatherbacks in waters of the eastern Pacific and Gulf of California originate from nesting beaches in the western Pacific, in Indonesia, Papua New Guinea, and the Solomon Islands, some

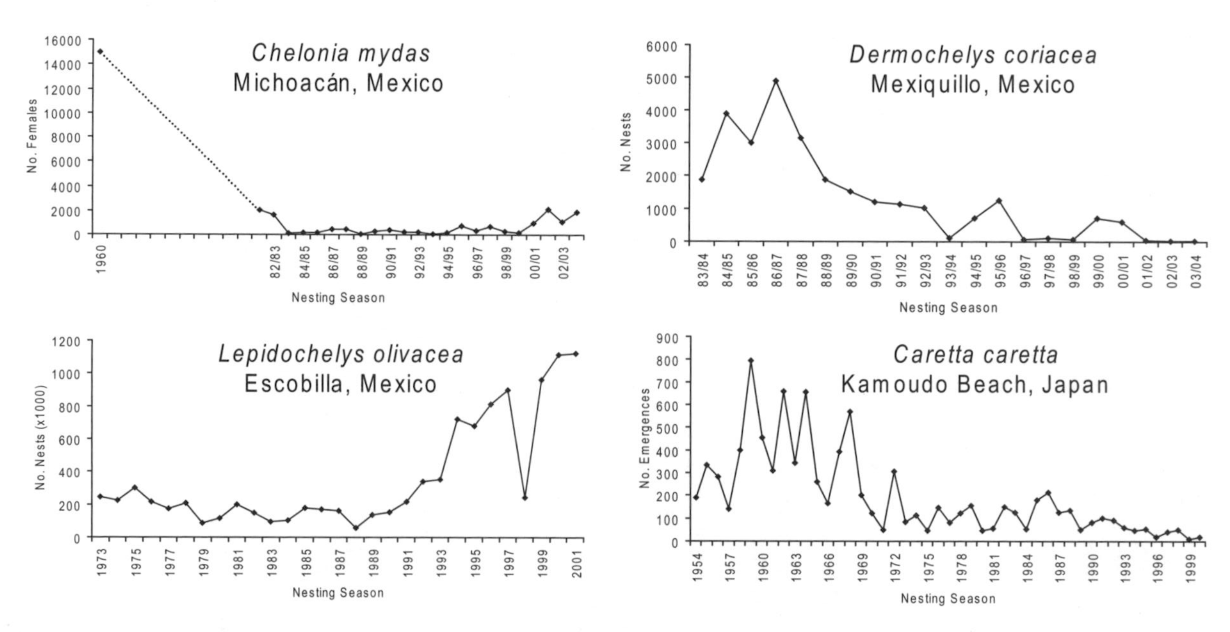

7.4. Nesting abundance trends at four sites. Data for Mexiquillo is from Sarti Martínez et al. 2007; data for Michoacán is from J. Alvarado and C. Delgado, unpublished data; data from Kamouda Beach is from Kamezaki et al. 2003; data from Escobilla is from Peñaflores et al. 2000 and Márquez et al. 2002.

15,000 km away (Benson et al. 2008; P. Dutton and colleagues, unpublished data). In the Gulf of California, leatherback distribution is poorly understood, although stranded turtles have been found as far north as the Colorado River delta. Seminoff and Dutton (2007) review all the recorded leatherback sightings and strandings in the Gulf of California, which amount to fewer than forty accounts. This low abundance is consistent with Seri traditional knowledge that suggests leatherbacks have never been particularly common in the central Gulf (Felger and Moser 1985). A telling sign of their uncommon nature is that, since 1943, the Comcáac have had only eleven leatherback ceremonies (Nabhan 2003; Seminoff and Dutton 2007).

Today, leatherback turtles are in serious decline in the Pacific Ocean. By-catch in commercial and artisanal long-line and drift-net fisheries—coupled with the impacts from decades of egg poaching—have decimated populations throughout the Pacific to the extent that some researchers fear that this species will become extinct there within the next two decades (Spotila et al. 2000). These declines are particularly evident in the western Pacific Ocean, where the nesting colony in Terengganu, Malaysia (once the Pacific's largest rookery) has vanished. The situation is similar in the eastern Pacific, where all of the main nesting beaches (in Mexico and Costa Rica) have either collapsed or are in serious decline (e.g., Eckert and Sarti 1997; Sarti et al. 2000; Spotila et al. 2000; figure 7.4). For example, during the 2002/03 nesting season only 120 leatherback nests were recorded on all the Mexican index beaches combined (Sarti-Martínez et al. 2007), which equates to roughly 14 individual turtles. Recall that this was once the most important nesting complex in the entire eastern Pacific Ocean, with thousands of females nesting each year (Pritchard 1982). A similar situation is present at Playa Grande, the largest nesting aggregation in Costa Rica, where the 2004/05 nesting season had the lowest number of nests on record since beach monitoring commenced nearly twenty years ago (F. Paladino, pers. comm.).

Green Turtle (*Chelonia mydas*)

Green turtles are by far the most abundant sea turtle in the Gulf of California. Locally known as *la negra* or *la prieta*, green turtles may attain a shell length of slightly over a meter and weigh up to 180 kg. They are present throughout coastal waters of the eastern Pacific from southern California to Chile, where they forage along islands and in coastal bays, estuaries, and lagoons. Their diet has been described as primarily herbivorous, consisting of marine algae and seagrass with occasional consumption of mangrove leaves (Seminoff et al. 2002b; López-Mendilaharsu et al. 2005). Green turtles in this region augment their diet with animal protein, consuming greater amounts of invertebrates than green turtles in any other part of the world (Bjorndal 1997; Seminoff et al. 2002c). Still, the morphology of green turtles suggests that they evolved to eat primarily seagrasses and marine algae. The sharp, incisor-like upper and lower portions of their beak are perfectly shaped to crop marine algae and plants that are anchored in the seafloor.

Green turtles in the eastern Pacific (also called east Pacific green turtles or black turtles) have been the focus of considerable taxonomic debate in recent years. Although some scientists believe their unique morphology and behavior warrant distinction as a separate species, *Chelonia agassizii* (Pritchard 1999), other scientists cite genetic data to argue that it is at most a subspecies of the pantropical green turtle (i.e., *Chelonia mydas agassizii*) and perhaps nothing more than a unique population (Karl and Bowen 1999). Notwithstanding this disagreement, there is no doubt that the Pacific Mexican green turtle population is severely depleted and in need of long-term conservation.

Although green turtles were once thought to carry out their entire life cycle in the Gulf (e.g., the northeastern Pacific green turtle; Caldwell 1962), it is now clear that green turtles in the Gulf of California originate from the rookeries in Michoacán and the Revillagigedos Islands, Mexico (Nichols 2003a; Peter Dutton, unpublished data). The initial connection came from an adult female captured in 1985 near Piedra San Bernabe in the Alto Golfo that bore flipper tags applied in April 1984 in Michoacán, a site over 1,750 km to the south (Alvarado and Figueroa 1992; Figueroa et al. 1993). Links between nesting beaches and foraging areas have since been established through satellite telemetry (Nichols et al. 1999), genetics (Nichols 2003a; Peter Dutton, unpublished data), and additional flipper-tag recoveries (Seminoff et al. 2002a; A. Resendiz, unpublished data).

Hawksbill Turtle (*Eretmochelys imbricata*)

Hawksbills reach 90 cm in length and 100 kg in weight. Locally known as *la carey,* the species is most easily identified by the imbricate, coffee- and caramel-colored scutes on the carapace and plastron, especially during juvenile and subadult life stages. The beauty and workability of these keratinous plates—known as tortoiseshell, *carey, penca,* or *bekko*—have made the hawksbill turtle the target of an exhaustive harvest for artisanal uses (Groombridge and Luxmoore 1989). The material is fashioned into combs, pendants, and fine jewelry.

In decades past, the species was once common throughout much of the Gulf (Cliffton et al. 1982). Seri elders explain that large hawksbills were abundant as recently as the 1950s in their territory (Felger and Moser 1985). Today, however, hawksbills are seen regularly only in more southern waters: near La Paz, Cabo Pulmo, and especially Farrallón Island (M. Paxson and T. Pfister, pers. comm.). In these areas, they forage along rocky shores and coral reefs. As their name suggests, hawksbill turtles possess an unusual hawk-like beak, a morphology that is functionally beneficial for their spongivorous diet (Witzell 1983). Few dietary data are available, although Seminoff et al. (2003c) report that two hawksbills from the Gulf of California had stomachs filled with fragments of the sponge *Haliclona* spp.

Hawksbills are the only sea turtle species present in the eastern Pacific for which we have virtually no information on nesting beach origins. In Mexico, a large population was reported in the 1950s and 1960s from the Tres Marías Islands, which may have been a major breeding ground (Parsons 1962). Cornelius (1982) reported infrequent nesting along the Pacific coast of Central America. By the 1980s, however, no major hawksbill nesting rookeries remained in the eastern Pacific (Cliffton et al. 1982). Hawksbill turtles continue to nest, albeit rarely, along the Pacific coast of Mexico. Accounts of sporadic nesting in the Mexican coastal states of Jalisco and Nayarit indicate potential nesting beach origins within 500 km of the Gulf (Raquel Briseño, pers. comm.). Adult-sized hawksbills are present near Cabo Pulmo in the Cape region and there is speculation that the species may nest in the area on occasion, although this has not actually been detected so far (M. Paxton, pers. comm.).

Loggerhead Turtle (*Caretta caretta*)

The loggerhead turtle reaches 100 cm in length and up to 200 kg in weight, although most individuals in the Gulf are substantially smaller. There are a variety of names for this turtle in northwestern Mexico, including *la cahuama, la javelina, la perica,* and *la amarilla.* As their name suggests, loggerheads have a disproportionately large head relative to other sea turtle species. The large skull is thought be functionally beneficial by supporting the larger musculature needed to crush their hard-shelled foods such as molluscs and crustaceans—especially the gooseneck barnacle (*Lepas anatifera*) and the pelagic red crab (*Pleuroncodes planipes*) (Peckham and Nichols 2003; Seminoff, unpublished data), which are regularly consumed by loggerheads in the eastern Pacific.

The loggerhead is an enigmatic species in the eastern Pacific Ocean. It is seen occasionally in foraging habitats of the central Gulf (Seminoff 2000; Seminoff et al. 2004), but its distribution in the northern and southern Gulf is poorly known. Although this turtle can be found in foraging habitats along the Pacific coast of the Baja California peninsula and in the Gulf of California, the closest nesting beaches are across the Pacific in Japan and Australia. After years of speculation, the first link between these foraging and nesting populations came in the early 1990s via genetic analysis (Bowen et al. 1995). Direct evidence has since been gathered through the recovery of a flipper-tagged turtle in Japan that was initially released in Baja California (Resendiz et al. 1998) and through satellite telemetry of the famed loggerhead named "Adelita" that was released in Baja and tracked as she swam back to Japan—a journey that spanned 11,500 km and took 368 days between August 1996 and August 1997 (Nichols et al. 2000). The trans-Pacific movement of the loggerhead turtle is one of the most amazing examples of animal migration. After spending years foraging in the eastern Pacific, once loggerheads return to their natal Japanese nesting beaches for reproduction they remain in the western Pacific for the remainder of their life cycle.

Relative to historic abundance, today's loggerhead turtle population in the Pacific is substantially reduced. However, loggerhead population trends in the North Pacific (a distinct stock from those in the South Pacific) are showing early signs of stabilizing. This promising news is likely related to near-complete protection of Japan's remaining nesting beaches, the closure of the North Pacific high-seas drift-net fishery, and increased management of the remaining U.S.-based fisheries in the North Pacific. Nowadays, as with the leatherback, loggerhead populations continue to be affected by commercial fisheries (in particular, those that use long lines) as they move across the Pacific Ocean. Loggerhead turtles also face additional threats of entanglement in nearshore artisanal fisheries as they enter Baja Californian waters (Hoyt Peckham, unpublished data).

Olive Ridley (*Lepidochelys olivacea*)

The olive Ridley, locally called *la golfina,* is the smallest turtle in the Pacific Ocean, reaching 70 cm in length and 60 kg in weight. Ridleys forage on a variety of invertebrates as they inhabit both neritic and pelagic habitats of the eastern Pacific (Kopitsky et al. 2005). Whereas olive Ridleys have been seen feeding in waters as shallow as 3 m in the Gulf (Seminoff, unpublished data), Bob Pitman (1990; pers. comm.) has reported seeing thousands of Ridleys in pelagic waters hundreds of miles from the Mexican coast during his numerous cruises on NOAA vessels. When this pelagic behavior is compared with the presence of olive Ridleys in shallow nearshore habitats of the Gulf (Felger and Moser 1985; Seminoff 2000), it is apparent that the olive Ridley's habitat preferences are highly flexible. However, as much as this may allow the species to occupy varied habitats throughout the eastern Pacific, it also increases the susceptibility of olive Ridleys to impacts from nearshore set-net fisheries as well as offshore long-line, drift-net, and trawl fisheries. The recent increase in artisanal long-line fishing in the Gulf is particularly alarming given recent field observations that suggest this sea turtle is regularly captured and then usually landed rather than released alive (John Brakey, pers. comm.).

Despite their lower abundance in the Gulf relative to pelagic waters of the eastern Pacific—as indicated by Seri knowledge and as evidenced by recent scientific research (Felger and Moser 1985; Seminoff 2000)—the olive Ridley is by far the most common sea turtle species in the eastern Pacific Ocean. It has been the focus of considerable attention since the recovery of some nesting populations along the Pacific coast (Márquez et al. 2002). For example, at the largest Mexican rookery (Escobilla, Oaxaca), olive Ridleys were nearly wiped out by decades of egg harvest and killing of adults for meat and leather products. But since the 1990 federal moratorium went into effect, this population has rebounded to what many believe is pre-exploitation abundance: over 400,000 turtles nested during an *arribada* (a mass nesting of turtles) in 2001 (Rene Márquez, pers. comm.; fig. 7.4). This is one of the greatest success stories in the history of sea turtle conservation, and it is perhaps the best example of how conservation legislation—such as the 1990 closure of legal sea turtle fisheries in Mexico—can, when coupled with nesting beach protection, pave the way for population recovery.

Sea Turtle Conservation in Mexico

The Gulf of California has long been considered a sea turtle hot spot in the Pacific Ocean. Renowned for its diversity of marine habitats and abundance of algae, seagrass, and invertebrate food resources, this region once hosted abundant sea turtle populations. Today, however, the populations of all five species in the area are a small fraction of their former levels. Consequently, the International Union for the Conservation of Nature (IUCN 2008) lists leatherbacks and hawksbills as critically endangered, green turtles and loggerheads as endangered, and olive Ridleys as vulnerable. All are also listed in the U.S. Endangered Species Act and included in Appendix 1 of CITES, the Convention on International Trade in Endangered Species of Wild Fauna and Flora (see table 7.1).

Recognition of the plight of sea turtles is a relatively recent phenomenon in many countries. However, sea turtle conservation in Mexico has been ongoing since 1966, when the federal government instigated on-the-ground protection at several of the primary sea turtle nesting areas along the Pacific, Caribbean, and Gulf of Mexico coasts (PESCA 1990; Seminoff 1994). Along the Pacific coast, the green turtle nesting beaches in Michoacán, the most important site for the species in the eastern Pacific, were afforded protection in 1979 (Figueroa et al. 1993). With such early conservation efforts, Mexico was one of the world's leaders when it came to protecting the nesting beaches of sea turtles.

Today, many nesting beaches of primary and secondary importance are protected throughout the country by various federal, state, nongovernmental, and academic groups. Many areas have served as testing grounds for innovative approaches to sea turtle conservation. In Baja California Sur, for example, a unique effort commenced in 2004 that for the first time partnered the commercial fishing industry with the U.S. government to provide assistance to one of Baja's local sea turtle conservation organizations. In this effort, the Federation of Independent Seafood Harvesters (FISH), an association of California (U.S.) drift gill-net swordfishers, teamed with the National Marine Fisheries Service (U.S.) to provide financial assistance to ongoing leatherback recovery efforts in Baja by the Mexican NGO Asociación Sudcaliforniana de Proteción al Medio Ambiente y a la Tortuga Marina (ASUPMATOMA) (Janisse et al. 2009). This was the first time an industry association had supported mitigation measures for sea turtle population recovery, and its genesis can be traced to the implementation of swordfish fishery regulations enacted for the purpose of protecting Pacific leatherback sea turtles under the U.S. Endangered Species Act. Regardless of whether this proves to be an isolated example of investment in marine conservation by commercial fishing entities, the program provides a unique example of crafting progressive multinational approaches to sea turtle conservation.

Unfortunately, although this program and other efforts have had substantial success in reducing egg collection and mortality of nesting females, sea turtles continue to suffer substantial human impacts that include directed capture and incidental mortality in marine habitats. In Mexico, the first effort to reduce fishing pressure on "in-water" sea turtle stocks occurred in the early 1970s when a federal mandate was passed that created size limits for turtles landed by the then-legal sea turtle fisheries. In 1972, the Mexican government even went so far as to establish a closed season for sea turtle fisheries of the Pacific. These efforts certainly helped to slow the decline in sea turtle populations, but it wasn't until the 1990 presidential decree by Carlos Salinas de Gortari, which outlawed the use and trade of all sea turtle products in Mexico (*Diario Oficial* 1990), that sea turtle conservation had a legal and permanent mandate. A year later, Mexico became a signatory of CITES, paving the way for collaboration between Mexico and neighboring countries to limit illegal trade and to reduce the incidental capture of sea turtles.

Among the first results of this enhanced sea turtle conservation awareness was the presidential decree mandating the use of turtle excluder devices (TEDs), a contraption that prevents turtle entrapment in shrimp trawling nets. This law was enacted in March 1996 on an emergency basis, and in December 1999 it was adopted permanently (*Diario Oficial* 1999). However, although the implementation of TEDs was a promising step toward reducing sea turtle by-catch in marine fisheries, their integration into daily shrimping practices still lags. This is due both to lack of enforcement and to the simple fact that fishers are not keen on legislators telling them how to do their job. Although TED enforcement may not be perfect, enforcement in other facets of sea turtle conservation has dramatically improved in recent years.

Beginning in the late 1990s, the Mexican government's wildlife enforcement branch, Procuraduría Federal de Protección al Ambiente (PROFEPA), became particularly active in sea turtle protection. For the first time there have been publicized cases in which turtle poachers have been incarcerated in Baja California and Sonora (Nichols and Safina 2004; L. Fueyo, pers. comm.).

In addition to the Mexican government, there are several nongovernmental organizations (NGOs) whose efforts are raising awareness about sea turtle conservation. In northwest Mexico, perhaps the most active NGO is Grupo Tortuguero, a grassroots community-based conservation network within the coastal communities lining the Gulf of California and Baja's Pacific coast. Initiated in 1999 with the leadership of Dr. Wallace J. Nichols, this network now enlists a cross section of marine resource stakeholders that includes commercial fishermen, sport fishermen, biologists, educators, students, resource managers, and conservation practitioners. Sea turtles are the flagship species for the group's broadening efforts to address the social and ecological roots of marine vertebrate exploitation. By promoting a new conservation ethic throughout the region, Grupo Tortuguero is helping transform turtles from a food and economic resource to a cultural and ecological resource. This progression has been promoted in a variety of ways, including research and protection of turtles in their foraging areas and on their nesting beaches as well as efforts that involve and enlist community members. Community-based environmental education, socioeconomic studies, and turtle "festivals" are just a few of the many innovative activities by the group. In addition, Grupo Tortuguero organizes monitoring efforts in foraging areas in Baja California, Baja California Sur, Sonora, and Sinaloa (fig. 7.5). Now, for the first time, changes in sea turtle population

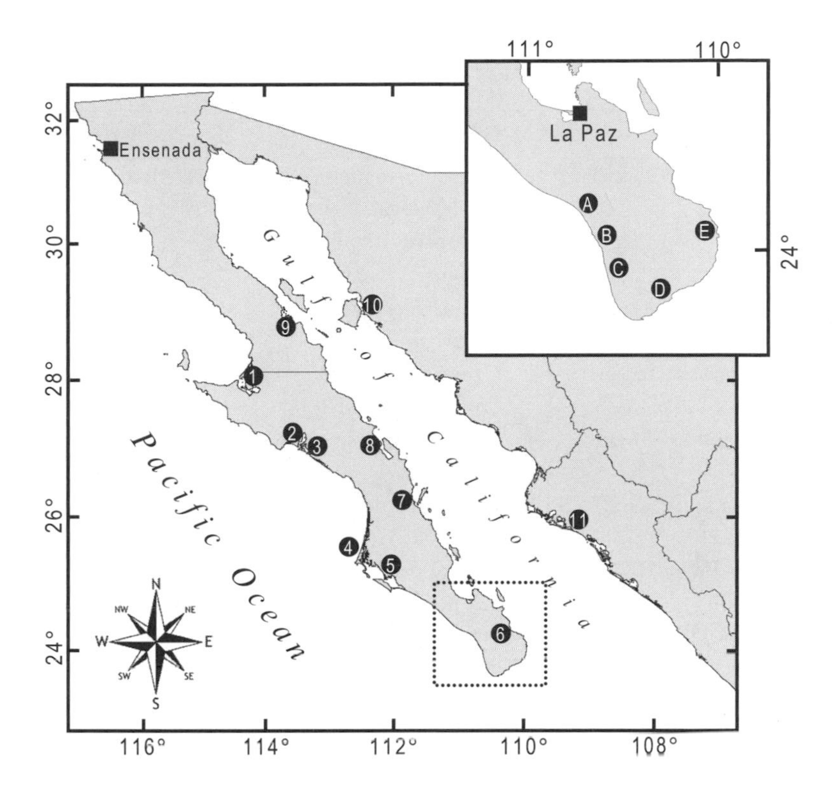

7.5. Map showing locations of ongoing sea turtle research and monitoring efforts in northwestern Mexico. Numbered circles depict activities in foraging areas; lettered circles show nesting beach study sites. See table 7.3 for a summary of each site.

size—both positive and negative—can be quantified over time. This will allow us to identify conservation victories and will also show where additional protection efforts are most needed. It is essential that the data generated from these efforts be promptly shared with federal and state authorities as well as with community members in each of the towns where monitoring occurs. Much of the information exchange occurs during the annual Grupo Tortuguero meeting, a 3-day event (always over the last weekend of January and usually in Loreto) that provides a forum for dialogue among those who are monitoring and enforcing protective legislation (Pesenti and Nichols 2002). Since the first Grupo Tortuguero meeting in 1999, this event has grown in size each year and has continued to draw in more people from throughout the region. In 2004, for example, nearly 300 people attended!

TABLE 7.3. Summary of sea turtle research and monitoring sites in northwest Mexico.

ID	Location	Organization(s)	Species/Activity
1	Laguna Ojo de Liebre, BCS	Grupo Tortuguero, Consejo Nacional de Areas Protegidas, Exportador de Sal	CM/Monitoring site
2	Punta Abreojos, BCS	Grupo Tortuguero, Cadernab	CM/Monitoring site
3	Laguna San Ignacio, BCS	Grupo Tortuguero	CM/Monitoring site
4	Puerto Lopez Mateos, BCS	ProCaguama, Grupo Tortuguero	CC, LO/Pelagic research
5	Bahía Magdalena, BCS	Grupo Tortuguero, SFS CIBNOR, UABCS	CM/Monitoring site
6	Cabo Pulmo, BCS	Amigos del Cabo Pulmo, Grupo Tortuguero	LO, EI, CM/Monitoring site
7	Loreto, BCS	Grupo Ecologista de Antares, Grupo Tortuguero	CM/Monitoring site
8	Mulegé, BCS	Ecomujeres, Grupo Tortuguero	CM/Monitoring site, environmental education
9	Bahía de los Ángeles, BC	SEMARNAT, Pronatura, Grupo Marino, Grupo Tortuguero	CM/Monitoring site
10	Canal Infiernillo, SON	Equipo de Conservación de la Tortuga Marina – Nación Comcáac, Grupo Tortuguero	CM/Monitoring site
11	Bahía Navachiste, SIN	Instituto Politecnico Nacional de Sinaloa, Grupo Tortuguero	CM/Monitoring site
A	Agua Blanca, BCS	ASUPMATOMA	LO, DC/Nesting site
B	Todos Santos, BCS	Grupo Tortuguero de Todos Santos	LO, DC/Nesting site
C	Pescadero, BCS	Grupo Ecológico y Tortuguero del Pescadero	LO, DC/Nesting site
D	San José del Cabo, BCS	Município de los Cabos	LO, DC/Nesting site
E	Cabo Pulmo, BCS	Amigos de Cabo Pulmo	LO, DC/Nesting site

Notes: In "ID" column, numbers denote foraging area projects and letters denote nesting beach projects; see figure 7.5 for location of each site. Primary species codes: CM = green turtle, CC = loggerhead turtle, LO = olive Ridley, EI = hawksbill, DC = leatherback.

The Future of Sea Turtle Conservation in the Gulf of California

Despite strong policies and increasing efforts at monitoring and protection, shortcomings in our understanding of sea turtle life history—and the difficulty of ferreting out corruption—have slowed the progress of sea turtle conservation. Nevertheless, when contemplating the question of how to conserve the Gulf's sea turtles, we must recognize that it is not too late for recovery. From a biological standpoint, there are encouraging signs and reasons for hope. For example, since the 2001/02 nesting season in Michoacán, Mexico, researchers report the largest nesting abundance of green turtles since the early 1980s (Javier Alvarado, pers. comm.; fig. 7.4). Many areas in Mexico are also seeing an increase in the number of olive Ridley nestings, perhaps none more so than the arribada beach in Escobilla, Oaxaca (Márquez et al. 2002; fig. 7.4). However, as much as these are promising signs for sea turtle recovery, many challenges remain.

In response to the problems of illegal poaching, we must focus on ways to slow the take of turtles. One solution is increased enforcement of existing laws. This will require a concerted effort on the part of enforcement officers and conservationists to address both the supply and demand of sea turtles as well as the cultural context in which sea turtle conservation issues are embedded. Recovery efforts must focus not only on the subsistence harvesting of these beautiful animals but also on the culture that propels their value for the tortoiseshell industry and memento trade. We should applaud PROFEPA for its tremendous vigilance, but one group or one government branch clearly cannot do it alone. Instead, it is imperative that PROFEPA liaise with the grassroots conservation organizations that are increasing their activity in and around the Gulf: Amigos de Cabo Pulmo, ASUPMATOMA, Grupo Ecológico y Tortuguero del Pescadero (GETUP), Grupo Ecologista de Antares, ProCaguama, ProNatura, ProPeninsula, and WildCoast. Enlisting the participation of such groups in community-based conservation should be a priority (e.g., Nichols et al. 2000).

Recovery efforts will benefit from greater focus on habitat protection and restoration. Coastal seagrass beds and marine algae pastures should be protected. Existing algae harvest practices must be assessed to ensure that they are sustainable and have no direct impact on foraging turtles, particularly during their earlier life stages. Water quality standards should be established and enforced through coastal monitoring efforts. Marine protected areas—such as the Alto Golfo Biosphere Reserve, the Loreto Marine Park, and the Bahía de los Ángeles Marine Park—can provide the necessary framework within which management efforts and legal mechanisms can function. Even with these parks, however, controlling illegal capture may require increased community vigilance and better monitoring of highways and other human movement corridors used to transport turtle contraband.

Conservation efforts should promote on-the-ground protection at key nesting and foraging areas and should emphasize public outreach and education in the associated communities. When appropriate, these efforts should also provide economic alternatives that are carefully planned and implemented. Whenever possible, local community members should be included early in the planning and decision-making process. In areas where development is a foregone outcome, we need to develop and implement economic strategies that balance development and tourism activities with the needs of marine species and ecosystems. It is also essential that marine fisheries are managed in a way that reduces incidental sea turtle by-catch. Clearly the TED mandates were a step in the right direction, but it is increasingly apparent that the use of drift nets, set nets, and long lines must be managed in order to prevent local turtle populations from further decline. We should explore additional opportunities for involvement in sea turtle conservation by the fishing industry, as is ongoing with leatherback protection in southern Baja.

Critical to all of these recovery strategies is our ability to measure the performance of populations as conservation measures are implemented. This should include efforts to track subpopulations at both nesting beaches and foraging areas. Because of the slow growth rates, delayed maturity, and extended longevity of sea turtles, it is crucial that monitoring programs be long-term in nature. The regional green turtle monitoring program coordinated by Grupo Tortuguero is a fantastic example, and it should serve as a blueprint for similar monitoring of other species in marine habitats and of regional nesting sites.

Finally, on the front of international collaboration, we must realize that sea turtle migrations connect the coastal waters of northwestern Mexico to other regions of the Pacific Ocean—some near and some far away.

Conservation must therefore involve international partnerships and information exchange when developing protection strategies that encompass the entire life history of these ocean connectors.

ACKNOWLEDGMENTS

I thank the government and people of Mexico for the opportunity and privilege to study sea turtles in this spectacular country. In particular, I gratefully acknowledge Antonio Resendiz, one of my early mentors in the Gulf, for his guidance and insight over the years, and Dr. J. Nichols, my friend and colleague, who helped me blossom as a sea turtle scientist and with whom I've shared fruitful conversations about sea turtle conservation that resulted in many of the thoughts presented here. I thank my wife and best friend, Jennifer, for all her help and support since my initial turtle research in the early 1990s. Javier Alvarado, Ana Barragan, John Brakey, Carlos Delgado, Richard Felger, Susan Gardner, Carlos Mandujano, Rene Márquez, Wallace J. Nichols, Frank Paladino, Melissa Paxton, Hoyt Peckham, Chris Pesenti, Bob Pitman, Rodrigo Rangel, Amadeo Rea, Antonio Resendiz, Laura Sarti, and Graciela Tiburcios provided information included in this chapter, Tom MacFarland supplied the illustrations, and Rick Brusca provided comments that improved early versions of this manuscript. My research in Mexico was made possible by the generous assistance of friends, colleagues, and mentors such as Steve Collins, Richard Felger, Todd Jones, Sandra Lanham, Laurie Monti, Gary Nabhan, Tad Pfister, Cecil Schwalbe, William Shaw, Travis Smith, Donald Thomson, and many others. Much of the information presented in this chapter was gathered during field projects supported by Earthwatch Institute, National Geographic Society, National Marine Fisheries Service, PADI Foundation, Wallace Research Foundation, the University of Arizona, and the Archie Carr Center for Sea Turtle Research at the University of Florida.

CHAPTER 8

Ospreys of the Gulf of California

Ecology and Conservation Status

JEAN-LUC E. CARTRON, DANIEL W. ANDERSON,
CHARLES J. HENNY, AND ROBERTO CARMONA

Summary

The Gulf of California and Pacific coast of the Baja California peninsula harbor a large year-round resident population of ospreys (*Pandion haliaetus*). The eastern coast of the Gulf of California also corresponds to a fall migration route for ospreys breeding in the western United States, and from Sinaloa southward it receives an influx of winter residents. The species' regional diet consists almost exclusively of fish, although it seems ospreys occasionally prey on chuckwallas (Sauromelas) and some seabirds, including young cormorants (*Phalacrocorax* sp.) still in the nest. The diet of ospreys has been especially well documented in the Midriff Islands Region of the Gulf, showing that soft-bottom and estuarine fish species, such as needlefish (*Strongylura* and *Tylosurus*) and mullet (*Mugil cephalus*), form the bulk of the prey base along coastal waters. In contrast, ospreys nesting on offshore islands tend to feed on deeper-water and offshore species, including *Girella, Paralabrax, Calamus,* and *Scorpaena*. During the breeding season, ospreys of the Gulf of California typically nest on giant cacti and on boulders and pinnacles. Southward, ospreys also nest on mangroves. Ground nests have become increasingly rare during the last few decades, but ospreys also now nest readily on man-made structures such as utility poles. Gulf of California ospreys typically lay three eggs, which they then incubate for an average of 38 days. Young (post-fledging) ospreys from the Midriff Region may wander over large areas before reaching breeding age, but many if not most would be expected to become breeders in the Gulf of California, mainly in the Midriff Region. An estimated 751 osprey pairs nested in the

Gulf of California survey area in 1977 (all of the Gulf except south of Mazatlán on the mainland). By 1992–1993, the size of the nesting population had increased by approximately 50 percent, with all of that increase observed in the eastern half of the Gulf. With an estimated 651 (58 percent) of the total 1,116 nesting pairs, the Midriff Islands Region had what appeared to be the largest density of ospreys in the entire Gulf of California in 1992–1993. Some local osprey population declines have been observed more recently, as well as wide variation in reproductive success. Pesticide contamination in the southeastern Gulf, human disturbance, overfishing, shooting, and the risk of electrocution on concrete power poles all represent threats to ospreys in the Gulf of California. The species has also been proposed as an indicator of regional ecosystem health, further warranting the need for continued monitoring.

Introduction

A nearly cosmopolitan bird of prey, the osprey (*Pandion haliaetus*) is also widely and fairly uniformly distributed in the Gulf of California, where it is arguably the best studied raptor. It ranges from the head of the Gulf of California to its mouth, with nesting recorded at multiple locations along the coasts of both the Baja California peninsula and mainland Mexico and on many of the offshore islands and islets (e.g., Grayson 1871; Baird et al. 1874; Mailliard 1923; Bancroft 1927; Huey 1927; Van Rossem 1932; Stager 1957; Banks 1962b; Grant and McT. Cowan 1964; Boswall and Barrett 1978; Henny and Anderson 1979, 2004; Judge 1983; Escalante-Pliego 1988; Mellink and Palacios 1993; Carmona and Danemann 1994; Massey and Palacios 1994; Carmona et al. 1994, 1996; Cartron 2000). Howell and Webb (1995) give the Islas Tres Marías (21°30' N) as the southernmost point where the species breeds in the Gulf of California (and, more generally, along the western coast of North America). Although not mentioned by Howell and Webb (1995), ospreys also nest along the coast of mainland Mexico as far south as San Blas, Nayarit (Escalante-Pliego 1988; H. Drummond, pers. comm.), at the same latitude as the Islas Tres Marías (fig. 8.1).

Ospreys nesting in and along the Gulf of California are year-round residents and part of a larger population that also includes pairs nesting along the Pacific coast of the Baja California peninsula (Henny and

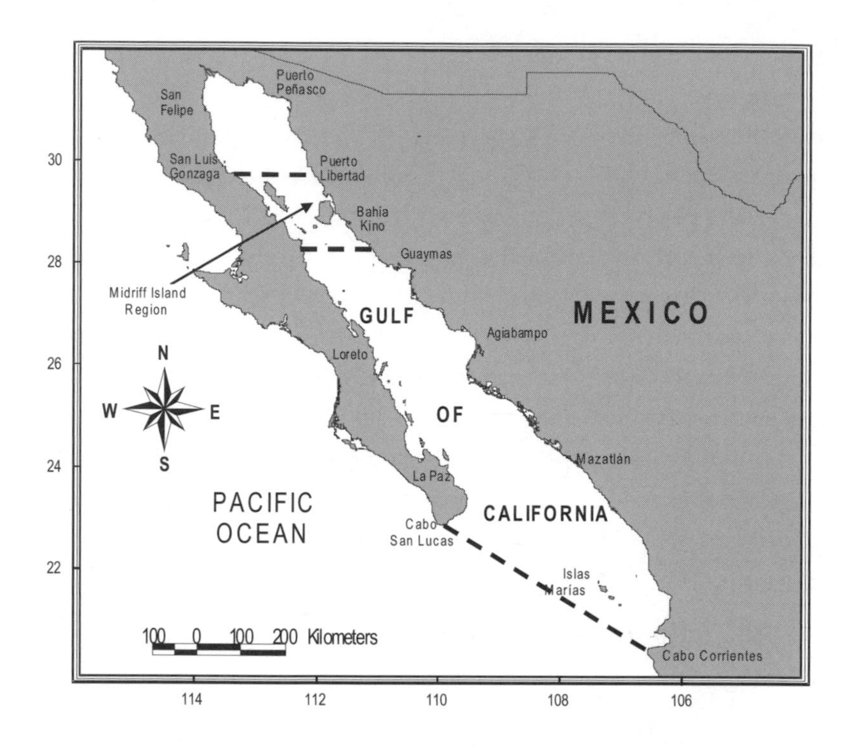

8.1. Gulf of California. The osprey is distributed from the head of the Gulf to its mouth, and it is a breeding year-round resident as far south as San Blas and the Islas Tres Marías. The eastern side of the Gulf is a fall migration route for ospreys nesting in the western United States, and from Sinaloa southward it harbors a wintering population of the species.

Anderson 1979, 2004; see below). The eastern side of the Gulf of California is also a fall migration route for ospreys nesting in the western United States, and from Sinaloa southward it receives an influx of winter migrants (Martell et al. 2001; fig. 8.1). Thus, the southeastern portion of the Gulf of California corresponds to an area of overlap in the distribution of resident and wintering ospreys. From San Blas to Cabo Corrientes, Jalisco, at the mouth of the Gulf of California, the osprey is only a winter migrant. On the Islas Marietas (the small archipelago lying 9.5 km to the southwest of Punta Mita along the coast of Nayarit) there are no nesting records for the osprey, whose occurrence there is only accidental (Rebón 1997; F. Rebón, pers. comm.).

The account presented here summarizes distributional and ecological information for the osprey for the entire Gulf of California (Cartron

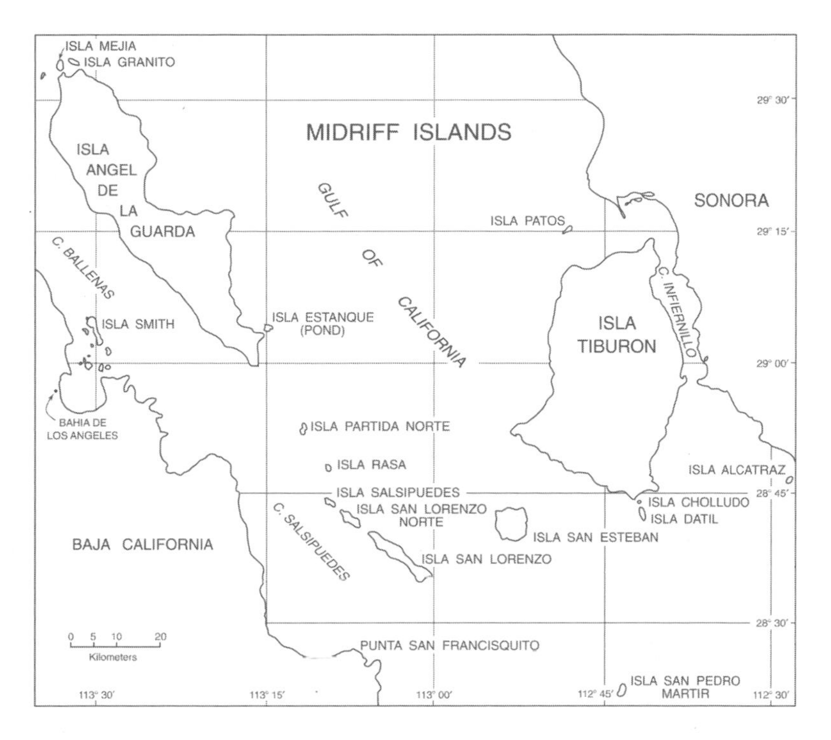

8.2. Midriff Region of the Gulf of California.

et al. 2006b provided a similar summary but only for coastal Sonora). We also discuss some of the human activities that may be a threat to the osprey, either directly or through changes in the Gulf of California ecosystem. Some of the information is new and based on detailed studies from 1971 through 1978 throughout the Midriff Islands Region (DWA, unpublished data; fig. 8.2) and recent local surveys along coastal Sonora (JLEC, unpublished data). Ospreys nesting along the Pacific coast of the Baja California peninsula are not included in this account.

History of Osprey Research in the Gulf of California

Beginning with the first published reports of ospreys in the 1870s and for nearly a century thereafter, incomplete knowledge on the distribution of the species in the Gulf of California was gained in small increments through local ornithological surveys, most of which focused on single islands or groups of islands. While residing in Mazatlán at the end of his life, the naturalist and artist Andrew Jackson Grayson visited the Islas Tres Marías in 1865, 1866, and 1867, noting nesting ospreys as part of the archipelago's avifauna (Grayson 1871; Lawrence 1874). John Xantus was a tidal observer for the U.S. Coast Survey stationed at Cabo San Lucas from April 1859 to mid-1861 (see Brusca 2004b; Brusca et al. 2005). His collecting work was the basis for Baird et al.'s (1874) report of the osprey from that location, the first published record of the species from the Baja California peninsula (Grinnell 1928).

Subsequent records of ospreys in the Gulf of California are too numerous to be cited exhaustively here. Early in the history of ornithological research in the region, nesting ospreys were reported from Isla San Luís and Isla Ángel de la Guarda by Mailliard (1923) and at San Felipe by Huey (1927). The San Luís Archipelago (seven islands and islets, including Islas San Luís and Cholluda) was again visited during the 1920s, this time by Bancroft (1927), who estimated a total of 60–75 nesting pairs at that location, including half of that total on Isla San Luís itself. Van Rossem (1932) described ospreys as common on Isla Tiburón, mentioning in particular one occupied nest. During the Belvedere Expedition of the San Diego Natural History Museum in March and April 1962, osprey nests were observed on Islas San Esteban, Salsipuedes, San Lorenzo Norte, and San Lorenzo (Banks 1962b). In the southern Gulf of California, several naturalists visited the Islas Tres Marías after Grayson's time. Among them, Nelson (1899), Stager (1957), and Grant and McT. Cowan (1964) all mentioned osprey.

The scope of osprey research expanded in the late 1970s and early 1980s. In 1977, Henny and Anderson (1979) conducted an aerial survey of the Gulf of California and Baja California region, thereby providing the first estimate (later revised) of osprey population size in the area. Judge (1981, 1983) detailed both the natural history and productivity of nesting pairs on islands of Bahía de los Ángeles, on the Baja California side of the Midriff Island Region (fig. 8.2). Similar information was provided from coastal Sonora by Cartron (2000), Cartron and Molles (2002), and Cartron et al. (2006b). Henny and Anderson (2004) repeated their survey of the Gulf of California and Baja California region in 1992 and 1993 and estimated population trends since their original survey. Several published ornithological surveys during the last three decades mention ospreys in the Gulf of California. Mellink and Palacios (1993) and Carmona and Danemann (1994) in particular provide new and important local data. Regrettably, there is a lack of information on ospreys in the southern Gulf of California. All of the ornithologists who visited the Islas Tres Marías gave little information other than distributional, and the last detailed published information (by Grant and McT. Cowan 1964), is now more than forty years old. The largest island of the archipelago, Isla María Madre, serves as a penal colony, and access to the other three islands is restricted by the Mexican government (see Velarde et al. 2005).

Natural History and Ecology

Diet and Foraging

The diet of the osprey throughout its distribution consists almost exclusively of fish (Poole et al. 2002). This is true in particular in the Gulf of California, where ospreys prey on fish from more than twenty families (Judge 1981; Cartron and Molles 2002; DWA, unpublished data; RC, unpublished data). Fairly intensive dietary studies of ospreys from the Midriff Islands Region of the Gulf of California (DWA and D. A. Thomson, in preparation) from 1973 to 1978, and less intensively since then, have generally indicated that coastal- and estuarine-nesting ospreys in the eastern Midriff Islands Region feed predominantly on estuarine and soft-bottom fish species (as also reported by Cartron and Molles 2002). Needlefish (*Strongylura* and *Tylosurus* spp.) and striped mullets (*Mugil cephalus*) are the primary prey along the coast of Sonora. Only at some locations do ospreys also catch large numbers of finescale triggerfish (*Balistes polylepis*), Pacific porgy (*Calamus brachysomus*), and several sea bass species (e.g., *Paralabrax maculatofasciatus*) (Cartron and Molles 2002). Occasionally, ospreys bring back bullseye puffers (*Sphoeroides annulatus*) to their nests but apparently recognize those fish as poisonous and drop them to the ground (Cartron and Molles 2002).

In contrast to mainland Sonora, ospreys nesting on the offshore islands of the Midriff Islands Region from Isla Turner (Isla Dátil) to Bahía de los Ángeles tend to feed primarily on rockfishes, other reef-fishes, and fish taxa that are usually considered to be more pelagic (e.g., *Sebastes, Girella, Paralabrax, Calamus, Scorpaena, Myctoperca, Hoplopagrus, Balistes, Anisotremus, Epinephelus, Paranthias, Oligoplites, Scarus,* and many others in lesser proportions). On islands of Bahía de los Ángeles in particular, Judge (1981) found the Gulf opaleye (*Girella simplicidens*) to be the species most frequently represented in collections of prey remains throughout the nesting season, followed by sea basses of the genera *Paralabrax* and *Epinephelus.*

The occasional occurrence of *Opistognathus* ("jawfish") in osprey diet in the western Midriff Islands Region is somewhat surprising, and DWA and colleagues believe that this presumably exclusive bottom-dweller becomes available to ospreys through sport fishing activities. In fact, anywhere in the Midriff Islands Region, one cannot discount the possible role of sport fishermen in making some species available to ospreys. For example, some fish pulled up from depths great enough to inflate their air bladders may be discarded and, unable to return to depth, remain alive but vulnerable to ospreys at the surface. Pelagic fishes also seen commonly in offshore-island osprey diet include: *Seriola, Sarda,* and *Euthyhnus* (DWA and D. A. Thomson, pers. obs.).

South of the Midriff Islands Region, the diet of ospreys has not been published. In Bahía de La Paz, ospreys have been observed eating mullets (*Mugil* spp.), mojarras (Gerreidae), needlefish (*Strongylura exilis*), Pacific sierra (*Scomberomorus sierra*), and Pacific mackerel (*Scomber japonicus*). In Sinaloa, ospreys catch sanddabs (Bothidae), mullets (*Mugil* spp.), spotted sand basses (*Paralabrax maculatofasciatus*), and mojarras (R. Carmona, unpublished data).

Finally, some birds and reptiles are also believed to serve as occasional prey of osprey when available, and these include pre-fledging cormorants (*Phalacrocorax* sp.), adult wintering eared grebes (*Podiceps nigricollis*), and chuckwallas (*Sauromelas*) (D. Anderson, unpublished data, has observed ospreys consuming these items on rare occasions). Boswall and Barrett (1978) documented the presence of partly consumed eared grebes in an osprey nest on Isla Rasa, and DWA has seen remains of this species in many nests in the Midriff Islands Region. Yet items such as these are also often scavenged from beaches near nests and then added to nests by osprey as nesting materials, thus making it difficult to determine their actual importance as osprey food.

Gulf of California ospreys usually forage on the wing, using patrols of shallow waters as well as high circling flights (Judge 1981; Cartron et al. 2006b). However, perch hunting has been observed in coastal Sonora, particularly along Bahía Sargento (Cartron et al. 2006b), and in Bahía de

Los Ángeles (DWA, pers. obs.). On the islands of Bahía de Los Ángeles, Judge (1981) found that 76 percent and 80 percent of foraging bouts resulted in the capture of a fish on the first dive in 1977 and 1978, respectively. This is in sharp contrast with coastal Sonora, where dive success for ospreys foraging on the wing ranged only between 14 percent and 31 percent (Cartron et al. 2006b). At that location also, 44 percent of completed dives from cliffs abutting the beach resulted in the capture of a fish. According to Swenson (1979), differences in dive success are largely the result of the ecology of prey species, with piscivorous pelagic fish likely being the most difficult to capture (see also Cartron et al. 2006b).

When an osprey seizes a fish (especially a large one), it may have to fend off other birds converging toward it in piracy attempts. Interspecific piracy involves the magnificent frigatebird (*Fregata magnificens*), yellow-footed gull (*Larus livens*), and common raven (*Corvus corax*) in addition to terns (Judge 1981; Cartron et al. 2006b; R. Carmona, pers. obs.). In fact, Henny (1988, p. 181) suggested that the magnificent frigatebird may limit the breeding distribution of ospreys in the tropics (e.g., southern Gulf). Individuals (i.e., wintering ospreys) could obtain adequate food for themselves, but would have difficulty consistently bringing food back to young in nests in the presence of the pirate species. Piracy may also be attempted by other ospreys, which can result in the fish being dropped in the water or on the ground (Judge 1981; Cartron et al. 2006b). Occasionally, large fish dropped by ospreys can be recovered by researchers, providing an opportunity to measure them. For example, a needlefish caught by an osprey weighed 450 g while a mullet measured 38 cm and weighed 800–900 g (Cartron et al. 2006b).

Nesting

The nesting season begins very early for ospreys, some pairs laying their eggs as early as December (see Cartron et al. 2006b). However, the degree of nesting asynchrony in the Gulf of California can be extremely high, and some pairs may be with eggs as late as May.

In most cases, the bulky nest is built in a cardón cactus or on a pinnacle or large boulder (Henny and Anderson 1979, 2004; Cartron et al. 2006b; color plate 6). Nests on man-made structures (e.g., utility poles, towers, buildings) are increasing in number and relative frequency (Henny and Anderson 2004). In contrast, ground nests are known from only a few localities and may be becoming increasingly rare. In the general area of Bahía de Santa Maria (25°00' N, 108°10' W), Sinaloa, R. Carmona and G. Danemann found most osprey nests to be in cardón (R. Carmona, unpublished data). However, they located three ground nests on two small mangrove islands, El Salero and Las Tunitas. On the tiny (0.4 km^2) island of Santa Ynez off the Gulf coast of Baja California Sur, Henny and Anderson (2004) found five occupied ground nests in 1977 but only one in 1992. In the Bahía de Los Angeles area, at least two ground nests were occupied in the mid-1970s, but no ground nests have been recorded during the most recent surveys (Henny and Anderson 2004).

In Sonora and the Baja California peninsula, osprey do not appear to use mangroves as nesting substrate (Henny and Anderson 1979, 2004; see also Whitmore et al. 2005). In coastal Sinaloa, however, about 10 percent of all osprey nests detected from the air were on mangroves (Henny and Anderson 2004). The northernmost mangrove nest in Sinaloa in 1993 was at 25°37.2' N, 109°12.2' W (near the town of Boca del Río). Farther south, Isla San Ignacio (near Los Mochis) also had some mangrove nests. To the south of Bahía de Santa María, in Dautillos, R. Carmona (unpublished data) observed two occupied nests on white mangrove (*Laguncularia racemosa,* Combretaceae), the rarest mangrove species in the area.

In the Gulf of California, clutch size is usually three but sometimes two or four (Cartron 2000). One-egg clutches have been reported by Judge (1981), Schaadt (1989), and Cartron (2000) but could represent incomplete clutches (Poole et al. 2002). Clutches of five and even seven eggs have been found along coastal Sonora, but these could be the result of two females laying their eggs in the same nest (Cartron 2000; Cartron et al. 2006b). Mean clutch size reported in the region has ranged from 2.38 to 2.90 (Judge 1981; Schaadt 1989; Cartron 2000) but depends on whether or not one-egg clutches are taken into account. The incubation period averages 38 days (Judge 1983). The young fledge about 58 days after hatching (Schaadt 1989). Fledging usually occurs in May, although it may be as early as April or as late as June. Nesting pairs can raise up to four nestlings to fledging (JLEC, unpublished data). Annual productivity (i.e., mean number of young fledged by nesting pairs during a year) levels reported from the Gulf of California have ranged from 0.10 to 1.14 (Judge 1983; Cartron 2000; Cartron et al. 2006b). In Bahía de Los Ángeles, mean annual productivity in 1977–1978 was 0.86 (0.94 in 1977 and 0.79 in 1978; Judge 1983). Along coastal mainland Sonora, the productivity of 42 pairs was 1.14 in 1995. Observed annual productivity in 1992 and 1993 was comparable to levels reported by Judge (1983) but ranged lower or much lower in 1994 and for 1996–1998 (see below).

Dispersal

Between 1971 and 1978, D. Anderson and students (unpublished data) banded and/or color-marked 489 pre-fledging ospreys throughout the Midriff Islands Region (from Bahía Kino, Sonora, west across the Midriff Islands to Bahía de los Ángeles and northern Isla Ángel de la Guarda). More than 75 re-sightings of banded individuals were entirely within the Gulf of California, and most were within the Midriff Islands Region itself. Of 46 band recoveries reported by others, nine (20 percent) were from the northern Gulf of California (Puerto Peñasco, San Felipe) and Río Colorado Delta area north to southern Arizona, five (11 percent) from southern California and the west coast of the Baja California peninsula (as far south as Bahía Magdalena), and one (2 percent) from Galena, Chihuahua. All 31 others (67 percent) were from the Midriff Islands Region and southward in the Gulf of California. The most distant osprey movement north was to about 34° N (coastal southern California), and south it was to about 23° N (in the Gulf of California). Band recoveries and re-sightings tended to be from less distant locations for adults than for immature birds. Altogether, these small samples suggested that ospreys originating in the Midriff Islands Region—although they wandered extensively before reaching breeding age—were basically nonmigratory, and many if not most ospreys would be expected to occur in the Midriff Islands Region itself year-round.

Population Status and Important Nesting Areas

Regional Population Levels and Trends

Two of us (DWA, CJH) conducted aerial surveys of the Gulf of California and Baja California region in 1977 and 1992–1993 (Henny and Anderson 1979, 2004). Our coverage along the eastern side of the Gulf of California extended south to Mazatlán, in coastal southern Sinaloa. It did not include the southernmost stretch of Sinaloan coast, coastal Nayarit, or any of the offshore islands of the southern Gulf of California (e.g., Isla Isabel, Islas Marietas, and Islas Marías). During the second survey of the region, coverage of the Baja California peninsula and western Midriff Islands Region was in 1992; coverage of the eastern Midriff Islands and coastal Mexican mainland was in 1993.

Based on adjustments for the lack of visibility of some nests from the air and for nesting asynchrony, an estimated 751 osprey pairs nested in the Gulf of California survey area in 1977. By 1992–1993, the size of the nesting population had increased by approximately 50 percent to an estimated 1,116 pairs (Henny and Anderson 2004). The increase was all observed in the eastern half of the Gulf of California. Along the coast of mainland Sonora and associated offshore islands, the osprey population approximately doubled between 1977 and 1993 (see Henny and Anderson 1979, 2004). A northward range extension was noted into the northernmost part of Sonora's coast, where newly erected utility poles provided osprey pairs with nesting substrates in the absence of cliffs or suitable cacti (Henny and Anderson 2004; see also Mellink and Palacios 1993). At the same time, those areas with nests in 1977 also experienced a large population increase. Along coastal Sinaloa, the nesting population appeared to grow by 150 percent between 1977 and 1993, with no obvious explanation for what was the highest rate of increase anywhere in the Gulf of California (Henny and Anderson 2004). In contrast to Sonora and Sinaloa, the Gulf coast of the Baja California peninsula saw either no increase or a slight but general decline in its number of nesting pairs (Henny and Anderson 2004).

Important Nesting Areas of the Gulf of California

With an estimated 651 (58 percent) of the total 1,116 numbers of pairs, the Midriff Islands Region stood out as the single most important area for ospreys in the Gulf of California in 1992–1993 (Henny and Anderson 2004). Henny and Anderson (2004) estimated a total of 401 pairs nesting on offshore islands of that region. The estimated total number of pairs nesting in the Midriff region along the mainlands of the Baja California peninsula and coastal Sonora were 53 and 197, respectively (Henny and Anderson 2004).

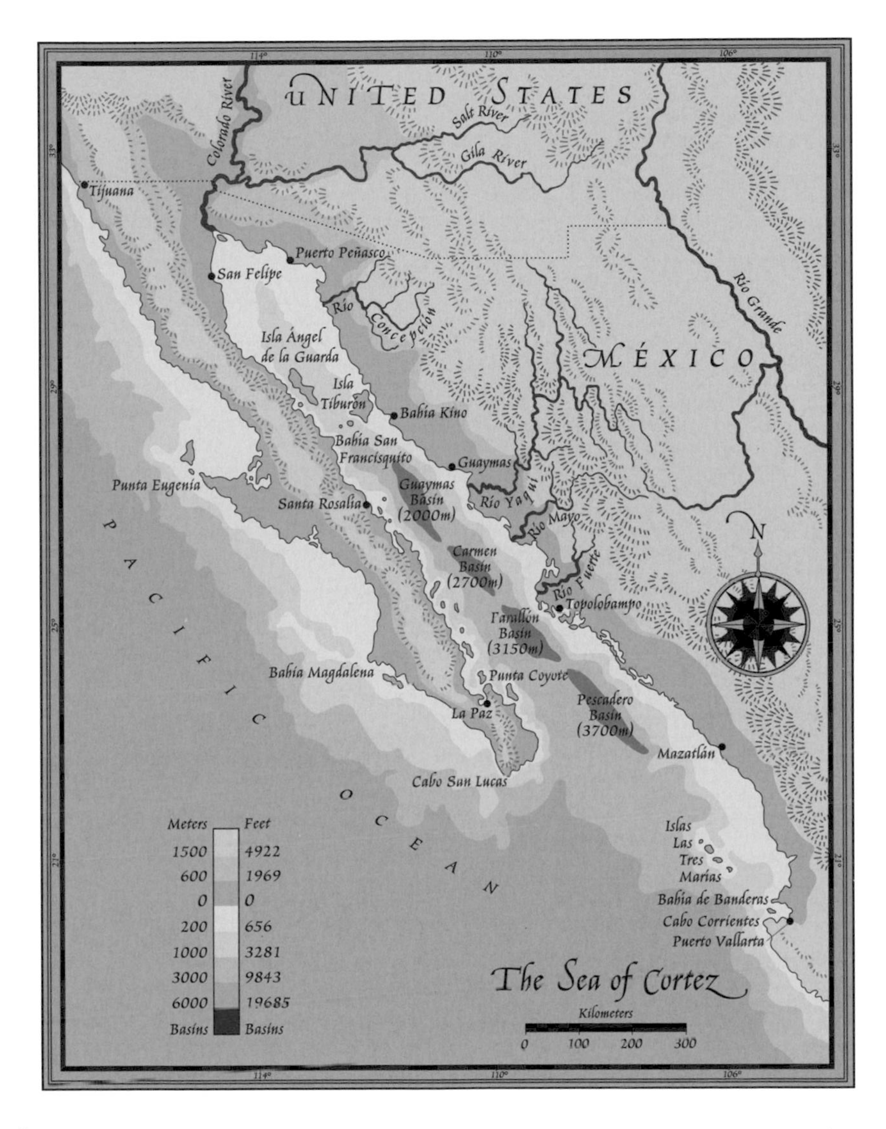

1. The Gulf of California (Sea of Cortez) with basins identified and surrounding lands marked with major rivers and indications of mountain ranges.

2. A rhodolith bed near Isla Requeson in Bahía Concepción. (A) Aerial photograph (arrows indicate bed; scale bar = 500 m). (B) Underwater view of the surface of the bed (scale bar = 10 cm for foreground and 1 m for background). (C) Clear core through the surface showing living rhodoliths grading into sediment with rhodolith fragments (scale bar = 5 cm). (D) A common sea star, *Phataria unifascialis* (Gray), in the rhodolith bed (scale bar = 5 cm). (E) Rhodolith bed at a time when the cover of primarily green algae and sponges was high (scale bar = 5 cm).

3. (on facing page) Some invertebrates of conservation concern (see text in chapter 4 for details). The gastropods (A) *Conus dalli* and (B) *Oliva incrassata;* like most other large and colorful molluscs in the Gulf, these species have become rare because of collecting by seashore visitors and for the commercial shell trade. (C and D) The sand dollar *Encope grandis;* collected by the casual seashore visitor and in shrimp trawls (and sold to curio shops), this species is becoming increasingly rare in the Gulf. A, B photos by A. Kerstitch; C, D photos by R. C. Brusca.

A

B

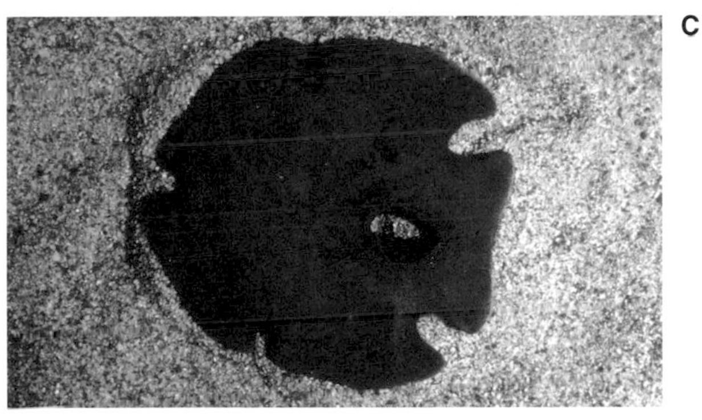

C

D

A

B
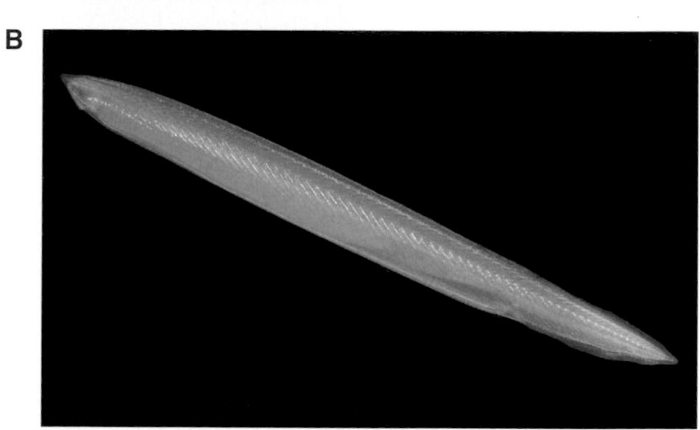

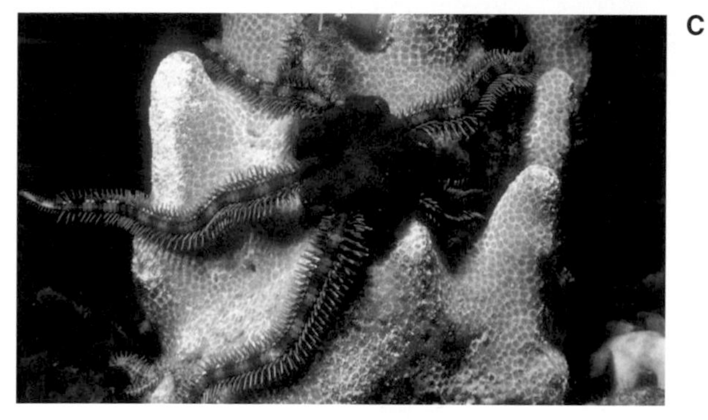
C

D

5. Panoramic of the five sea turtle species inhabiting the Gulf of California. Clockwise from top left: leatherback turtle (*Dermochelys coriacea*), loggerhead turtle (*Caretta caretta*), olive Ridley turtle (*Lepidochelys olivacea*), hawksbill turtle (*Eretmochelys imbricata*), and green (a.k.a. black) turtle (*Chelonia mydas*). Painting by Rachel Ivanyi.

4. (on facing page) Some invertebrates of conservation concern (see text in chapter 4 for details). (A) The polychaete *Aphrodita sonorae* is a Northern Gulf endemic inhabiting the seafloor at the same depths dragged by shrimp boats; trawling has driven it to rarity. (B) Amphioxus (Cephalochordata), *Branchiostoma californiense,* which lives in clean sandy habitats in tidal embayments, is threatened by pollution and coastal "development." (C) *Ophiocoma alexandri* and other large brittle stars and sea stars have become rare in the rocky intertidal zone of mainland shores because of collecting by tourists. (D) The giant sea pen, *Ptilosarcus undulatus,* has become rare because it lives at the same depths dragged by shrimp trawlers. Photos by A. Kerstitch.

6. Osprey nests and nesting substrates in the Gulf of California: (A) nest on a cardón (*Pachycereus pringlei*) near Bahía Sargento, Sonora (photo by J.-L. Cartron); (B) nest on a spire on a Midriff Region island (photo by J.-L. Cartron); (C) nests on utility poles in Bahía Kino, Sonora (photo by J.-L. Cartron); (D) nest on a tower at El Desemboque, Sonora (photo by J.-L. Cartron); (E) nest on a beached boat about 35 km north of Puertecitos, Baja California Norte (photo by D. Anderson).

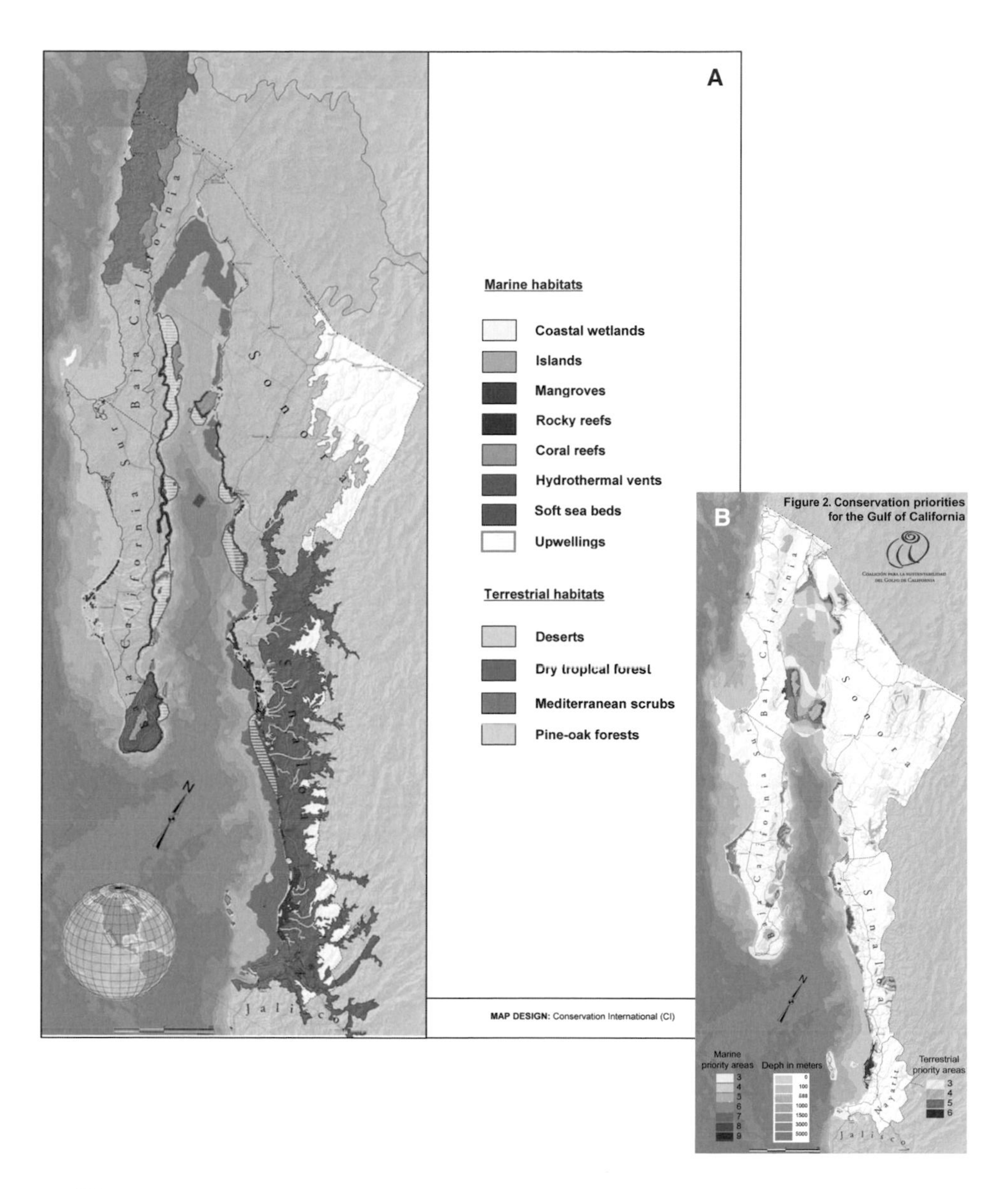

7A. Marine and terrestrial habitats in the Gulf of California. *Sources:* Conservation International, Gulf of California Geographic Information System (compiled from CEC 1997; CONABIO 1999; Sala et al. 2002; Garcillán and Ezcurra 2003; Riemann and Ezcurra 2005).

7B. Regional conservation agenda for the Gulf of California as developed by a group of conservationists and presented at the "Defying Ocean's End" (DOE) Conference. *Source:* DOE's Gulf of California Working Group; compiled by Carvajal et al. 2004).

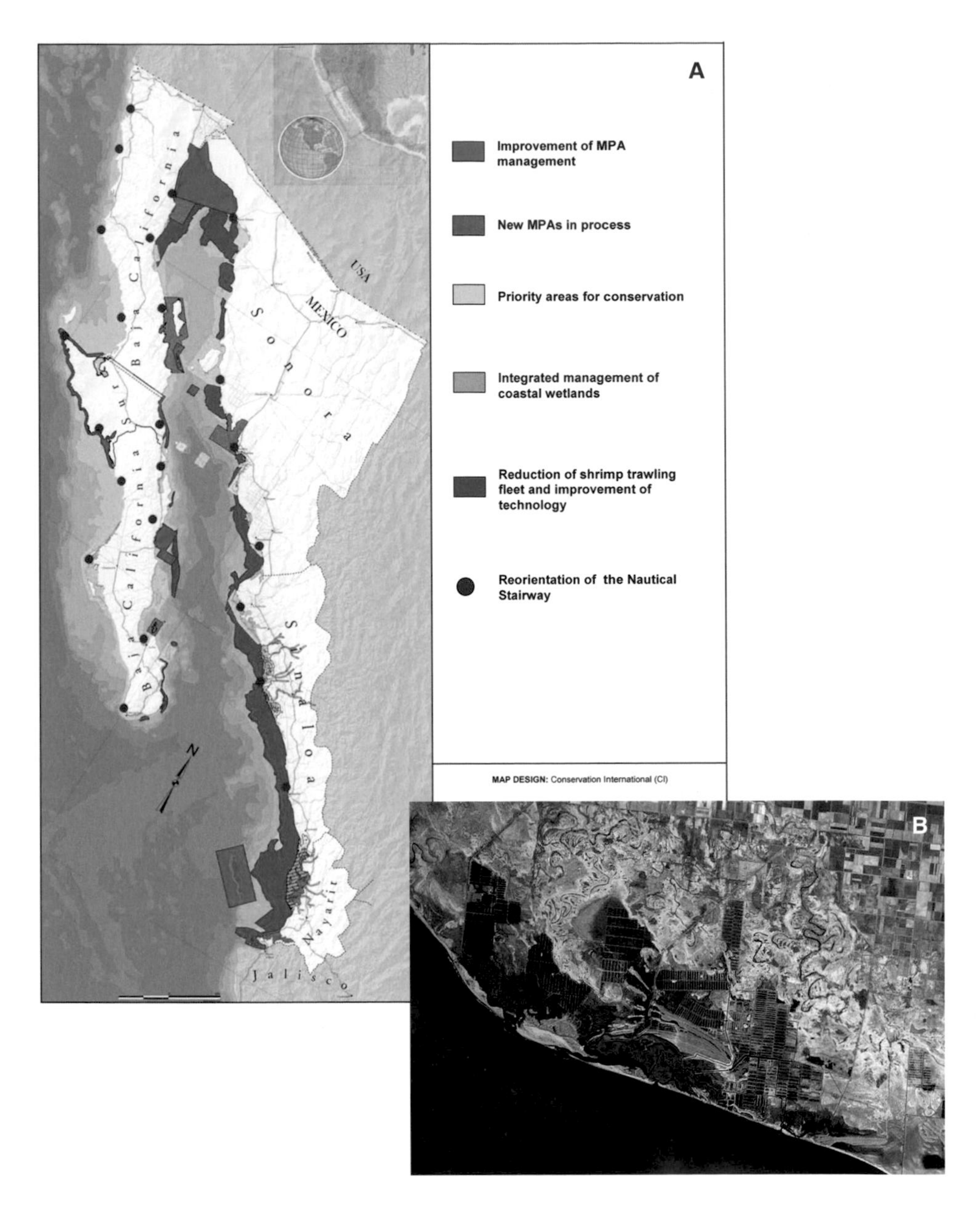

8A. Conservation priorities for the Gulf of California as developed by *Coalición para la Sustentabilidad del Golfo de California. Sources:* CSGC (2001, 2004) and Enríquez-Andrade et al. (2005).

8B. On the coast around the Yaqui Valley in southern Sonora, between Estero Lobos and Estero Tobari, most of the mangrove forests have been cut to develop shrimp farms, and the lush subtropical Sonora Desert scrub has been cleared for agriculture. Salty meanderings are all that remain of the salt-flat hinterland that was once one of the most luxuriant coastal ecosystems of the Gulf of California. *Source:* Google Earth image bank, accessed August 2005.

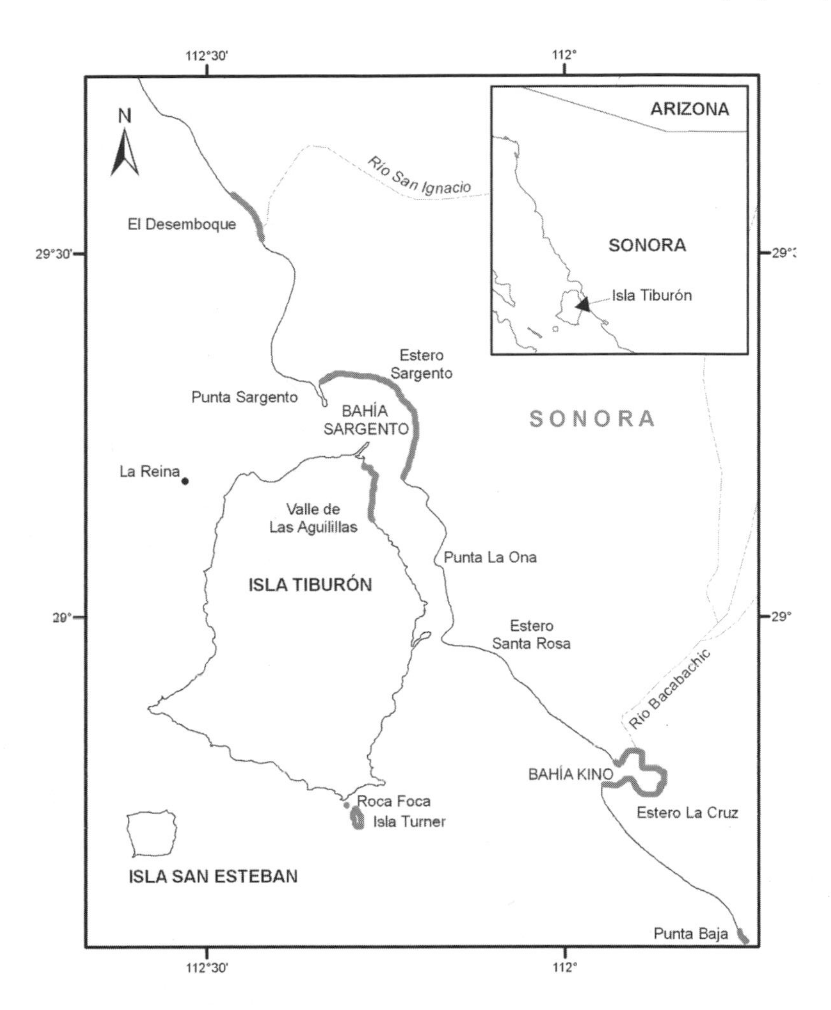

8.3. Coastal west-central Sonora (eastern Midriff Region), showing locations surveyed and monitored by Cartron (2000) and Cartron et al. (2006b) as well as important osprey nesting areas (thick gray lines).

Within the Midriff Islands Region, some areas had greater densities of nesting ospreys than others. In the eastern half of that region (see fig. 8.3), Isla Tiburón alone harbored an estimated total of 195 pairs in 1993, or more than one nesting pair (on average) for every kilometer of its 156-km-long coastline. However, the distribution of pairs was far from uniform on the island, and most pairs nested along the northern and northeastern shores. As recently as 1998, Cartron et al. (2006b) reported finding 12 nesting pairs along a 2-km transect in the Valle de las Aguilillas, a coastal plain near the northeastern corner of the island. Other islands with high densities of nesting ospreys in 1998 included Isla Turner (or Isla Dátil; 15 pairs) and Roca Foca (or Isla Cholludo; 5 pairs). Along coastal mainland Sonora, Cartron et al. (2006b) reported high densities of nesting ospreys in an area centered on Estero de la Cruz and Bahía Kino, along Bahía Sargento, and around the Seri village of El Desemboque (fig. 8.3). Farther south along the coast of mainland Sonora, Punta Baja also had high numbers of nesting ospreys (Schaadt 1989). In the western (Baja California) half of the Midriff Islands Region, Isla Partida, Isla Ángel de la Guarda, and Bahía de los Ángeles all had high densities of nesting ospreys as recently as 1992 and later. The numbers of nesting ospreys on Isla San Luís, described as high at the time of Bancroft's (1927) survey, was nowhere nearly as high in 1992 (Henny and Anderson 2004) or in any other year of the last three decades. During annual visits to Isla San Luís since 1971, DWA has found that the island consistently harbors about 4–6 nesting pairs, not 30 or more as in Bancroft's time (1927).

Outside the Midriff Islands Region, high densities of nesting ospreys were found in 1993 on some barrier islands just south of Los Mochis along the coast of Sinaloa (CJH and DWA, unpublished data). Farther south, Bahía de Santa María harbored about 40 nesting pairs in 1988, with some nests as close as 15–20 m from one another (Carmona and Danemann 1994).

Indications of Wide Population-Level Variation and Recent Local Declines

With an area of 99 km^2, Isla Espíritu Santo near Bahía de La Paz, Baja California Sur, is one of the larger islands in the Gulf of California (see Gastil et al. 1983). Its coastline consists of cliffs, shallow coves with sandy beaches, and mangrove estuaries; inland, the vegetation is typical of the Sonoran Desert (Carmona et al. 2005; JLEC, pers. obs.). The island was surveyed from the air in 1977 (Henny and Anderson 1979) but only one occupied nest was observed (CJH and DWA, unpublished data; fig. 8.4). Population variation since that time can be assessed via surveys of the island by R. Carmona and associates from 1984 to 1986 and in 1988 and 1999 (Carmona et al. 1994; RC, unpublished data) and via the 1992 aerial count

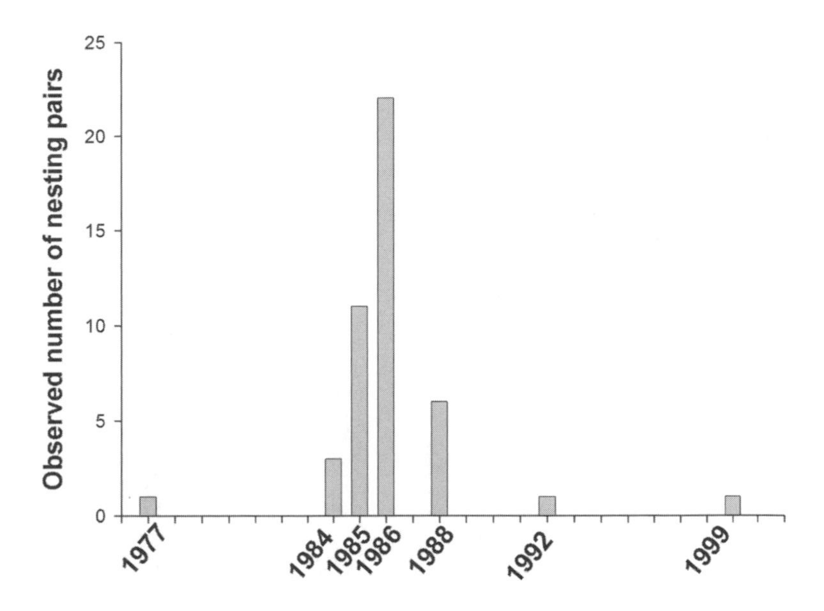

8.4. Observed number of nesting pairs in selected years during 1977–1999 on Isla Espíritu Santo, near Bahía de La Paz, Baja California Sur. Note the sharp and rapid increase in the observed number of nesting pairs beginning in 1984 and continuing through 1986, followed by an equally sharp and rapid decline. Data for 1977 and 1992 are unadjusted aerial counts (Charles J. Henny and Daniel W. Anderson, unpublished data); all other numbers are from Carmona et al. (1994) and Roberto Carmona (unpublished data).

from C. Henny and D. Anderson's second regional survey. Apparently, the number of nesting ospreys on the island increased sharply beginning in 1984 and peaked in 1986, with a total of 22 occupied nests. After 1986, the number of nesting pairs decreased and, by 1992, only one nesting pair again nested on the island (16 unoccupied nests were also observed)—the same count as in 1999. The patterns observed on Isla Espíritu Santo serve to illustrate that osprey population numbers and trends can change rapidly (perhaps with the help of anthropogenic factors; see below).

Along the coast of central mainland Sonora (fig. 8.3), J.-L. Cartron monitored the reproductive success of ospreys from 1992 through 1998 (Cartron 2000; Cartron et al. 2006b). During those years, the productivity of nesting pairs (i.e., the mean number of young fledged per occupied nest) varied elevenfold, with a low of only 0.1 in 1997 (only 2 of 40 nests were successful) and with an apparent overall downward trend. The observed

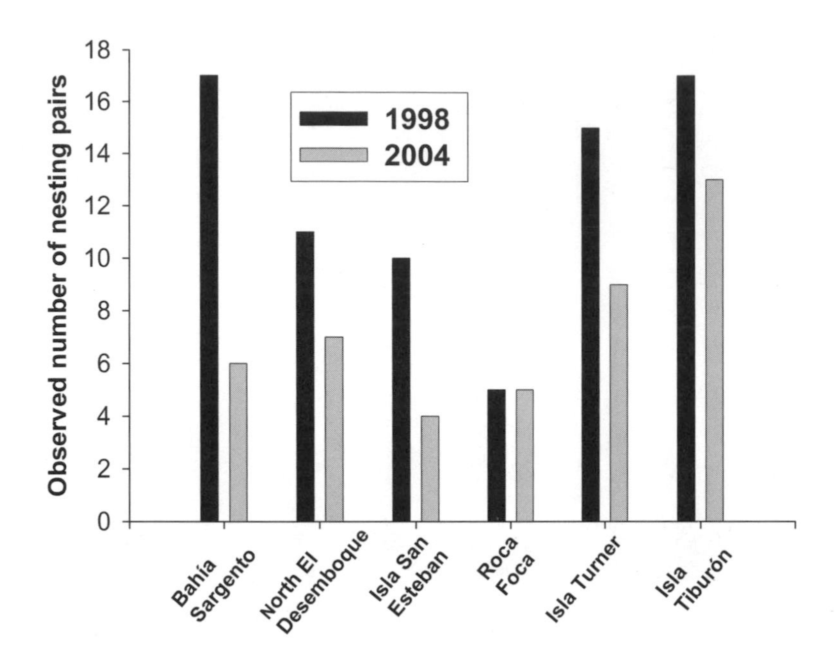

8.5. Numbers of nesting pairs observed during one-time surveys in 1998 versus 2004 at five different locations. Except for the 1998 survey of Isla San Esteban, which was conducted in April, all other surveys were in mid-March. The surveys of Isla Tiburón covered the southern shoreline and the western shoreline north to La Reina. Note the decline in the number of observed nesting pairs in 2004.

number of nesting osprey pairs increased through 1995 but subsequently began to decline. In March–April 1998, Cartron et al. (2006b) expanded the study area and conducted a survey of a 100-km stretch of coast between Estero de la Cruz and the vicinity of Cerro El Puerto (29°34.69 N, 112°27.08 W) on the mainland and along the coasts of Roca Foca, Isla Turner, Isla San Esteban, and Isla Tiburón (southern and western shoreline only). On Isla San Esteban, Isla Turner, and Roca Foca combined, the observed number of pairs in 1998 was comparable to or higher than the unadjusted count reported by Henny and Anderson (2004). A repeat of the 1998 survey in 2004 suggested significant declines in local osprey population (fig. 8.5). The decrease in the number of nesting pairs noted in 2004 was most severe along Bahía Sargento. At that location, nests occupied in previous years were more likely to be still occupied if they were located near the local mangrove estuary (JLEC, unpublished data; see below).

Actual and Potential Threats to Ospreys

At present, the impact of humans and human activities on the Gulf of California osprey population remains poorly documented and poorly understood. However, some threats have been identified, most of which are likely to intensify as human populations continue to grow in the region and place increased pressure on marine and coastal ecosystems.

Contamination

In the 1960s and 1970s, pesticide contamination caused osprey populations to plummet throughout much of North America (e.g., Ames and Mersereau 1964; Ames 1966; Hickey and Anderson 1968; Henny and Wight 1969; Wiemeyer et al. 1975, 1978, 1988; Spitzer et al. 1977). Since the ban of DDT and derivatives, osprey numbers have largely recovered, but residual contamination by organochlorine pesticides remains a problem in some areas (see Poole et al. 2002 and references therein).

No impact of contaminants has ever been documented on ospreys in the Gulf of California; in fact, over much of its area the Gulf may be one of the least contaminated ecosystems in western North America (Anderson, pers. obs.). Yet impacts of pesticides on ospreys cannot be ruled out, especially in the southeastern Gulf, where agricultural runoff to the coast is high and contaminants can more readily reach offshore and estuarine waters. Pesticide contamination does occur in coastal wetlands of Sinaloa, a state where agriculture is an especially important economic activity (Mora et al. 1987; Mora and Anderson 1991; Stoleson et al. 2005). Mora et al. (1987) reported benzene hexachloride and dieldrin residues in northern pintails (*Anas acuta*) wintering in the Culiacán area, Sinaloa. Although not reaching levels known to have detrimental effects on birds, residues were sufficient to indicate local sources of contamination. In the same general area, Carmona and Danemann (1994) did not detect any impact of pesticides on wildlife but noted the possibility of contamination of Bahía de Santa María from streams that discharge into it. At the northern end of the Gulf of California, selenium and organochlorine pesticide contamination has been documented in the Colorado River delta ecosystem (Mora and Anderson 1995; Mora et al. 2003). However, few ospreys nest in the northernmost Gulf of California.

Human Disturbance

Ospreys nesting in urban or semi-urban areas readily habituate to humans and their activities (see Poole et al. 2002). Elsewhere, however, ospreys do not have an opportunity to build up their tolerance to humans, so disturbance carries a greater risk of flushing adults from the nest and interrupting incubation of the eggs or shading of young nestlings. As noted by Henny and Anderson (2004) and many others, there has been a sharp increase in human activities in the Gulf of California. At the time of the 1992 aerial survey of Isla Espíritu Santo, during which only one occupied nest was found, C. Henny and D. Anderson also noted the presence of people camping in nearly every cove. Thus, the possibility exists that human disturbance played a part in reversing the growth of the osprey nesting population on that island in the mid-1980s. In the Bahía de Los Ángeles area, disturbances are indicated by trails leading directly to some osprey nests (on Islas Ventana, Bota, and Smith), the presence of many camps on the beaches near some osprey nests, and observations of people climbing up to nests. Monitoring of the osprey population in the Bahía de Los Ángeles area from 1971 to present (D. Anderson, pers. obs.) shows a local decline in the number of nesting pairs (from 30–35 in early years to about 25 in 2003), normal productivity of successful pairs, and a shift in the distribution of occupied nests toward less disturbed areas. All of these patterns point to an impact of human disturbance on nesting ospreys in the area.

Overfishing

Commercial trawlers and fishing boats with large gill nets have had a substantial negative impact on the Gulf of California marine ecosystem, and overfishing may now apply to nearly every fishery of the region (see Brusca et al. 2005; Cisneros-Mata, chapter 6 in this volume). As already mentioned, Cartron (2000) noted an apparent trend toward much lower productivity of ospreys beginning in 1996. For 1997, in particular, only 2 of 40 osprey pairs produced fledglings along an 80-km stretch of the coast of mainland Sonora. Low productivity was combined with lower clutch size and delayed onset of egg laying. Throughout most of the monitoring study (1992–1998), successful and unsuccessful nests alike tended to be clumped spatially, with no relationship between success and distance to the nearest potential source of disturbance (Cartron 2000; Cartron et al. 2006b). These patterns, together with the shift in distribution of occupied nests at Bahía Sargento (mangrove waters are generally considered to be very productive; see Brusca et al. 2005; Whitmore et al. 2005), suggest that gill nets, trawlers, or perhaps even crab traps may negatively affect local osprey reproductive success.

Concrete Power Poles

Concrete power poles fitted with steel cross-arms are responsible for high raptor mortality in parts of northwestern Mexico, including coastal Sonora (Cartron et al. 2000, 2005, 2006a; Manzano-Fischer et al. 2006). Raptors that perch on steel cross-arms become grounded and can be electrocuted by touching a single energized wire (see Cartron et al. 2005). One dead osprey was recovered under a concrete pole in northwestern Chihuahua. In coastal Sonora, ospreys nest on pole tops rather than cross-arms (JLEC, pers. obs.), which likely entails a reduced risk of electrocution. However, engineers from the Comisión Federal de Electricidad have reported occasional osprey electrocutions (Cartron et al. 2006b).

Shooting

Most Gulf of California osprey band recoveries have been the result of mortality caused by shooting (DWA, unpublished data). In the 1970s "sportsmen" on yachts in the Gulf of California bragged to DWA that they used ospreys as target practice for wing shooting, commenting that "those birds are hard to bring down." Although infrequent, shooting was also reported at El Desemboque in coastal Sonora, where young boys practiced with rifles on nesting ospreys (JLEC, unpublished data).

Conclusions and Recommendations

In the contiguous United States, estimated osprey numbers reached 16,000–19,000 pairs in 2001—a 25 percent increase since 1994 (Poole et al. 2002). The osprey experienced population increases and/or range extensions in most states, including those bordering northwestern Mexico: Arizona (Dodd and Vahle 1998), California (Henny and Anthony 1989; Poole et al. 2002), and New Mexico (D. Stahlecker, pers. comm.). Between 1977 and 1992, osprey numbers also increased along the Pacific coast of the Baja California peninsula, especially from Punta Eugenia south to Cabo San Lucas, where the increase was a tremendous 394 percent (Henny and Anderson 2004). Certainly the increase in the number of osprey pairs between 1977 and 1992–1993 in the Gulf of California accords with all of those other recent population trends. On the basis of the 1992–1993 osprey survey, it would seem that the outlook for the osprey in the Gulf of California is good.

Yet the data in Henny and Anderson's (2004) survey is now more than fifteen years old. The more recent numbers reported here suggest a possibly widespread decline of the osprey population along coastal Sonora and on islands of the eastern Midriff Islands Region (fig. 8.5). Another cause for concern is a possible downward population trend for osprey along the eastern coast of the Baja California peninsula between 1977 and 1992 as well as the population decline observed at Bahía de Los Ángeles, in particular since the 1970s. In addition, R. Rodríguez-Estrella (pers. comm.) and students have observed during the last few years a very sharp decrease in the number of osprey pairs at Laguna San Ignacio, along the Pacific coast of Baja California Sur. Has the osprey population of the Gulf of California and Pacific coast of the Baja California peninsula declined, perhaps even substantially, since the 1992–1993 survey? Do some local osprey populations in the region go through rapid cycles of growth and decline in response to density-dependent factors? The rapid, dramatic increase and then decrease in osprey population size on Isla Espíritu Santo could be part of a natural cycle, or it could have been caused by human disturbance (and/or some other anthropogenic factor). Despite the large body of information gathered on ospreys in the Gulf of California, population dynamics and human impacts remain poorly understood.

Among the priorities for future research are the following.

1. A detailed phylogenetic analysis to determine the degree of differentiation between the resident osprey population of the Gulf of California (and Baja California) region and migratory populations farther north. Gulf of California resident ospreys are presumed to be the subspecies *P. haliaetus carolinensis,* also found farther north in North America. However, they show similarities with the subspecies found in the Caribbean and along the coast of southeastern Mexico and Belize, *P. haliaetus ridgwayi* (see Poole et al. 2002).

2. Continued monitoring of the Gulf of California and Baja California osprey population, at a regional level and at intervals of no more than 10–15 years.
3. Multiyear local studies to better understand population dynamics and document the impact of anthropogenic impacts.
4. Through more extensive marking and banding, a demographic study to estimate minimum replacement rates (i.e., minimum annual productivity necessary for maintaining population size) for ospreys in the Gulf of California (and Baja California) region.
5. Also through more marking and banding, determining the possible role and importance of the Gulf of California and Baja California ospreys in expanding the population into northwestern Baja California and southern California offshore islands, where they nested historically (Kiff 1980).
6. Along mainland Sonora, monitoring human disturbance and prey delivery at nests; also, testing the hypothesis of food shortage by conducting a controlled study in which randomly selected nests are supplemented daily with fish.

Management and protection of the osprey in the Gulf may require the following measures.

1. Where camping is allowed in areas with osprey nests, it would be useful and educational to erect signs describing the osprey and its sensitivity to disturbance during the nesting season. If signs are already in place but focus on other species, some mention of the osprey could be added.
2. Continue to educate and collaborate with the Comisión Federal de Electricidad on the issue of raptor electrocutions on concrete power poles. "Suggested Practices for Avian Protection on Power Lines" (APLIC 2006) addresses the issue of conductive poles, such as those made of concrete and fitted with steel cross-arms.

Note added in proof. Between the time this chapter was finalized and the book published, another osprey survey was conducted in the Gulf of California and Baja California region. See Henny et al. (2008) for more information.

CHAPTER 9

Marine Mammals of the Gulf of California

An Overview of Diversity and Conservation Status

JORGE URBÁN R.

Summary

Interest in the study of Mexican marine mammals dates from the mid-nineteenth century, a result of their commercial exploitation. Nevertheless, until the 1970s scientific studies of the marine mammals of Mexico were infrequent and mainly limited to occasional observations by American and a few Mexican scientists. Beginning in the late 1970s, the number of reports published on individual species or groups of species in the Gulf of California began to increase substantially.

The marine mammal fauna of the Gulf is surprisingly diverse, with 36 species representing 31 cetaceans in eight families (Balaenidae, Eschrichtiidae, Balaenopteridae, Physeteridae, Kogiidae, Ziphiidae, Delphinidae, Phocoenidae), four pinnipeds in two families (Phocidae, and Otariidae), and one bat in the family Vespertilionidae. The Gulf of California has 39 percent of the world's 83 cetacean species. The Odontoceti (toothed whales) are represented by 23 species, 33 percent of the extant species, and the Mysticeti (baleen whales) by 8 species, 61 percent of the extant species.

The notable diversity (i.e., species richness) and abundance of marine mammals in the Gulf can be explained by three main factors. First, the Gulf has an exceptionally high rate of primary productivity that supports complex and productive food webs. Second, the complex topography and oceanography of the Gulf presents a high diversity of habitats. And third, the warm and relatively calm waters found in the Gulf during winter and spring are exploited by several migratory species to give birth, nurse, and care for their newborn.

According to the IUCN, the Gulf of California includes one species that is *critically endangered* (vaquita), four that are *endangered* (blue whale, fin whale, sei whale, northern right whale), and two that are *vulnerable* (humpback whale, sperm whale). The Mexican Government's Official Standard (NOM-59-ECOL-2001) includes all of the cetacean species entering Mexico's waters. The North Pacific right whale and the vaquita are listed as "in danger of extinction" and all others as "subject to special protection."

The main human-caused mortalities of cetaceans in the Gulf of California are related to fisheries. The cetaceans most frequently involved in incidental mortality from fisheries are the short-beaked common dolphin, long-beaked common dolphin, common bottlenose dolphin, pantropical spotted dolphin, and vaquita. Yet large cetaceans, including sperm, gray, and humpback whales, also have been affected by fisheries.

There is no single body of legislation enacted for the sole benefit of cetaceans in the Gulf. Instead there are several different laws relevant to their conservation and management that apply to all of Mexico. Additionally, in May 2002, Mexico established the Mexican Whale Refuge (Refugio Ballenero Mexicano), consisting of its entire Exclusive Economic Zone (about 3 million km^2).

Introduction

Interest in the study of Mexican marine mammals dates from the mid-nineteenth century, mainly between 1850 and 1870, and was a direct result of their commercial exploitation. Whalers, principally American, made numerous voyages to hunt large cetaceans in the Mexican Pacific. From sailing ships and whaling dories they sought mostly gray whales, *Eschrichtius robustus,* but also took humpback whales, *Megaptera novaeangliae* (Scammon 1874/1968; Henderson 1972, 1984; Urbán R. et al. 2003b), and sperm whales, *Physeter macrocephalus* (Rice 1974). By the early twentieth century, modern mechanized whalers had expanded the catch to include some of the faster rorquals that were previously uncatchable by the "Yankee" whalers: blue whales, *Balaenoptera musculus,* Bryde's whales, *B. edeni,* and minke whales, *B. acurostrata* (Rice 1974). The California sea lion, *Zalophus californianus,* Guadalupe fur seal, *Arctocephalus townsendi,* and northern elephant seal, *Mirounga angustirostris,* were also heavily exploited on islands off the Pacific coast of Baja California (Bartholomew and Hubbs

1960; Le Boeuf et al. 1983; Zavala-Gonzalez and Mellink 2000). Before the 1970s, scientific studies of the marine mammals of Mexico were infrequent and mainly limited to occasional observations by American and a few Mexican scientists (e.g., Bartholomew and Hubbs 1952; Berdegué 1956a; Norris and McFarland 1958; Gilmore 1960; van Gelder 1960; Norris and Prescott 1961; Gilmore et al. 1967; Lluch 1969; Barham 1970; Rice 1974). Beginning in the early 1970s, an increasing number of reports began to be published on individual species or groups of species in the Gulf of California. Vidal, Findley, and Leatherwood published the first comprehensive account of this sea's diverse and rich marine mammal fauna in 1993. It was largely based on important unpublished information existing in difficult-to-obtain "gray literature": researchers' field notes, osteological materials from scientific collections, and strandings and sightings documented with photographs. The present paper is an update of that pioneering publication.

The marine mammal fauna of the gulf is surprisingly diverse (Aurioles-Gamboa 1993; Vidal et al. 1993; Urbán R. and Rojas 1999; Brusca et al. 2005; Urbán R. et al. 2005; Lluch-Cota et al. 2007), with 36 species representing 31 cetaceans, 4 pinnipeds, and 1 bat. With 31 species (21 genera) of whales and dolphins recorded, this sea has 39 percent of the world's 83 cetacean species. The Odontoceti (toothed whales) are represented by 23 species, 33 percent of the extant species, and the Misticeti (baleen whales) by 8 species, 61 percent of the extant species.

This notable diversity (species richness) and abundance of marine mammals in the Gulf can be explained by three main factors (Urbán R. et al. 2005). First, the Gulf has an exceptionally high rate of primary productivity (Zeitzschel 1969) that supports complex and productive food webs. Toothed whales (Odontoceti)—including fish-eating (ichthyophagi), squid-eating (teuthophagi), and (otherwise) flesh-eating (sarcophagi) species—all find abundant and diverse prey to support generally large populations year-round. For baleen whales (Mysticeti), the great abundance of euphausiid shrimps (krill) as a prey resource allows for the presence of the blue whale (*Balaenoptera musculus*) and the fin whale (*B. physalus*), among others. A second major factor is the great diversity of habitats that reflect the complex topography and oceanography of the Gulf (Brusca 2004a; Brusca et al. 2005; Lluch-Cota et al. 2007). This allows for the presence of both coastal and oceanic species as well as species from both tropical and temperate waters. Finally, the warm and relatively calm waters found in the Gulf during winter and spring are exploited by several migratory species. For example, humpback whales (*Megaptera novaeangliae*) give birth, nurse, and care for their newborn in the Gulf, and until recently so did part of the migrating population of the gray whale (*Eschrichtius robustus*) (Findley and Vidal 2002).

Species Accounts

The classification given in this section follows Rice (1998) except for the family Balaenidae, where I use the classification of the International Whaling Commission (2001).

Order Cetacea Brisson, 1762—Whales, dolphins, and porpoises; ballenas, delfines, y marsopas

Suborder Odontoceti Flower, 1867—Toothed whales; cetáceos con dientes

- **Superfamily Physeteroidea (Gray, 1821)**
  - Family Physeteridae (Gray, 1821)
    - **Genus *Physeter* Linnaeus, 1758**

PHYSETER MACROCEPHALUS LINNAEUS, 1758. SPERM WHALE; *CACHALOTE*. The sperm whale has one of the largest distributions of any marine mammal, ranging throughout all the world oceans from the equator to the edges of the polar ice (Rice 1989). Since the eighteenth century, sperm whales have been opportunistically sighted in the Gulf of California (Townsend 1935; Vidal et al. 1993). Adult males, groups of subadult males, and groups of females and their dependant young have been recorded here, suggesting that the Gulf is an important breeding ground for this species (Jaquet and Gendron 2002; Jaquet et al. 2003; Rubio C. et al. 2006). Their distribution is tied to the presence of one of their most important prey resources, the jumbo squid, *Dosidiscus gigas* (Jaquet and Gendron 2002; Ruíz-Cooley et al. 2004; Davis et al. 2007). Sperm whale abundance in the Gulf of California is estimated at 417 (95 percent confidence interval = 164–1,144) individuals, a figure that is based on line transect methods (Gerrodette and Palacios 1996). Movements are poorly understood, but photo identification data have so far identified seven females that moved into the Gulf of California from the Galapagos Islands, traveling up to 3,803 km (Jaquet et al. 2003).

- Family Kogiidae Gill, 1871
  - **Genus *Kogia* Gray, 1846**

KOGIA BREVICEPS (BLAINVILLE, 1838). PYGMY SPERM WHALE; CACHALOTE PIGMEO. The pygmy sperm whale is a cosmopolitan species recorded in nearly all temperate, subtropical, and tropical waters. Preliminary stomach content studies indicate a distribution off the continental shelf (Leatherwood and Reeves 1983; Rice 1998; Gendron 2000). Stranding data reviewed by Vidal et al. (1987; Vidal et al. 1993) indicated that pygmy sperm whales are distributed throughout the entire Gulf, from San Felipe and Puerto Peñasco in the north to La Paz and Mazatlán in the south. There are no abundance estimates for this species in the Gulf.

KOGIA SIMA (OWEN, 1866). DWARF SPERM WHALE; CACHALOTE ENANO. The distribution of dwarf sperm whales includes both temperate and tropical waters of the world (Caldwell and Caldwell 1989). Numerous strandings and sightings, especially along the west coast of the Gulf of California, indicate an important population of these whales in this region. During numerous cetacean surveys in the southwest Gulf of California between 2004 and 2006, *Kogia sima* was the odontocete most frequently observed, second among cetaceans only to bottlenosed dolphins, *Tursiops truncatus* (Cárdenas 2008).

- **Superfamily Ziphioidea (Gray, 1865)**
  - Family Ziphiidae Gray, 1850—Beaked whales; zífidos y mesoplodontes
    - **Genus *Ziphius* G. Cuvier, 1823**

ZIPHIUS CAVIROSTRIS G. CUVIER, 1823. CUVIER'S BEAKED WHALE; ZÍFIDO DE CUVIER. This species is one of the world's most widely distributed cetaceans and the most common beaked whale in the eastern North Pacific (Leatherwood et al. 1988; Ferguson et al. 2006). Their distribution in the Gulf of California is related to the limit of the continental shelf. Two main areas of concentration have been detected: southwest of the Los Frailes area (the east cape region); and in the central Gulf, around Isla San Pedro Mártir (Barlow et al. 1997). Cárdenas (2008) registered *Ziphius cavorostris* as the most common beaked whale in the southwestern Gulf, and Barlow et al. (2006) claimed that the highest density of Cuvier's beaked whales in all the Pacific (38 animals per 1,000 km^2) occurs in the southern Gulf of California.

• Genus *Berardius* Duvernoy, 1851

BERARDIUS BAIRDII STEJNEGER, 1883. BAIRD'S BEAKED WHALE; ZÍFIDO DE BAIRD. Baird's beaked whale (*Berardius bairdii*) is endemic to the North Pacific Ocean and adjacent seas (Bering Sea, Okhotsk Sea, Sea of Japan, and the southern Gulf of California) (Leatherwood et al. 1988; Balcomb 1989). Seasonal movements are poorly understood in the eastern North Pacific. California catches suggest two peaks of abundance, July and October (Rice 1974). In the Gulf of California there are seven records: four strandings (two single and two mass strandings) and three sightings. All occurred in summer months and, except for one skull collected at Isla San Esteban, all were in the southwestern Gulf (Vidal et al. 1993; Urbán R. et al. 2007; Urbán R. and colleagues, unpublished data). The last mass stranding occurred on July 31, 2006, near the southern end of Isla San José, north of Bahía de La Paz. This stranding included ten animals, all sexually mature males ranging in size from 9.97 m to 11.05 m and with estimated ages (based on dental layers) from 8 to 40 years (Urbán R. et al. 2007).

• Genus *Indopacetus* Moore, 1968

INDOPACETUS PACÍFICUS (LOGMAN, 1926). LOGMAN'S BEAKED WHALE; ZÍFIDO DE LOGMAN. Before the publication by Delebout et al. (2003), the records of this species in the North Pacific were identified as an undescribed species of the genus *Hyperoodon* (see, e.g., Leatherwood et al. 1988). There are only two sightings of this species in the Gulf of California, both in the southwestern area (Urbán R. et al. 1994; Urbán R., unpublished data).

• Genus *Mesoplodon* Gervais, 1850

MESOPLODON PERUVIANUS REYES, MEAD, AND VAN WAEREBEEK, 1991. PYGMY BEAKED WHALE; MESOPLODONTE PIGMEO. This species of *Mesoplodon* has more records in the Gulf of California than any other. Records include: five strandings, four in the area of Bahía de la Paz, Baja California Sur (B.C.S; Urbán R. and Aurioles 1992; Urbán R. and Pérez-Cortés 2000; Urbán R., unpublished data), and one in Bahía de Banderas, Jalisco (unpublished data); one skull collected at Bahía de San Carlos, Sonora (Vidal et al. 1993); and at least seven live sightings, including at Bahía de Banderas (Barlow et al. 1997; Romo 2004), Los Barriles, B.C.S. (Cárdenas 2008), and Bahía Concepción, B.C.S. (unpublished data). There have also been a few sightings in Bahía de Banderas and in the central and southern Gulf of California of an unknown *Mesoplodon* described as *Mesoplodon* sp. A (Pitman et al. 1987). Based on the distribution and pigmentation patterns, these have been thought to be *Mesoplodon peruvianus* (Pitman and Lynn 2001), but confirmation is needed.

MESOPLODON DENSIROSTRIS (BLAINVILLE, 1817). BLAINVILLE'S BEAKED WHALE; MESOPLODONTE DE BLAINVILLE. Blainville's beaked whale is widespread in tropical and warm-temperate waters of all oceans, where it is regularly sighted over slope-depth waters (500–1,000 m; Reeves et al. 2002). Only one sighting is confirmed in the Gulf of California, and that was in the central Gulf region (Pitman and Lynn 2001; Macleod et al. 2006).

MESOPLODON GINKGODENS NISHIWAKI AND KAMIYA, 1958. GINKGO-TOOTHED BEAKED WHALE; MESOPLODONTE DE DIENTES DE GINKGO. The ginkgo-toothed beaked whale is found in tropical and warm temperate waters of the Indo-Pacific (Reeves et al. 2002). In Mexico there is only one confirmed record based on a skull collected at Malarrimo Beach in the west coast of the Baja California peninsula (Leatherwood et al. 1988). The only records for this species within the Gulf of California come from two photographs of one skull found in the fishing village of El Golfo de Santa Clara in the uppermost Gulf (unpublished data). Because positive identification is difficult when based only on skull photographs, the presence of *Mesoplodon ginkgodens* in the Gulf should be taken as possible but not confirmed.

- **Superfamily Delphinoidea (Gray, 1821)**
  - Family Delphinidae Gray, 1821—Dolphins; delfines
    - **Genus *Steno* Gray, 1846**

STENO BREDANENSIS (G. CUVIER IN LESSON, 1828). ROUGH-TOOTHED DOLPHIN; DELFÍN DE DIENTES RUGOSOS. The rough-toothed dolphin exists worldwide in tropical and warm-temperate waters. Although it is not a frequently observed species, there are sightings throughout the Gulf of California—particularly in the lower Gulf, including the Bahía de Banderas (Jalisco), and Bahía de La Paz (B.C.S.), where four single strandings have been documented (Urbán R. et al. 1997). There are an estimated 6,341 (95 percent C.I. = 2,853–14,757, based on line transect methods) such animals in the Gulf of California (Gerrodette and Palacios 1996).

- **Genus *Tursiops* Gervais, 1855**

TURSIOPS TRUNCATUS (MONTAGUE, 1821). COMMON BOTTLENOSE DOLPHIN; TONINA, TURSIÓN. Bottlenose dolphins exist worldwide in temperate and tropical waters, with populations occurring in coastal waters, around islands and atolls, and over shallow banks as well as offshore in deep water (Rice 1998). Differences in morphology, feeding habits, and parasite loads suggest there are at least two distinctive forms, coastal and offshore, in the eastern North Pacific and the Gulf of California (Walker 1981; Scott and Chivers 1990; Díaz-Gamboa 2003). This species is distributed in the entire Gulf, where it is one of the most frequently seen dolphins. There are an estimated 33,799 (95 percent C.I. = 20,500–58,358, based on line transect methods) individuals in the Gulf (Gerrodette and Palacios 1996).

• Genus *Stenella* Gray, 1866

STENELLA ATTENUATA (GRAY, 1846). PANTROPICAL SPOTTED DOLPHIN; DELFÍN MANCHADO PANTROPICAL. Pantropical spotted dolphins are found throughout the tropical and subtropical (and some warm-temperate) regions of the world, between about 40° N and 40° S (Perrin 2002). It is the most abundant dolphin occurring in the mouth of the Gulf of California, especially off the coasts of Sinaloa and Nayarit (Urbán R. 1983). Two subspecies of the pantropical spotted dolphin (the offshore species, *S. attenuata attenuata*, and the coastal *S. attenuata graffmi*) have been identified in the Eastern Tropical Pacific. Although the coastal species is most common, records of both species have been documented in the southern Gulf (Urbán R. 1983; Leatherwood et al. 1988). The abundance estimation of this species in the Gulf of California is 23,734 (95 percent C.I. = 14,419–40,913, based on line transect methods; Gerrodette and Palacios 1996).

STENELLA LONGIROSTRIS (GRAY, 1828). SPINNER DOLPHIN; DELFÍN GIRADOR, DELFÍN TORNILLO. Spinner dolphins are found around the world in tropical waters, and they occur in the southern Gulf. All recorded observations of this species in the Mexican Pacific, including the Gulf of California, belong to one of the three subspecies described in the Eastern Tropical Pacific, *Stenella longirostris orientalis* (Urbán R. 1983; Perrin 1998). There was a mass stranding of a hundred or so spinner dolphins in the Bahía de La Paz in August 1993; about half of them were released alive back into the sea (Urbán R. et al. 1997). Abundance estimates of this species in the Gulf are 22,724 individuals (95 percent C.I. = 12,411–43,572, based on line transect methods; Gerrodette and Palacios 1996).

STENELLA COERULEOALBA (MEYEN, 1833). STRIPED DOLPHIN; DELFÍN LISTADO. Striped dolphins are found in warm-temperate and tropical waters worldwide (Perrin et al. 1994). This species is not common in the

Gulf of California and is distributed mainly in its southern half (Mangels and Gerrodette 1994; Urbán R. et al. 1997). There was a mass stranding of seventeen individuals (fifteen adult females, one calf, and one adult male) of this species in the Bahía de La Paz on January 9, 2003 (unpublished data). The striped dolphin's abundance estimate in the Gulf is 8,642 individuals (95 percent C.I. = 3,314–23,603, based on line transect methods; Gerrodette and Palacios 1996).

• Genus *Delphinus* Linnaeus, 1758

DELPHINUS DELPHIS LINNAEUS, 1758. SHORT-BEAKED COMMON DOLPHIN; DELFÍN COMÚN DE ROSTRO CORTO. Although common dolphins exist worldwide in warm-temperate and tropical waters, published records do not always specify which of the two species (*D. delphis* and the long-beaked common dolphin *D. capensis*) were seen. A means of distinguishing these species was recently published by Heyning and Perrin (1994). In addition to its noticeably longer beak, the long-beaked dolphin has a longer and narrower head, and its melon is less rounded and thus flatter in profile; the overall impression is of a more slender, longer-bodied animal. The color differences are subtle, but the "thoracic patch" on the long-beaked dolphin tends to be darker and to contrast less with the dark back color (Reeves et al. 2002). In the Eastern Tropical Pacific, the habitat preference of the short-beaked common dolphin is offshore in equatorial and subtropical waters (Best 2007). There are records of this species throughout the Gulf, but their main distribution is south of the Midriff Islands. The abundance of short-beaked common dolphin in the Gulf of California is estimated at 28,681 individuals (95 percent C.I. = 14,287–72,316, based on line transect methods; Gerrodette and Palacios 1996).

DELPHINUS CAPENSIS GRAY, 1828. LONGBEAKED COMMON DOLPHIN; DELFÍN COMÚN DE ROSTRO LARGO. Common dolphins are found worldwide in warm-temperate and tropical waters, although the long-beaked common dolphins seem to be distributed as a series of disjunct populations in nearshore waters off California, Mexico, Peru, Venezuela south to Argentina, West Africa, South Africa, Korea, southern Japan, and Taiwan (Heyning and Perrin 1994). School sizes can range from fewer than ten to as many as several thousand individuals (Reeves et al. 2002). This species is distributed throughout the Gulf and is the most abundant cetacean; its estimated population size is 61,976 individuals (95 percent C.I. = 31,295–154,153, based on line transect methods; Gerrodette and Palacios 1996).

• Genus *Lagenorhynchus* Gray, 1846

LAGENORYNCHUS OBLIQUIDENS GILL, 1865. PACIFIC WHITE-SIDED DOLPHIN, DELFÍN DE COSTADOS BLANCOS DEL PACÍFICO. In the Eastern North Pacific, white-sided dolphins are common from the northern Gulf of Alaska to about the tip of Baja California. In general, white-sided dolphin populations shift southward and inshore during winter, when the water is colder, and then shift northward from spring to autumn, when the water is warmer (Leatherwood et al. 1988). In the Gulf of California their presence had been recorded on the southwest coast between Loreto and Cabo San Lucas from November to May, although there are few sightings during the summer (Vidal et al. 1993; Urbán R. et al. 1997). There are no abundance estimates for this species in the Gulf.

• Genus *Grampus* Gray, 1828

GRAMPUS GRISEUS (G. CUVIER, 1812). RISSO'S DOLPHIN; DELFÍN DE RISSO. Risso's dolphin has an extensive distribution in tropical and warm-temperate waters of all oceans (Reeves et al. 2002). Risso's dolphin is a pelagic species that is most commonly seen seaward of the continental shelf and above rugged seafloor topography (Leatherwood et al. 1980). There are relatively few records of sightings or strandings in the Gulf, and all but one of them occurred south of the Midriff Islands (Vidal et al. 1993; Urbán R. et al. 1997). Risso's dolphin abundance in the Gulf of California is estimated at 16,918 individuals (95 percent C.I. = 9,027–33,205, based on line transect methods; Gerrodette and Palacios 1996).

• Genus *Peponocephala* Nishiwaki and Norris, 1966

PEPONOCEPHALA ELECTRA (GRAY, 1846). MELON-HEADED WHALE; DELFÍN CABEZA DE MELÓN, CALDERÓN PIGMEO. The melon-headed whale is oceanic and pantropical, distributed mainly between 20° N and 20° S (Reeves et al. 2002). The only record known for the Gulf of California is of a stranded individual collected at Isla Espíritu Santo (Vidal et al. 1993).

• Genus *Pseudorca* Reinhardt, 1862

PSEUDORCA CRASSIDENS (OWEN, 1846). FALSE KILLER WHALE; ORCA FALSA. False killer whales are found worldwide from tropical to temperate waters; their latitudinal limits are generally between 50° N and 50° S (Odell and McClune 1999). Although they are uncommon in the Gulf of California, there are about ninety records of sightings and strandings distributed throughout all but the uppermost region of the Gulf (Vidal et al. 1993; Urbán R. et al. 1997; unpublished records). There are no abundance estimates in the Gulf for this species.

• Genus *Orcinus* Fitzinger, 1860

ORCINUS ORCA (LINNAEUS, 1758). KILLER WHALE; ORCA. Killer whales are said to be the world's most widespread mammal, found from the tropics to high latitudes in both hemispheres and including both oceanic and coastal waters (Best 2007). Between 1972 and 2006, more than 230 sightings were reported throughout the Gulf of California. Group size ranged from 1 to 45 animals, with an average of 5.6 individuals per group (Guerrero-Ruíz et al. 2006a). Four individual strandings of killer whales have been documented in the Gulf: (1) a "killer whale dead on the beach" in the 1950s (Dahlheim et al. 1982); (2) a "dead animal" in 1992 (Delgado-Estrella et al. 1994); (3) a "skull" in 2002 (Guerrero-Ruíz et al. 2006a); and (4) most recently (April 10, 2007) a 2.47-m live female stranded at San Blas, Nayarit. In addition to these, two live mass strandings have also recently taken place in the Gulf of California. In the first, on July 31, 2000, eight killer whales stranded themselves in Bahía de La Paz. All the individuals died despite attempts by local fishermen to return them to the sea. The group consisted of an undetermined number of females, immature males, and two calves (Guerrero-Ruíz et al. 2006a). The second live mass stranding, also in Bahía de La Paz, occurred on March 3, 2006, and involved three killer whales (no adult males and calves were present). In this second incident, all animals were successfully released back into the sea thanks to the efforts of tourists on the beach (unpublished data). There are no abundance estimates of this species, but the number of photo-identified individuals suggests that their abundance could be a few hundred.

• Genus *Globicephala* Lesson, 1828

GLOBICEPHALA MACRORYNCHUS GRAY, 1846. SHORT-FINNED PILOT WHALE; CALDERÓN DE ALETAS CORTAS. The short-finned pilot whale is widespread and abundant throughout the world's tropical and warm-temperate marine waters (Reeves et al. 2002). In the Gulf of California, this species is distributed in all regions except the upper Gulf. Three mass strandings have been reported: 14 individuals in the early 1980s at Bahía Guadalupe, north of Bahía de los Ángeles (Vidal et al. 1993); 30 at Bahía de San Rafael, Baja California (Vidal et al. 1993); and 31 at Bahía de La Paz, Baja California Sur, which were successfully driven offshore (Urbán R. 1993).

In the 1930s and 1940s, local fishermen hunted short-finned pilot whales in the waters off Loreto and La Paz for their oil (to be used in a well-known tannery in La Paz). There is no information about the numbers of animals taken, although today older fishermen still remember looking for large males (unpublished data). The abundance of short-finned pilot whales in the Gulf of California is estimated at 3,923 individuals (95 percent C.I. = 1,591–9,829, based on line transect methods; Gerrodette and Palacios 1996).

◦ Family Phocoenidae (Gray, 1825)—Porpoises; marsopas

• **Genus *Phocoena* G. Cuvier, 1816**

PHOCOENA SINUS NORRIS AND MCFARLAND, 1958. VAQUITA; GULF OF CALIFORNIA HARBOR PORPOISE. The vaquita is endemic to the upper Gulf of California (north of a line connecting San Felipe, Baja California, and Puerto Peñasco, Sonora). It is the most endangered cetacean species in the world (Rojas-Bracho and Jaramillo-Legorreta 2002; Rojas-Bracho et al. 2006; Jaramillo-Legorreta et al. 2007). Abundance estimates based on line transect surveys indicate that 567 (95 percent C.I. = 177–1,073) vaquitas were present in the upper Gulf in 1997 (Jaramillo et al. 1999). The population size in 2007 was estimated at just 150 individuals (Jaramillo-Legorreta et al. 2007).

Mortality from commercial fisheries, especially entanglement net fishing (e.g., gill nets), is the greatest risk today for vaquitas (Rojas-Bracho and Taylor 1999). It will take time to plan and raise money to buy out entanglement net fisheries, so immediate action must be taken to establish a total moratorium on all entanglement nets. The situation is urgent, given the little time remaining before the population becomes so small that it's vulnerable to stochastic events or processes that can lead to complete extinction (ecological, genetic, and demographic) (Jaramillo-Legorreta et al. 2007).

Suborder Mysticeti Flower, 1864—Baleen or whalebone whales; cetáceos o ballenas con barbas

◦ Family Balaenidae (Gray, 1821)

• **Genus *Eubalaena* (Gray, 1864)**

EUBALAENA JAPONICA (LACEPEDE, 1818). NORTH PACIFIC RIGHT WHALE; BALLENA FRANCA DEL PACÍFICO NORTE. The North Pacific right whale is distributed from the Sea of Okhotsk east to the Bering Sea and the Gulf of Alaska. There have been occasional sightings off the west coast of the United States and Baja California (Reeves et al. 2002). There is only one record in the Gulf of California, at 23°02' N, 109°30' W (Gendron et al. 1999). North Pacific right whales are among the world's most critically endangered mammals. The Eastern Pacific population was reduced by illegal Soviet whaling during the 1960s and currently may comprise fewer than 100 animals (Reeves et al. 2002).

- Family Eschrichtiidae Ellerman and Morrison-Scott, 1951
    - **Genus *Eschrichtius* Gray, 1864**

ESCHRICHTIUS ROBUSTUS (LILLJEBORG, 1861). GRAY WHALE; BALLENA GRIS. Gray whales occur most frequently in shallow coastal waters. The Eastern Pacific population migrates from summer feeding grounds in the Bering, Chukchi, and Western Beaufort Sea to its winter breeding and calving areas off the west coast of the Baja California peninsula (Reeves et al. 2002). The main wintering areas on the peninsula are Laguna Ojo de Liebre, Laguna San Ignacio, and the Bahía Magdalena lagoon complex (Gilmore 1960; Rice et al. 1981; Norris et al. 1983; Jones and Swartz 1984; Urbán R. et al. 2003b). Though less widely known, smaller numbers of gray whales are also found throughout the Gulf of California (Vidal et al. 1993; Findley and Vidal 2002). Until about 1984, gray whales still visited the two winter calving sites most distant from the summer range: the areas of Yavaros-Tijahui, Sonora, and Bahía Santa Maria (Reforma), Sinaloa. A few mothers with calves and occasional solitary subadults could be seen at these two sites during the winter. However, with the accelerated economic development of these areas and especially with increased fishing and shipping, it appears that gray whales no longer visit these sites or do so only sporadically (Vidal et al. 1993; Findley and Vidal 2002). Gray whales have been seen in small groups along both coasts of the Gulf; the sightings are of mothers with calves or, most commonly, of solitary animals. Their presence in the Gulf is greater when the sea-surface temperature is colder than normal—for example, during La Niña events (Le Boeuf et al. 2000; Urbán R. et al. 2003a).

- Family Balaenopetridae Gray, 1864
    - **Genus *Megaptera* Gray, 1846**

MEGAPTERA NOVAEANGLIAE (BOROWSKI, 1781). HUMPBACK WHALE; BALLENA JOROBADA. Humpback whales are distributed throughout the world's oceans. They are a highly migratory species, spending spring through fall in feeding areas in mid- or high-latitude waters and then wintering in calving grounds in the tropics (Clapham 2002). In the North Pacific there are at least five separate breeding grounds—in Japan, Hawaii, coastal Mexico, offshore Mexico (Archipiélago de Revillagigedo), and Central America. Humpback whales from these wintering areas migrate primarily to the Kamchatka Peninsula, Bering Sea, Gulf of Alaska, and the west coasts of Canada and the United States. Although some migratory connections have been established (e.g., Central America–California, Hawaii–Gulf of Alaska), the structure of the populations in the North Pacific is still unclear (Baker et al. 1998; Urbán R. et al. 2000; Calambokidis et al. 2001). In the Gulf of California, humpback whales aggregate around the tip of the Baja California peninsula, particularly between Cabo Pulmo and Cabo San Lucas (around the Islas Tres Marías and Isla Isabel) and off the mainland coast south of Mazatlán, with the highest density in the Bahía de Banderas region, Nayarit (Urbán R. et al. 1999). Whereas the humpback whales from the mainland coast (including Islas Tres Marías and Isabel) migrate mainly to California, Oregon, and Washington, whales from the coast of Baja California are considered a mixture of groups in transit from the mainland and the Revillagigedo archipelago and others that use that area as a wintering ground (González-Peral 2006). The abundance of the whales in these congregations has been estimated at 1,813 individuals (95% C.I. = 918–2,505, based on mark-recapture methods; Urbán R. et al. 1999).

• Genus *Balaenoptera* Lacepede, 1804

BALAENOPTERA ACUTOROSTRATA LACEPEDE, 1804. MINKE WHALE; BALLENA MINKE. Minke whales are distributed throughout all the world's oceans. Although it is not a frequently observed species, there are records throughout the entire Gulf of California—from San Felipe in the north to Bahía de La Paz in the south—particularly during the winter and spring (Urbán R. et al. 1997; Urbán R. 2000). There are no abundance estimates for this species in the Gulf.

BALAENOPTERA EDENI ANDERSON, 1879. BRYDE'S WHALE; BALLENA DE BRYDE, RORCUAL TROPICAL. Bryde's whales are found worldwide in tropical to temperate waters and rarely above latitudes of about 35° (Reeves et al. 2002). Annually, Bryde's whales are frequently observed throughout the Gulf of California, particularly in coastal waters (Urbán R. and Flores 1996). Molecular analyses suggest the possibility that two stocks may be present in the southern Gulf, one resident and another from the Eastern Tropical Pacific (Dizon et al. 1995). Bryde's whale abundance in the Gulf of California is estimated at 564 individuals (95 percent C.I. = 453–2,085, based on line transect methods; Gerrodette and Palacios 1996).

BALAENOPTERA BOREALIS LESSON, 1828. SEI WHALE; BALLENA SEI. The sei whale is distributed worldwide from subtropical and tropical waters to high latitudes, and it inhabits both shelf and oceanic waters (Reeves et al. 2002). Fewer than twenty sightings have been recorded in the Gulf of California, all of them on the southwest coast between Loreto and Isla Cerralvo (Gendron and Chavez 1996; Urbán R. 2000). There are no abundance estimates for the sei whale in the Gulf of California.

BALAENOPTERA PHYSALUS (LINNAEUS, 1758). FIN WHALE; BALLENA DE ALETA, RORCUAL COMÚN. Fin whales are found in all the world's oceans from the equator to polar regions; the largest concentrations are in temperate and cold waters. There is a resident population of fin whales in the Gulf of California (Berubé et al. 2002; Urbán R. et al. 2005), where they exhibit a generalized pattern of distribution: in the west from Canal de Ballenas to Bahía de La Paz in winter and spring and in the Midriff Islands and northern Gulf in summer and fall. The fin whale population size in the Gulf of California has been estimated at 820 individuals (95 percent C.I. = 594–3,229, based on line transect methods; Gerrodette and Palacios 1996) and 656 individuals (95 percent C.I. = 374–938, based on mark-recapture methods; Díaz-Guzmán 2006).

BALAENOPTERA MUSCULUS (LINNAEUS, 1758). BLUE WHALE; BALLENA AZUL. The blue whale is a wide-ranging species that is distributed throughout the world's oceans in coastal, shelf, and oceanic waters (Reeves et al. 2002). Blue whales concentrate in the Gulf of California during winter and spring, especially on the southwest coast from Loreto to Los Cabos, but there are records of sightings during all seasons of the year as far north as the Midriff Islands. Blue whales from the Gulf of California spend the summer feeding in the waters off California (Gendron 2002). Gendron and Gerrodette (2003) estimate a population size of 362 individuals (coefficient of variation = 47.5 percent, based on aerial line transect methods).

Order Carnivora Bowdich, 1821

◦ Family Otariidae (Gray, 1825)

• **Genus *Arctocephalus* E. Geoffroy Saint-Hilaire and F. Cuvier, 1826**

ARCTOCEPHALUS TOWNSENDI MERRIAM, 1897. GUADALUPE FUR-SEAL; LOBO FINO DE GUADALUPE. The Guadalupe fur seal's current breeding range is limited almost exclusively to Isla Guadalupe off the Pacific coast of Baja California. Researchers discovered another small colony at Isla San Benito, off central western Baja California in 1997 (Maravilla and Lowry 1999), and the birth of one pup at San Miguel Island, in the Channel Islands off southern California, has also been reported (Melin and DeLong 1999). In the Gulf of California there are only three records of individual animals: the first at Los Islotes in Bahía de La Paz; the second in Guaymas, Sonora (Aurioles et al. 1993); and the third at Isla Lobos in the northern Gulf of California (Maravilla and Gallo 2000). The total population was estimated at about 6,000 individuals in 1987 and at 10,000 in the late 1990s (Reeves et al. 2002).

• **Genus *Zalophus* Gill, 1866**

ZALOPHUS CALIFORNIANUS (LESSON, 1828). CALIFORNIA SEA LION; LOBO MARINO DE CALIFORNIA. California sea lions breed in the Channel Islands, islands along the Pacific coast of the Baja California peninsula (including the offshore Isla Guadalupe), and in the Gulf of California from Rocas Consag (310°03' N, 1140°28' W) to Los Islotes (240°33' N, 1100°26' W) (Le Boeuf et al. 1983; Aurioles-Gamboa and Zavala 1999). The abundance estimate of California sea lions in the Gulf of California ranges from 25,000 to 30,000 animals (Maravilla and Gallo 2000).

◦ Family Phocidae (Gray, 1821)

• **Genus *Phoca* Linnaeus, 1758**

PHOCA VITULINA LINNAEUS, 1758. HARBOR SEAL, COMMON SEAL; FOCA COMÚN. Harbor seals range widely in the inner coastal areas of the North Pacific and North Atlantic. Five subspecies are recognized; *P. vitulina richardii* range from the eastern Aleutians southward along the west coast of the Baja California peninsula. In the Gulf of California there are only two records of individual animals, one at Los Islotes in the Bahía de La Paz and the other at Los Frailes (230°23' N, 1090°25' W) (Gallo and Aurioles 1984).

• **Genus *Mirounga* Linnaeus, 1758**

MIROUNGA ANGUSTIROSTRIS GRAY, 1827. NORTHERN ELEPHANT SEAL; ELEFANTE MARINO DEL NORTE. The largest breeding colonies of northern elephant seals occur at the Channel Islands (California), and their distribution extends south to Coronado, Guadalupe, San Benito, Cedros, and the Natividad Islands along the Pacific coast of Baja California (Reeves et al. 2002). At least eight sightings of individual elephant seals have been recorded for the Gulf of California, all but one in the Midriff Islands (the other, at Los Islotes in Bahía de La Paz, consisted of sightings of the same individual during several years in the 1990s) (Aurioles et al. 1993; unpublished data).

Order Chiroptera Blumenbach, 1779

○ Family Vespertilionidae Rafinesque, 1815

• **Genus *Myotis* Kaup, 1829**

MYOTIS VIVESI MENGAUX, 1901. COASTAL FISHING BAT OR MEXICAN FISHING BAT; MURCIÉLAGO PESCADOR. The Mexican fishing bat is endemic to the islands and coastal areas on both sides of the Gulf of California and also to the west-central coast of the Baja California peninsula (Patten and Findley 1970; Arita and Ortega 1998; Bogan 1999). These bats forage over water, plucking fish from the surface. They roost in caves or rock crevices and have also been found living under large flat rocks or boulders along the beach (Altringham 1996). Arita and Ortega (1998) have evaluated the Mexican fishing bat as a species of special concern. The bat's chances for survival are threatened because it is endemic to the Gulf of California and Baja California peninsula, it has a specialized diet and habitat, and its habitat is rapidly being transformed by human activities.

Conservation Issues

The 2000 edition of the International Union for the Conservation of Nature (IUCN)'s Red List of Threatened Species listed one species of marine mammal in the Gulf of California as *critically endangered* (vaquita), four species as *endangered* (blue whale, fin whale, sei whale, and northern right whale), and three as *vulnerable* (humpback whale, sperm whale, and Guadalupe fur seal) (Hilton-Taylor 2000; see table 9.1).

The Convention on International Trade in Endangered Species of Wild Fauna and Flora (CITES) includes all of the baleen whales as well as the sperm whale, bottlenose whale, Baird's beaked whale, vaquita, and Guadalupe fur seal in its Appendix I (species most endangered among CITES-listed animals and plants, with commercial international trade in specimens of these species prohibited). All the other toothed whales in the Gulf of California are included in its Appendix II (species for which commercial trade is strictly regulated with special permits delivered by the exporting country; table 9.1).

The Mexican Government's Official Standard NOM-59-ECOL-2001 identifies several "at risk" ("en riesgo") categories of species and subspecies for Mexico's terrestrial and aquatic flora and fauna. It considers all of the marine mammal species entering Mexico's waters to be "at risk" (*Diario Oficial* 2002a). The North Pacific right whale, the vaquita, and the Guadalupe fur seal are listed as being "in danger of extinction," the northern elephant seal as "threatened," and all others as "subject to special protection" (table 9.1).

Conservation Threats

The main human-caused mortalities of cetaceans in the Gulf of California are related to fisheries activities, especially coastal ("inshore") fisheries. Zavala-González et al. (1994) mention more than 125 specimens recovered from artisanal fisheries and killed by gillnet entanglement or even deliberately by harpoons or firearms. The cetaceans most frequently killed by fishing activities are the short-beaked common dolphin, long-beaked common dolphin, common bottlenose dolphin, pantropical spotted dolphin, and vaquita. Dolphins that are killed are frequently used as bait for shark fishing. Large cetaceans also have been affected by fisheries. At least seven cases of gray whale entanglements in gill nets have been documented in the Gulf, with five of those whales being released (Vidal et al. 1994; Urbán R. et al. 2003b; Guerrero-Ruíz et al. 2006b). From January 2003 to January 2008, thirty entangled humpback whales were reported in the Gulf of California, at least seven of which were dead (Urbán R. et al. 2007; unpublished data).

Another apparently human-caused mortality event occurred in the

TABLE 9.1. Marine mammals of the Gulf of California and their conservation status.

		Conservation Status				
Scientific Name[a]	Common Name[a]	IUCN[b]	CITES[c]	NOM-59[d]	ESA[e]	MMPA[e]
Family Balaenidae						
Eubalaena japonica	North Pacific right whale	EN	I	P	E	D
Family Eschrichtiidae						
Eschrichtius robustus	Gray whale	LC	I	Pr	—	—
Family Balaenopteridae						
Balaenoptera musculus	Blue whale	LR	I	Pr	E	D
Balaenoptera physalus	Fin whale	EN	I	Pr	E	D
Balaenoptera borealis	Sei whale	EN	I	Pr	E	D
Balaenoptera edeni	Bryde's whale	DD	I	Pr	—	—
Balaenoptera acutorostrata	Minke whale	LC	I	Pr	—	—
Megaptera novaeangliae	Humpback whale	LC	I	Pr	E	D
Family Physeteridae						
Physeter macrocephalus	Sperm whale	VU	I	Pr	E	D
Family Kogiidae						
Kogia breviceps	Pygmy sperm whale	DD	—	Pr	—	—
Kogia sima	Dwarf sperm whale	DD	—	Pr	—	—
Family Ziphiidae						
Ziphius cavirostris	Cuvier's beaked whale	LC	—	Pr	—	—
Berardius bairdii	Baird's beaked whale	DD	I	Pr	—	—
Indopacetus pacíficus	Logman's beaked whale	DD	—	—	—	—
Mesoplodon peruvianus	Pygmy beaked whale	DD	—	Pr	—	—
Mesoplodon densirostris	Blainville's beaked whale	DD	—	Pr	—	—
Mesoplodon ginkgodens	Ginkgo-toothed whale	DD	—	Pr	—	—
Mesoplodon sp. A	Beaked whale sp. A	—	—	—	—	—
Family Delphinidae						
Steno bredanensis	Rough-toothed dolphin	LC	—	Pr	—	—

Tursiops truncatus	Common bottlenose dolphin	LC	—	Pr	—	—
Stenella attenuata	Pantropical spotted dolphin	LC	—	Pr	—	D
Stenella longirostris	Spinner dolphin	DD	—	Pr	—	D
Stenella coeruleoalba	Striped dolphin	LC	—	Pr	—	—
Delphinus delphis	Short-beaked common dolphin	LC	—	Pr	—	—
Delphinus capensis	Long-beaked common dolphin	DD	—	Pr	—	—
Lagenorynchus obliquidens	Pacific white-sided dolphin	LC	—	Pr	—	—
Grampus griseus	Risso's dolphin	LC	—	Pr	—	—
Peponocephala electra	Melon-headed whale	LC	—	Pr	—	—
Pseudorca crassidens	False killer whale	DD	—	Pr	—	—
Orcinus orca	Killer whale	DD	—	Pr	—	—
Globicephala macrorhynchus	Short-finned pilot whale	DD	—	Pr	—	—
Family Phocoenidae						
Phocoena sinus	Vaquita	CR	I	P	E	D
Family Otariidae						
Arctocephalus townsendi	Guadalupe fur seal	VU	I	P	T	D
Zalphus californianus	California sea lion	LR	—	Pr	—	—
Family Phocidae						
Phoca vitulina	Harbor seal	LR	—	Pr	—	—
Mirounga angustirostris	Northern elephant seal	LR	—	A	—	—
Family Vespertilionidae						
Myotis vivesi	Mexican fishing bat	VU	—	P	—	—

Notes: International Union for the Conservation of Nature (*IUCN*): CR = critically endangered, EN = endangered, VU = vulnerable, LR = lower risk, LC = least concern, DD = data deficient. Convention on International Trade in Endangered Species of Wild Fauna and Flora (*CITES*): I = Appendix I (most endangered CITES-listed species), II = Appendix II (species not currently threatened with extinction but at risk of becoming so unless trade is strictly regulated). NOM-59-ECOL-2001 (*NOM-59*): P = danger of extinction, A = threatened, Pr = special protection. Endangered Species Act (*ESA*): E = endangered, T = threatened. Marine Mammal Protection Act (*MMPA*): D = depleted.

[a] Scientific and common names from Rice (1998) and International Whaling Commission (2001).

[b] http://cmsdata.iucn.org/downloads/cetacean_table_for_website.pdf (accessed August 2008).

[c] http://www.cites.org/eng/app/appendices.shtml (accessed May 2009).

[d] From NOM-59-ECOL-2001 (*Diario Oficial* 2002a).

[e] Marine Mammal Commission (2007).

winter of 1995 in the upper Gulf of California when 367 dolphins (including long-beaked common dolphins, common bottlenose dolphins, and striped dolphins), 8 baleen whales (including fin, common minke, and Bryde's whales), 51 California sea lions, and 215 sea birds (mostly Pacific loons, eared grebes, brown pelicans, and double-crested cormorants) were found dead—possibly due to sea contamination by NK-19, a fluorescent cyanide compound used by narcotraffickers to mark drop areas for unloading drugs (PROFEPA-SEMARNAP 1995; Brusca et al. 2005). Although discounting that specific agent as the proximal cause of mortality, Vidal and Gallo-Reynoso (1996) agreed that the die-off was likely caused by an unknown toxic substance in the water or in prey ingested by the affected animals.

Formal Protection of Cetaceans in the Gulf of California

There is no single body of legislation enacted for the sole benefit of cetaceans in the Gulf. Instead, there are several different laws relevant to their conservation and management that apply to all of Mexico. The General Law of Ecological Balance and Environmental Protection (Ley General del Equilibrio Ecológico y la Protección al Ambiente), enacted in 1988, is currently the responsibility of the Secretariat of the Environment and Natural Resources (SEMARNAT). Articles 15 through 19 of the law provide SEMARNAT with a broad mandate to formulate policy and planning initiatives and to implement management actions for the protection of the nation's natural resources (Estados Unidos Mexicanos 1993). The Fishing Law (Ley de Pesca) authorizes government agencies dealing with fisheries to "establish measures aimed at the protection of . . . marine mammals" (Secretaría de Pesca 1992, p. 10). Another piece of legislation, a 1991 addition to the Mexican Penal Code (Article 254 Bis), prohibits unauthorized capture of or injury to marine mammals and sea turtles. A prison term of 3–6 years is prescribed as the penalty (*Diario Oficial* 1991).

The General Law of Wildlife (Ley General de Vida Silvestre), under the responsibility of SEMARNAT, was approved on April 27, 2000 (*Diario Oficial,* 2000b). This is the first pertinent Mexican law related to wildlife that confronts the challenges of balancing protection of the country's mega-diversity with the need for socioeconomic development. Article 60

Bis was added on January 10, 2002; it states that no specimen of any marine mammal can be the subject of subsistence or commercial use. Exceptions for scientific research and educational purposes require prior approval of the authorities (*Diario Oficial* 2002b).

The Mexican Government's Official Standard NOM-131-ECOL-1998 provides specific guidelines for whale-watching activities that are compatible with their conservation and habitat (*Diario Oficial,* 2000c).

Additionally, in May 2002, Mexico established the Mexican Whale Refuge (Refugio Ballenero Mexicano) consisting of its entire Exclusive Economic Zone (about 3 million km^2). The decree stipulates that environmental conditions required for biological functions of whales (e.g., breeding, calving, growth, migration, learning, and feeding) must be maintained. Species that are protected include all members of the families Balaenidae, Balaenopteridae, Eschrichtiidae, Physeteridae, Kogiidae, and Ziphiidae in addition to the killer whale, short-finned pilot whale, false killer whale (*Pseudorca crassidens*), pigmy killer whale (*Feresa attenuata*), and melon-headed whale in the family Delphinidae (*Diario Oficial,* 2002c).

CHAPTER 10

A Brief Natural History of Algae in the Gulf of California

RICHARD MCCOURT

Summary

The marine algae of the Gulf of California are a diverse assemblage of more than 300 subtropical, tropical, and temperate species that are subject to a diverse range of physical circumstances, including: a wide range of seasonal temperatures, extreme tidal fluctuations (especially pronounced in the Northern Gulf), localized upwelling and other variations in water/nutrient chemistry, and substrate types (e.g., granite, basalt, sandstone, beachrock/coquina, plus a range of sandy habitats). The biogeographic source areas of the Gulf algal flora include the cooler waters of the Pacific coast of North America and warmer regions to the south. The high summer temperatures are presumably responsible for the absence of kelps, which leaves the Sargassaceae as the predominant group of large brown algae in the Gulf of California. Even after nearly two centuries of study, much remains to be discovered about the taxonomy and biology of this ecologically important group of organisms.

Introduction

Most plants on land are green flowering plants, ferns, and mosses, but most of the plants of the sea are algae—also called seaweeds. Among these are three main groups, which are only distantly related to each other in an evolutionary sense. These algal groups share the features of autotrophy (they are photosynthetic) and habitat (they grow in the sunlit ocean or in such habitats as the rocky intertidal or sandy coastal wetlands, *esteros,* which are inundated by seawater for long periods). Algae exhibit a less-derived form of sexual reproduction than so-called higher plants (embryophytes), and they produce no flowers, fruits, or seeds. Some algae produce structures that are superficially similar to the roots, stems, and leaves of land plants, but on closer inspection the fine structure of algae lacks the complexity of land plants.

Although most algae live in the oceans or freshwater, several evolutionary lineages have invaded the land (Lewis and McCourt 2004) and occupy aerial or subaerial habitats, including exposed rocky and sandy substrates in the intertidal zone. One group (the green algae) contains members most closely related to freshwater aquatic ancestors of land plants (Karol et al. 2001; McCourt et al. 2004). These relatives of land plants are, from an evolutionary perspective, only distantly related to most green seaweeds.

The major groups of algae—termed *phyla,* or divisions—are named for their colors: green algae, brown algae, and red algae. All three groups possess the grass-green pigment chlorophyll *a,* but the combination of additional pigments in brown and red algae often imbues them with the masking colors responsible for their common names. Generally, green algae are green and brown algae range from tan to deep brown. Red algae are most often red; however, some reds are bright green, purple, or even black, depending on the mix of pigments or on whether the plant is in a bleached-out condition.

Algae, the plural term for a single alga, provide food and shelter for marine animals. Snails and limpets, crustaceans, fish, and sea urchins graze upon algae. Crustaceans, bryozoans, and other invertebrates find refuge from predators and from the desiccating heat of low tides among the wet fronds of larger red and brown algae. Some marine invertebrate lineages (e.g., amphipods and isopods) have apparently undergone a lengthy co-evolution with coastal seaweeds that is not unlike the co-evolution between certain insects and land plants (Brusca and Wallerstein 1979a,b; Brusca 1984, 2000).

Throughout this chapter, "marine algae" refers to larger, or macroscopic, forms that can be viewed with the naked eye or a hand lens. This excludes most of the unicellular algae, such as diatoms and cyanobacteria, which are important components of the phytoplankton or surface scums on rocks and sediments but which require a microscope to view in any detail.

The morphology of algae can be quite complex and the terminology equally so. For details, refer to basic textbooks on algae (Bold and Wynne 1985; Graham and Wilcox 2000; Lee 2008). The algal body is called a thallus, which may consist of a single cell or, for the algae covered here, filaments (long thin chains of cells), colonies (clusters of cells in clumps or balls), of parenchymatous forms (with a tissuelike structure). Some larger algae, or seaweeds, bear leaflike structures that, as noted previously, lack the internal anatomy of true leaves; likewise, rootlike and stemlike structures on algae are not the same as these features of land plants. So for algae, the terms blade, holdfast, and stipe are used (respectively) to identify the analogues of leafs, roots, and stems.

Reproduction in algae may be asexual or sexual. Asexual reproduction occurs through vegetative growth of crusts and rhizoids of the holdfast. Fragmentation of the thallus can lead to dispersal and establishment of new thalli. Asexual spores may be produced in special cells or in pockets on the thallus. Many marine green and brown algae produce zoospores, single cells that swim by means of flagella, in specialized cavities or sacs. In contrast, red algae do not produce flagella on any cells.

Sexual reproduction is simpler than in higher plants because it occurs without flowers, fruits, or seeds. However, what algal sex lacks in complexity it makes up for in the diversity of processes involved. In all algae, sex cells (or gametes) unite to form a zygote, but beyond that the processes are so diverse that it is beyond the scope of this chapter to describe them. Gametes may be the same or nearly the same in size and form, or one cell (termed "male") is much smaller and mobile while the other is larger and does not move (termed "female"). Some marine algae have separate spore-producing and gamete-producing vegetative phases in their life cycle. These phases may look the same (as with, e.g., *Ulva* or sea lettuce), but in some algae (e.g., large brown seaweeds such as kelps) they are very different and occupy different habitats. Other brown algae (e.g., *Sargassum,* a common Gulf brown seaweed) have one large free-living growth phase that produces gametes. The basic phycology texts cited before should be consulted for details.

Green algae belong to the Chlorophyta and all possess chlorophylls *a* and *b* as well as other accessory pigments for light absorption (carotene and xanthophyll), and they store carbohydrates produced by photosynthesis as starch. Green algae are usually grass-green in color. They range from unicellular to relatively large blades or spongy tissues made of compacted filaments. A group of tropical green algae accumulate calcium carbonate in their thalli and are crispy in texture. Green algae are most abundant in freshwater, but certain groups are abundant in the ocean.

Brown algae (Phaeophyta) are among the most common seaweeds, and they are abundant in the intertidal and subtidal of the Gulf of California. Most are bladelike forms, although some are nearly microscopic filaments or crusts that look like brown lichens. The brown color comes from the accessory pigment fucoxanthin, which masks the green chlorophylls *a* and *c*. The simplest are nearly microscopic filaments, whereas the largest (kelps) have stipes as thick as small trees and more than 30 m long from the anchoring holdfast on the rocky bottom to the blades floating on the surface. Kelps (e.g., *Macrocystis*) are absent from the Gulf of California, although they occur on the outer northwest coast of the Baja California peninsula at least as far south as Bahía Tortola (Brusca and Wallerstein 1979b). Brown algae are found almost exclusively in marine waters.

Red algae (Rhodophyta) possess only chlorophyll *a*, which is masked by reddish pigments such as phycocyanin and phycoerythrin. This group includes some of the most delicate and lovely forms, ranging from microscopic filaments to larger gelatinous fronds and coralline crusts (or nodules, e.g., rhodoliths covered in chapter 3 in this volume) as hard as pavement. Most red algae are marine, although a few groups are common in freshwater fast-flowing streams.

History of Algal Exploration in the Gulf of California

It was not until the early twentieth century that scientists turned serious attention to the marine algae of the Gulf. Early collectors were usually from outside Mexico, and they sent their specimens back to phycologists (specialists in algae) in California and elsewhere, where the first monographs were written. Among the most important workers were William Albert Setchell and Nathanial Lyon Gardner (Setchell and Gardner 1924). Elements of the history of algal exploration in the Gulf are provided by Abbott and Hollenberg (1976) in their comprehensive treatment of marine algae of California. The proximity of the Gulf led to the exploration, or at least examination, of Gulf algae by many of the same phycologists who studied the western coast of the United States and Mexico. Mexican phycologists have contributed a growing body of taxonomic and ecological information in the last few decades. There have been few monographs on marine algae in the Gulf, and no comprehensive treatment of the flora of the entire Gulf has been published. Most attention has focused on the Northern Gulf of California. Undoubtedly, much remains to be discovered, especially in the less-explored areas of the Southern Gulf.

The study of marine algae in this region accelerated rapidly in the 1940s with the arrival of E. Yale Dawson, who spent most of his unfortunately brief professional career exploring the Gulf. Dawson was a prolific writer and produced a large body of work prior to his untimely death in 1966 in a drowning accident in the Red Sea (Abbott 1966). Among the synoptic works are several published from 1944 to 1966 (Dawson 1944, 1953, 1954b, 1960a–c, 1961a–c, 1966a,b).

Since Dawson, James N. Norris and colleagues have published on the taxonomy of Gulf algae (e.g., Norris and Abbott 1972; Norris and Bucher 1976), and over the past thirty years Luis E. Aguilar-Rosas, Raúl Aguilar-Rosas, and Mexican co-workers have published a series of papers on the Gulf and outer coast of Baja California (e.g., Aguilar-Rosas et al. 1984, 2000a, 2002).

Biogeography

Dawson (1960a) reviewed the biogeography of Gulf algal flora and drew a number of conclusions on the endemicity and patterns of diversity in the region that will be summarized here. He estimated that about a third of the species are endemic to the Gulf of California. Of the remainder, there are significant overlaps in species found in California. In addition, Dawson noted a number of disjunct Indo-Pacific distributions—that is, species that occur on both sides of the Pacific. For example, *Ishige foliacea* occurs in Asia and the Northern Gulf of California but not in between. Presumably the Gulf populations are the descendants of a relatively recent introduction. Other elements of the flora appear to have achieved disjunct distributions without human-aided dispersal.

Dawson (1960a) noted that the northern parts of the Gulf are dominated by species found in more temperate areas whereas the southern portions exhibit pantropical or pantropical-Pacific species. The flora of the Gulf has been described as smaller in size than the flora of the nearby California coastline; this generalization is drawn largely from the absence of larger brown algae known as kelps, which are so abundant on the temperate shores of California and the outer coast of the northern Baja California peninsula. Their absence is attributed to the higher summer temperatures inside the Gulf, which are presumably lethal to the microscopic gamete-producing stages of kelps. Though smaller in stature, the marine algal flora of the Gulf of California is nevertheless quite diverse and comprises more than 300 species. The most distinctive differences in the flora are the presence of tropical brown and red algae in the Gulf, in particular *Sargassum* species (McCourt 1984b; Pacheco-Ruiz et al. 1998; Rivera and Scrosati 2006).

Dawson (1960a) pointed out that the Gulf exhibits several gradients in habitat type and temperature that afford marine algae with a diverse set of growing conditions that support a diverse flora. These gradients include a north–south gradient in climate from temperate to subtropical. The far northern end experiences a marked difference in seasonal temperature, with winter sea temperatures averaging 15°C and summer averaging above 30°C. Intertidal areas experience an even wider range of seawater temperatures because of exposure to chilly winter air temperatures in early morning (~10°C) and hot summer temperatures at mid-day.

Dawson (1960a) noted that upwelling areas near various islands and mainland parts of the Gulf supported algal floras typical of colder regions outside the Gulf. He also pointed out the marked heterogeneity of habitats due to upwelling, oxygen content, and other factors that can lead to a patchy distribution of species. He emphasized that, because of this, limited sampling could lead to a biased picture of the flora.

The cooler waters of the Northern Gulf are isolated from the outer coast by the 900+ km–long Baja California peninsula and by the warm waters at the mouth of the Gulf. Thus, the peninsula sustains a number of disjunct species occurring in the northern reaches of this inland sea and on the southern California coast, but not along the shorelines connecting them.

Brusca et al. (2004) summarized biodiversity distributions of marine animals in the Gulf of California, and much of the discussion is applicable to marine algae as well.

Algal Habitats of the Northern Gulf of California

The marine algal flora of the Northern Gulf is distinguished by a distinct seasonality and a gradient in composition from the cooler (winter) climatic region of the Northern Gulf to the warmer southern section. The range of habitats where marine algae are found is typical for the marine coastal realm: subtidal, intertidal, tide pools, hard substrates of various degrees of porosity and surface heterogeneity, and coastal tidal flats. In addition, the dramatically large tidal range in the Northern Gulf—exposing tide pools and emergent substrates—offers another axis of habitat variation along with zonation patterns that sort algae into visually discernable communities. Add to this the marked seasonality of water temperatures in the Northern Gulf, and the result is a region that offers a wide range of algal habitats supporting a flora rich in numbers and variety that is distributed in time and space along various dimensions of habitat diversity.

Marine algae grow attached to hard substrates, of which there are several types in the Gulf of California. Three types of substrate are most common in the intertidal region of the Northern Gulf. Solid, relatively smooth basaltic or granitic surfaces are impermeable to water and are home to many species emergent at low tide, generally in the more shaded areas or the low intertidal. Punta Pelicano and Punta Peñasco are areas that support this habitat. Coastal beachrock, or coquina, is a soft limestone material composed of agglomerated sand and seashells. Coquina is an important though limited hard-bottom substrate in the Northern Gulf, and it provides a rough and friable surface that supports a large number of turf species (Stewart 1982) in addition to larger algae. This rough rock is made from shells and sand cemented together by calcium carbonate precipitated from seawater, and it weathers to form tidal pools and small depressions that harbor a distinctive flora of permanently submerged marine algae. Finally, esteros and mud flats are home to species able to attach to small points of rock or shell and to withstand the scour of sand moving with the tides.

In each of these habitats are many microhabitats—for example, under ledges at the edge of tide pools, at the water line of partially submerged basalt boulders, and in the protective moist understory beneath larger fronds of algae at low tide. There is some overlap in species occurrence in these habitats, but many species are most common in one or two.

Aguilar-Rosas et al. (2002) noted that the epiphytic habit is an important one in the Gulf of California, especially for delicate, filamentous forms growing on larger algae. The authors list several red and green algae that commonly grow on larger algae in the Northern Gulf, particularly species of brown algae such as *Sargassum johnstonii, Padina durvellei,*

Spyridia filamentosa, and *Dictyota flabellata.* Red algae (*Gelidium crinale, Gracilaria subsecundata*) and occasionally green algae (*Cladophora microcladioides, Chaetomorpha linum*) are also host species for many epiphytes. The habit of epiphytism is particularly common in sandy areas such as tidal flats and bays, where solid substrates are otherwise scarce. Likewise, the algal turf that carpets coquina contains anchor species, as described for southern California algal turfs (Stewart 1982), that occupy a significant amount of space and provide a substrate for epiphytes that persist throughout the year, despite annual fluctuations in abundance.

Rhodoliths (Foster 2001) are unattached spherules up to 10 cm in diameter that occur primarily on the southeast coast of the Baja California peninsula and on the mainland Sonoran coast. Riosmena-Rodriguez and colleagues (Reyes-Bonilla et al. 1997; Riosmena-Rodríguez et al. 1999; chapter 3 in this volume) have studied the distinctive red algae that grow in the habit of rhodoliths.

Seasonality of Northern Gulf Algae

One of the more pronounced features of marine algae in the Gulf is the seasonal change in abundance and species composition in intertidal communities. This pattern is exhibited even by the more perennial types of algae such as *Sargassum,* for which species may exhibit differences in abundance and phenology (McCourt 1984a,b).

Tidal pattern can ameliorate the exposure of intertidal algae to the highest air temperatures that occur during high-tide periods, but water temperatures nevertheless reach 30°C (air temperatures can surpass 35°C). Therefore, the two extreme seasons of winter and summer are marked by different dominant species (Dawson 1960a; Aguilar-Rosas et al. 2002). Even within genera, species can show a temporal succession of species within a season (e.g., *Colpomenia;* Wynne and Norris 1976) or differences in winter and summer species abundances (e.g., *Sargassum,* McCourt 1984b, 1985).

Zonation

Shoreline zonation of algae (globally) is common in areas subject to tidal changes and, although often less noticeable, in areas of less-steep gradient or with small tide ranges. In the Gulf, the tidal range can be ten times greater in the north than in the south, with a 7–9-m vertical difference between extreme low and high tides near the delta of the Colorado River. The upper intertidal rocks and boulders (e.g., at Pelican Point, near Puerto Peñasco, Sonora) are often colored black by a thin, slippery layer of cyanobacteria. This black zone is bordered below by macroalgae in bands dominated by color (e.g., green *Ulvas* or brown *Colpomenias).* The colored bands are visually dramatic, yet the distribution of algae that makes the pattern obvious is set by a number of different factors (Ricketts et al. 1985).

From a classical perspective, the upper limits of habitat zones are set by the tolerance of individual species to desiccation or temperature while the lower limits are determined by herbivores or competition. Disturbance and more complex interactions between algae, barnacles, and grazing gastropods—as well as their spatial and seasonal environment—may also have a significant effect on intertidal algal distributions. Zonation of intertidal animals in the Gulf was summarized in Brusca (1980).

Zonation mirroring that found on exposed shores has been documented on a smaller scale within tide pools. McCourt (1984a,b) reported on three *Sargassum* species in the Northern Gulf that show differences in distribution along an intertidal gradient of tens of meters. The same sequence of vertical zonation also occurs on a smaller vertical gradient of just a few centimeters at the edges of tide pools.

ACKNOWLEDGMENTS

I thank Richard C. Brusca for reviewing the manuscript, and I thank Ruth Hoshaw, wife of the late Dr. Robert W. Hoshaw, for access to Hoshaw's literature and information on E. Y. Dawson and Gulf of California phycology.

CHAPTER 11

Ecological Conservation in the Gulf of California

MARÍA DE LOS ÁNGELES CARVAJAL,
ALEJANDRO ROBLES, AND EXEQUIEL EZCURRA

Summary

With an immense biological richness and high marine productivity, the Gulf of California (*Mar de Cortés,* Sea of Cortez) is both a large marine ecosystem of high global conservation priority and a region that faces growing threats—mostly as a result of overfishing and significant degradation of coastal habitats—with 39 marine species listed in the International Union for the Conservation of Nature (IUCN) Red List.

The Gulf coastline is a sparsely populated and comparatively affluent region of Mexico. In recent decades, new economic opportunities and depletion of natural resources have led the economy away from its traditional reliance on agriculture and fisheries, shifting it toward tourism and manufacturing. Rapid economic growth has brought accelerated immigration with growing pressures on the environment. The implications of these socioeconomic changes for biodiversity conservation are significant.

In addressing the growing challenges in the region, the conservation movement has relied on three fundamental approaches: (1) research data from the scientific community, (2) ecosystem conservation through the establishment of protected areas, and (3) building capacity for conservation at local and regional levels. Conservation efforts began in the 1950s and early 1960s, when the Upper Gulf and Isla Rasa were officially declared protected sanctuaries for the reproduction of fisheries and seabirds, respectively.

Rapid progress was achieved in the 1990s when conservation organizations began working closely with academic and governmental institutions to better manage protected areas in the region. As a result of these efforts there are now ten terrestrial and thirteen terrestrial and marine protected areas in the Gulf that cover, in total, more than 9 million hectares. The National Commission for Protected Natural Areas (CONANP; see

TABLE 11.1. Acronyms used in this chapter.

ALCOSTA	*Alianza para la Sustentabilidad del Noroeste Costero Mexicano,* an alliance of several environmental organizations promoting coastal conservation
CES	*Centro Ecológico de Sonora,* a research center in Sonora
CIDESON	*Centro de Investigación y Desarrollo de los Recursos Naturales de Sonora,* a research center in Sonora
CIRVA	International Committee for the Recovery of the Vaquita
CONANP	*Comisión Nacional de Áreas Naturales Protegidas,* Mexico's National Commission for Protected Natural Areas
FONATUR	*Fondo Nacional de Fomento al Turismo,* Mexico's National Fund for the Promotion of Tourism
GEF	Global Environmental Facility, an international funding organization created as a result of the Rio Summit (UNCED)
INE	*Instituto Nacional de Ecología,* Mexico's National Institute of Ecology
IUCN	International Union for the Conservation of Nature
MAB	UNESCO's Man and the Biosphere Program
NAFTA	North American Free Trade Agreement
NGO	Nongovernmental organization
NOS	*Noroeste Sustentable* (Sustainable Northwest), a Mexican conservation initiative
PROFEPA	*Procuraduría Federal de Protección al Ambiente,* Mexico's Environmental Attorney General
SAGARPA	*Secretaría de Agricultura, Ganadería, Desarrollo Rural, Pesca y Alimentación,* Mexico's Ministry of Agriculture
SDNHM	San Diego Natural History Museum
SEDUE	*Secretaría de Desarrollo Urbano y Ecología,* Mexico's first Environmental Ministry (1983–1992)
SEMARNAP	*Secretaría de Medio Ambiente, Recursos Naturales y Pesca;* Mexico's Ministry of the Environment (1994–2000)
UNAM	*Universidad Nacional Autónoma de México,* Mexico's National University
UNCED	United Nations Conference on Environment and Development (also known as the Rio Summit)
UNESCO	United Nations Educational, Scientific, and Cultural Organization

table 11.1 for a summary list of acronyms) was created during that time. Also, professional staff was hired and assigned to protected areas, advisory groups were established, management plans were developed, legislation was updated, and the budgets for managing protected areas rose significantly. Yet in spite of increased legal protection, problems have continued to grow in many areas. This highlights the importance of finding alternative modes

of conservation, especially by learning from the many grassroots and regional success stories of conservation objectives achieved by means that are different from, but complementary to, those of federal reserves.

For conservation initiatives to succeed in the Gulf of California, they must be able to win the hearts and minds of local people, lead toward generation of alternative livelihoods, and increase local community capacity for stewardship of resources.

Introduction

The Gulf of California is a large, semi-enclosed sea covering approximately 260,000 square kilometers. It contains some outstanding natural features such as deep ocean basins with hydrothermal vents in its central and lower portions, expansive tides in its upper reaches, over a hundred large islands, myriad islets and offshore rocks, and strong upwellings of cold, nutrient-rich waters that make it extraordinarily productive. In the surrounding coastal deserts, rainfall and terrestrial productivity are tightly coupled to oceanographic variations (Polis et al. 1997; Velarde and Ezcurra 2002). The great diversity of topographic and bathymetric features has produced a variety of habitats for marine life and island species (Case and Cody 1983; Robles and Carvajal 2001; Case et al. 2002). The Gulf is a sort of "marine peninsula," isolated from the rest of the Pacific by the 1,500 km of land of Baja California. Habitat heterogeneity, geographic isolation, endemism, and rarity have been driving forces of evolution in the fragmented coastal lagoons and wetlands as well as in the oceanic islands, reefs, underwater seamounts, and other areas of the Gulf (color plate 7A).

Fragmentation also yielded unique human cultures. Separated from the rest of Mesoamerica, the Cochimí Indians of Baja California developed one of the most incredible prehistoric assemblages of cave paintings in the world. Later, during the Spanish colonial period, Jesuit fathers established a system of missions in Baja California that evolved in complete independence from the harsh rules of the mainland conquistadores. Other coastal indigenous nations—such as the Seri, the Yaqui, and the Cucapá—also developed unique lifestyles as fishers and sailors, with cultures finely adapted to their coastal resources (Felger and Moser 1985; Bowen 2000; Bahre and Bourillón 2002; Nabhan 2002).

Biologically, the Gulf of California is one of the most productive and diverse seas in the world (Alvarez-Borrego 1983, 2002; Brusca 2004a; Brusca and Bryner 2004; Brusca et al. 2005; Lluch-Cota et al. 2007). Its high biodiversity levels, biological productivity, and 857 endemic species of fish, seabirds, marine mammals, and macroinvertebrates give it one of the highest coastal ecosystem conservation priorities on the planet. Unfortunately, it also faces growing threats that result mostly from overfishing and degradation of the coastal habitats: 39 of its marine species are listed in the IUCN Red List as threatened or vulnerable (Thomson and Eger 1966; Brusca 1980; Thomson et al. 2000; Thomson and Gilligan 2002; Brusca 2004a; Brusca and Bryner 2004; Carvajal et al. 2004; Brusca et al. 2005; Findley et al. in press).

This vast natural wealth is not only of biological and conservation interest; it also yields some 30–60 percent of the national catch of Mexican fisheries and provides the socioeconomic sustenance of the inhabitants of the region, who have developed systems of natural resource access, use, and appropriation that often put the long-term sustainability of the resources in peril. The most important threats to biodiversity are driven by the growth of economic activities in the region that have caused the deterioration of coastal marine ecosystems due to decreasing freshwater flows, pollution by agrochemicals and urban waste, sedimentation, overfishing, and the use of inappropriate fishing technologies such as bottom trawling and gill nets. Critical habitat of mangrove forests is being lost at an annual rate of 9 percent from sedimentation, eutrophication, and changes in water flows caused by the construction of shrimp ponds, marinas, inland channels, and deforestation (color plate 8B). Invasion of exotic plant and animal species is putting at risk the native and endemic species of the Gulf's islands and the Sonora and Baja California desert regions (Carvajal et al. 2004).

In the 1930s, outboard motors and gill nets came into use in the Gulf, permanently transforming the once-abundant regional fisheries. The highly prized totoaba—an endemic fish—saw significant population declines (Cisneros-Mata et al. 1995a, 1997; Román-Rodríguez and Hammann 1997), while inshore fishing efforts began to have a strong impact on estuaries and lagoons. In 1933, the shrimp fishing industry introduced the use of bottom-trawling gear on soft seabeds, devices that have since been sweeping the seabed clean on a regular basis. Everything in the path of the trawling dragnet (fish, octopus, conch, sponges, starfish, etc.) is harvested, and the by-catch, if judged to be of lesser economic value, is returned dead to the sea. As the twentieth century progressed, the shrimp fishery became the most important economic activity in the fishing industry. In 1997 the five states of the Gulf of California produced 57,000 tons of shrimp (approximately 70 percent of the Mexican national shrimp production) while destroying the Gulf's soft seabeds and wiping out hundreds of thousands of tons of by-catch. The use of bottom trawlers had become one of the greatest threats to marine biodiversity in the Gulf (García-Caudillo et al. 2000).

Today, the Gulf region represents a large and still sparsely populated area of Mexico, with human densities only a third of the national average. It is also a relatively wealthy region of the country: the per capita income of Baja California and the State of Sonora is 22 percent higher than the national average (Ezcurra 2003; Carvajal et al. 2004). The region produces around 70 percent of the value of national fisheries, and 40 percent of the national agricultural production is harvested in its high-technology irrigated fields. In recent decades, however, depletion of natural resources and new macroeconomic opportunities have led to major shifts in the economic structure, leading the economy away from its traditional base on the primary sector—particularly agriculture, fisheries, and mining—and toward export-oriented manufacturing/industry and services such as tourism.

Regional Environmental Challenges

The rapid growth of the regional economy has brought a large demographic increase from immigration and ensuing pressures on its natural resources. Open-access, extractive use of natural resources seems to have reached a limit in the region, and rapid development of the manufacturing and services sectors is putting an additional strain on natural resources. The implications of this economic change for biodiversity conservation are significant. Decision makers have an important opportunity to reduce the pressure on natural resources in the region by supporting a reorientation of the economy away from primary-sector activities associated with overexploitation of natural resources and by promoting the establishment of a sustainable industry of low-impact, environmentally minded tourism. In this section we discuss some of the major regional environmental challenges brought about by agriculture, industrial and small-scale fisheries, and massive tourism in the environmentally sensitive areas of the Gulf of California.

Fisheries

With 39 species listed on IUCN's Red List, it seems clear that ecological degradation has already taken a major toll on the Gulf's biodiversity. The endemic vaquita porpoise (*Phocoena sinus*) is already near extinction, the fate of the totoaba (*Totoaba macdonaldi*) is questionable, and populations of five species of sea turtles have all but disappeared from the Gulf. The International Committee for the Recovery of the Vaquita (CIRVA) estimated the current vaquita population in the range of 268–464 for 2004, as compared with a previous estimate of 567 vaquitas for 1997. The new estimate took into consideration an annual growth rate of 4 percent and the loss of 39 to 78 individuals per year (D'Agrosa et al. 2000), which make the vaquita the most endangered marine cetacean in the world. More recent work suggested that the vaquita population might be as low as 150 (Jaramillo-Legorreta et al. 2007).

Overharvesting of the fishing stocks has become a strongly limiting factor for the success of the regional fisheries. Twenty years ago, there was a positive correlation between catch and effort in most of the regional fisheries—the more days the fleets fished, the higher the catch they achieved. Now that correlation is largely gone: the total landings in most fisheries are largely independent of the (generally excessive) fishing efforts, and the catch per unit effort has decreased severely for many species. In short, the fishers of the Gulf are often overharvesting, and in some cases even depleting, their stocks (Sala et al. 2004; Velarde et al. 2004). Additionally, there is clear evidence that coastal food webs in the Gulf of California have been "fished down" during the last thirty years (i.e., fisheries have shifted from large, long-lived species belonging to high trophic levels to small, short-lived species from lower trophic levels) and that the individual fish length for landings has decreased significantly (by about 45 cm) in only twenty years (Sala et al. 2004).

In some fisheries, the tragedy of common-access resources is hitting the Gulf hard. For example, thirty years ago, the shrimp-trawling fleet in the Gulf consisted of about 700 boats, each of which captured about 50 tons of shrimp per season. Today, the fleet is almost 1,500 boats but the annual catch scarcely surpasses 10 tons per boat. In spite of governmental transfers of about $30 million (U.S.) each year—provided in the form of discounted fuel prices—many boats of the fleet are facing economic collapse. Decrease in catch per unit effort combined with low wholesale prices for farmed shrimp led, in the early 2000s, to many shrimp trawlers going out of business or switching largely to nonshrimp target species.

The situation is also discouraging from an environmental standpoint, especially regarding the shrimp trawlers. The bottom trawlers exterminate some 200 thousand tons of by-catch every year for a meager annual catch of some 30 thousand tons of shrimp. The dragnets destroy some 30–60 thousand square kilometers of sea bottom (García-Caudillo et al. 2000), much of which lies within the Upper Gulf Biosphere Reserve, and the boats emit collectively some 30–40 thousand tons of greenhouse gases due to the cheap fuel that keeps their inefficient business going. Seabeds have been so depleted in some areas that local artisanal fishers in places such as Loreto Bay and Bahía de los Ángeles have been demanding the establishment of no-take zones and marine protected areas. In open conflict with the local communities, the larger fleets want no protected areas and demand permits to trawl inside federal reserves—such as the Upper Gulf Biosphere Reserve—to increase their scanty earnings. In the Gulf, conflict between sectors and between particular interests has increasingly become the rule.

However, not all stories of common resource use in the Gulf of California are despairing tales of unsustainability and collapse. There are also many success stories, and learning from them is fundamental for future conservation efforts. For example, local artisanal fishers have started to work with local researchers in the Gulf to understand the phenomenon of spawning aggregations so that they can identify and protect reproductive areas. As a result of pressures from these local resource users, the Bay of Loreto is now a marine park (*Diario Oficial* 1996a), and the fishers of Bahía de Los Ángeles are supporting the creation of a similar marine protected area. In San Ignacio Lagoon, fishers from a previously unsustainable fishery have organized to preserve the environment and train their people in basic natural history to organize whale-watching tours.

The abalone and lobster cooperatives of the Pacific coast of Baja provide yet another example of long-term sustainable use. With no support from the federal government, these fisheries have established strict rules for resource extraction and have developed their own law enforcement system. Many cooperatives generate their own electricity, run their own canneries, and finance their own schools. More than forty years after their establishment, their productivity is still high and their resources seem to be used in a sustained manner.

Not only small communities and conservationists are critical of some of the region's unsustainable modes of development: a growing number of entrepreneurs and business people are also becoming committed supporters of the environmental cause. As a result of increasing environmental concerns, a cluster of committed business leaders have organized an action and opinion group called grupo NOS (for Noroeste Sustentable, or Sustainable Northwest) to promote the appropriate use of resources in the Gulf region. Even large fishing fleets can maintain sustainable landings when their leaders and operators work in cooperation. In contrast with the continuing crises of the shrimp bottom-trawling fleet, the sardine fishery has proven capable of controlling its own fishing effort and—after a past collapse—is now healthy and productive (Cisneros-Mata 1995b, 1996; Lluch-Belda et al. 1986). In short, although the Gulf of California is under extreme overfishing pressures in many parts and has seen the consequent collapse of some of its resources, it also harbors a number of successful and encouraging experiences of communities that are trying to keep their resources healthy and productive now and in the future.

Coastal Tourism and Recreational Activities

The magnificent landscapes and the amazing density of marine wildlife make the Gulf of California a superb place for visitors. The first tourists coming into the Gulf in the first half of the twentieth century were driven by the extraordinary catches of sport fish. Nautical tourism developed later, partly driven by the success of sport fishing and partly by the beauty of the area. Soon the idea of connecting the region through a series of marinas with the states of California, Oregon, and Washington in the United States became a vision for developers. During President Vicente Fox's administration this idea materialized in the form of a regional project, the Escalera Náutica, or Nautical Stairway, whose goal was greatly increasing boat tourism in the Gulf. The project was developed by the National Fund for the Promotion of Tourism (FONATUR, a federally funded development agency).

Following its announcement in 2001 and in the years that followed, the Escalera Náutica became one of the most debated projects in the region. On the one hand, the project tried to generate a shift in the regional economy from unsustainable fisheries and water-intensive agriculture toward the services sector in the form of boat tourism. Given the irrefutable evidence that the economy's primary sector has reached its limits in the Gulf (and in some cases is even facing collapse), this shift seems like a desirable move in principle. On the other hand, experiences in Mexico with resource-intensive, unsustainable tourism (and its sequel of failed and abandoned projects, dredged mangrove swamps, and exhausted aquifers) have left a deep scar of mistrust in local groups and communities. Thus, the Escalera project presented a dilemma for conservation groups: it signified a positive shift in the local economy by moving it away from consumptive resource use, but it failed to assure the ecological sustainability of the project and opened the door for environmentally disruptive development. For regional conservation groups, the big challenge in the Gulf of California is to promote environmentally sustainable tourism while ensuring the preservation of the natural beauty and biodiversity of the region—the very attributes that initially triggered tourism in the region.

ALCOSTA (Alianza para la Sustentabilidad del Noroeste Costero Mexicano), an alliance of several environmental organizations), was born in the wake of the Escalera Náutica project. Together with other players, ALCOSTA brought a voice of concern into the development plan. The group spearheaded critical efforts to reduce the environmental impacts of the project and to reform the initiative by making it more accommodating to issues of environmental conservation. As a consequence of these efforts, FONATUR was compelled to prepare a regional Environmental Impact Statement and public hearings were held about the initiative—the first hearings ever held in Mexico about a development project of regional dimensions. ALCOSTA developed a critical analysis of the project and presented it to the federal authorities during the hearing. All the environmental conditions proposed by ALCOSTA for the authorization of the project were taken into account in the final resolution. Additionally, the discussions drove President Fox to designate the Gulf of California as a joint priority for tourism development and for conservation. As a result, the Ordenamiento Ecológico Marino (Marine Habitat Use Plan) program was initiated together with a promise to enlarge the protected areas in the Gulf islands to include surrounding waters.

The governments of states surrounding the Gulf have adopted aggressive plans to promote tourism by creating infrastructure. In their vision, tourism is an opportunity to create much-needed jobs and to foster economic development, thereby compensating for job losses in other sectors and perhaps also easing the growing demands for subsidies by farmers and fishers. The crisis of the primary sector makes tourism an increasingly attractive alternative for regional governments. This, in turn, will demand growing attention from conservation groups to address the issue on a regional scale.

Freshwater Resources

Water is a vital development resource in desert regions such as the Baja California peninsula and the State of Sonora. However, the rapid growth of activities such as agriculture, industry, tourism, and urban development (and its associated demographic growth) have placed increasing pressure on regional water resources during the last decades, not only from higher demand but also from the increase in pollutants that results from unplanned economic growth and from growing pressures on the deficient sanitary infrastructure and limited water treatment facilities. Thus, the rapid expansion of the more successful sectors of the regional economy has been mostly done at the expense of depleting underground aquifers, disrupting watershed flows, and destroying the natural wetlands around large urban conglomerates.

At the end of the nineteenth century, all regional rivers still ran free into the Gulf waters. In the early decades of the twentieth century, major agricultural developments were established in the lower basins and deltas of the Fuerte, Mayo, and Yaqui Rivers of Sinaloa and Sonora and also in the Colorado River valley along the Mexico–U.S. border, giving birth to the fast growth of settlements along the coastal plains and permanently modifying the ecological landscapes and the services that the deltas provided.

A good illustration of this dynamic is the Colorado River, which until the 1930s was the largest river flowing into the Gulf of California and included a vast delta covering 300 km^2 of wetlands (Sykes 1937; Fradkin 1984; Ezcurra et al. 1988; Felger 2000; Brusca and Bryner 2004). The development of steamboat traffic on the Colorado River during the nineteenth century, from the Gulf into the Yuma trail, was made possible partly because of the dense cottonwood forests on the river banks, which were intensely logged for charcoal. The devastation of these forests brought the first significant environmental impacts on the delta, which were merely a prelude to the devastation that the estuary would later see. In 1905, major infrastructure projects that diverted the waters of the Colorado gave birth to the Imperial and Mexicali agricultural valleys. In the early 1940s, the Hoover Dam and the International Water Treaty signed between the United States and Mexico brought about the irremediable demise of the great delta. In the same manner and in the same decade, dams and channel works were initiated along the Fuerte, Mayo, and Yaqui Rivers as part of a regional project for agricultural development that brought a dramatic reduction in the supply of freshwater to downstream ecosystems and thus led to degradation of the riparian corridors, estuaries, wetlands, and coastal lagoons (Robles et al. 1999). Economically, all these irrigation projects were extremely successful during their first decades. At present, however, the once-fertile soils of these man-made agricultural valleys are showing clear signs of exhaustion. Soil salinization and decreasing profit margins have forced farmers to abandon many fields, leaving behind a substantial ecological footprint of salinized and vegetation-denuded coastal plains and thus further compounding problems in the already heavily affected areas.

The decline of water and soil resources for agriculture has encouraged a reorientation toward high-value horticultural crops that provide better returns. The outlook also seems to be improving for restoration of the once-lush coastal wetlands: the abandonment of lands because of salinization, together with changes on the Mexican Law of National Waters (Ley de Aguas Nacionales), has created the possibility of buying water rights to restore the deltas.

The Need for Effective Conservation

If effective conservation in the region is to be achieved, then a strategy that addresses these critical regional environmental challenges must be developed. Such a strategy should allow for the protection of endangered species, spawning aggregation areas, and endangered ecosystems such as seamounts, coastal lagoons, coral reefs, estuaries, and marine mammal habitats (Sala et al. 2002). Expansion or establishment of protected areas is an approach that can be taken to ensure long-lasting conservation efforts, but alternative approaches are also needed. In order to be effective, any successful conservation effort will require the support of the local communities and regional stakeholders. If conservation initiatives are not able to win the hearts and minds of local people by generating alternative livelihoods, then future efforts are not likely to succeed.

The degradation of coastal wetlands is one of the Gulf's most serious threats. With little consideration given to the ecological services they provide, mangrove forests are being cut for the development of aquaculture (mostly shrimp farms) and tourism projects. Moreover, coastal wetlands in general are threatened by consumptive water use upstream and by the pollution of rivers and waterways. Because water is a mobile resource, the protection and the restoration of coastal wetlands demands comprehensive large-scale plans addressing both upstream and downstream habitats. The ecological services provided by estuaries and lagoons are critical for the survival of the Gulf of California fisheries and for the health of the large marine ecosystem as a whole. Comprehensive plans for effective protection of coastal wetlands, halting mangrove deforestation, and maintaining the ecological services of coastal lagoons and estuaries must be developed and implemented for the long-term survival of the entire Gulf.

A Brief History of Conservation Efforts

Possibly the first efforts to protect the ecosystems of the Gulf of California started in 1951 with the publication of Lewis Wayne Walker's popular paper on the seabirds of Isla Rasa in the *National Geographic* magazine. Walker was then a researcher at the San Diego Natural History Museum and later became associate director of the Arizona-Sonora Desert Museum. He was knowledgeable about the natural history of the region and possessed firsthand field experience in Baja California and on the islands of the Gulf, especially Isla Rasa. Walker wrote many popular articles on the natural history of the region and through these publications popularized the plight of Rasa (Walker 1951, 1965).

In the early 1950s the Audubon Society donated $5,000 for the preservation of Rasa. This started Walker's research on the island, which was later also supported by a grant from the Belvedere Scientific Fund from San Francisco. Part of this financial support reached Dr. Bernardo Villa's laboratory at the Institute of Biology in the National University of Mexico (Universidad Nacional Autónoma de México, or UNAM). The funds were used to maintain a biologist and a field station on the island. The work at Rasa was later supported by donations from the Roy Chapman Andrews Fund to the Arizona-Sonora Desert Museum.

The results of these investigations soon reached the Direction of Forestry and Wildlife in the Mexican federal government, which in the late 1950s was headed by Dr. Enrique Beltrán, an eminent Mexican conservationist. Beltrán's own interest in the issue—and the public notoriety that Isla Rasa had achieved through popular publications and the field trips of many biologists—helped to prepare the way for the first federal decree protecting the insular ecosystems of the Gulf of California: in 1964, the official governmental federal register (*Diario Oficial de la Federación*) published a decree declaring Isla Rasa a nature reserve and a refuge of migratory birds (*Diario Oficial* 1964).

In 1973, a natural history expedition was organized, using a chartered Catalina flying boat to traverse the Gulf and visit small, remote islands. The group included George Lindsay, then director of the California Academy of Sciences; Charles Lindbergh, the legendary aviator; Joseph Wood Krutch, a celebrated American nature writer; and Kenneth Bechtel, a philanthropist from San Francisco. Lindbergh had become a committed conservationist and Krutch had written *The Forgotten Peninsula*, a sparkling natural history description of Baja California. George Lindsay had previously organized a series of scientific explorations into the Gulf of California and the Baja California peninsula (Banks 1962a,b; Lindsay 1962, 1964, 1966, 1970). Bechtel had given financial support to the Audubon Society in the 1960s to study the seabird rookery at Isla Rasa. Two or three months later, both Lindbergh and Lindsay traveled to Mexico City and met Presidents Echeverría's top-level cabinet to promote the conservation of the Gulf. Although no cause and effect has been established, the fact is that four years after Lindbergh's appearance in Mexico City a decree was issued protecting all of the islands of the Gulf of California (SDNHM 1996; *Diario Oficial* 1978).

The Seri People and the Protection of Tiburón Island

Isla Tiburón was actually the first part of the Gulf of California to receive official status as a protected area (through a decree published a year before that concerning Isla Rasa's ecology). The largest island in the Gulf, Tiburón occupies 120,756 hectares. In pre-Hispanic times it was an important part of the territory of the Comcáac or Seri Indians (Felger and Moser 1985). As a result, the island not only is an important natural site but also harbors historic, archeological, and cultural artifacts. Although the twentieth-century Seri have not lived permanently on the island, they have always used it as their main fishing camp, hunting ground, and plant-collecting territory and have always considered it part of their tribal land.

On March 15, 1963, in response to an initiative by Enrique Beltrán, Tiburón was decreed a wildlife refuge and nature reserve by President Adolfo López Mateos (*Diario Oficial* 1963). However, the declaration was based solely on biological and ecological grounds, and it failed to take into consideration the needs and demands of the Comcáac themselves. Twelve years later, in 1975, the Secretary of the Agrarian Reform gave the Seri formal possession of Tiburón Island as part of an *ejido* (communal land) allotment for the tribe, thus recognizing for the first time the Seri's right to their ancestral homeland. On February 11, 1975, a decree was issued by President Luis Echeverría that restored Tiburón Island to the Seri People as part of their communal property. Although this decree was issued as part of a series of governmental actions to empower native peoples within their traditional lands, it also had conservation implications for the island as well as for the mainland coast. The decree established that the coastal waters of the island could be only used by the Seri and by their fishing cooperative, the Sociedad Cooperativa de la Producción Pesquera Seri (INE 1994); these waters were declared off-limits to other fishers.

The Development of Mexican Conservation Efforts

The Mexican government and civil society have shown increasing concern for these regional environmental issues. New legal and institutional frameworks have been adopted, and innovative institutional arrangements have been created among various governmental agencies and stakeholders to face the mounting problems. In the second half of the twentieth century, several regulations were adopted to avoid the overexploitation of a number of species or to prevent their extinction. Perhaps the most notable effort was the declaration in 1955 of a protection zone in the delta of the Colorado to protect the breeding habitat of many marine species, including the totoaba. In the 1960s, efforts were made to protect the seabird colonies of the islands. The failure of the protection zone in the Colorado River delta—coupled with the persistent decline in the totoaba population—forced the government to implement even more drastic measures, including a total ban on the totoaba fishery by 1975.

Over the years, modest but continued funding was provided to Bernardo Villa's laboratory at UNAM by the Audubon Society, the Roy Chapman Andrews Fund of the Arizona-Sonora Desert Museum, Conservation International, and others who contributed to maintaining the presence of Mexican researchers and students on Isla Rasa. Many of these students later became leading conservationists in the Gulf of California. Dr. Villa's work in the early 1980s effectively combined research with conservation. One of his young students at the time, Dr. Enriqueta Velarde, decided to extend this idea to other islands of the Gulf. With scientific support from George Lindsay of Cal Academy and Daniel Anderson of the University of California at Davis, together with financial support from Spencer Beebe of the Nature Conservancy, Enriqueta Velarde launched the first conservation project for the islands. The project produced, among many other applied results, the book *Islas del Golfo de California* (Bourillón et al. 1988), which was extremely influential in bringing attention to the islands and their conservation problems.

Many of the biologists who participated in this early team are now crucial players in conservation efforts throughout the Gulf. The team included, among others, Alfredo Zavala, now Baja California's regional director for the Protected Area of the Islands of the Gulf of California; the late Jesús Ramírez Ruíz, who in the early 1990s eradicated introduced rodents from Isla Rasa; and Luis Bourillón and Antonio Cantú, who now head conservation-oriented nongovernmental organizations (NGOs) in the region. In many ways, the conservation work at Isla Rasa was the catalyst that started many of the more recent conservation efforts in the Gulf (Velarde and Anderson 1994).

The Biosphere Reserve Concept

In the early 1970s, many changes were occurring within the Mexican scientific and conservation groups that also helped to protect the ecosystems of the Gulf of California. In 1974 the Instituto de Ecología, a nonprofit research organization, was created in Mexico City and soon began to promote the concept of Biosphere Reserves in the country. Although widely accepted at present, the idea of Biosphere Reserves, which had been developed by UNESCO's Man and the Biosphere Program (MAB), was radically new in 1975. Biosphere Reserves were conceived as natural protected areas where the indigenous populations living inside the area or in the surrounding "buffer zones" were encouraged to use their natural resources in a sustainable manner. The new approach departed radically from the natural park concept, which basically advocates pristine areas free of human influence. Instead, Biosphere Reserves promote sustainable use as an effective tool for conservation.

Many of the concepts of global ecology and conservation were already operational in MAB's concept of Biosphere Reserves almost twenty years before. These concepts included: (a) a global approach to conserving biodiversity through a planetary network of protected areas; (b) the preservation of cultural diversity together with natural diversity; (c) the involvement of local populations in the protection of natural resources; and (d) the promotion of the sustainable use of nature. Although the islands of the Gulf of California were initially not conceived as a Biosphere Reserve but rather as a Wildlife Refuge (*Refugio de la Vida Silvestre*), it was in the wake that the decree protecting them was issued in 1978 (*Diario Oficial* 1978).

The Establishment of SEMARNAP

In December 1994, a revolutionary transformation occurred within the Mexican federal administration. The newly elected president, Ernesto Zedillo, decided to create a ministry of the environment in charge of pollution control and natural resources management. The creation of the Secretaría de Medio Ambiente, Recursos Naturales y Pesca (SEMARNAP) opened the door for many impressive changes in the establishment and management of protected natural areas. In the most important reserves—including the Upper Gulf of California and Colorado River Delta Biosphere Reserve—"paper parks" gave way to "hands on" resource management. Also, this decade saw a steady increase in funding from the Mexican government and in human resources dedicated to the management of nature reserves.

The establishment of some key marine protected areas in the Gulf shortly before or after the creation of SEMARNAP (namely, the Upper Gulf Biosphere Reserve and the Cabo Pulmo and Loreto Bay National Parks; see table 11.2) were of immense conceptual importance, as they opened the way for other marine reserves in Mexico where, because of a long-standing feud between conservationists and the fisheries authorities, no marine protected areas had previously been accepted. Specifically, it facilitated current efforts by various NGOs to extend decreed protection into waters adjacent to some important islands such as Islas Marías, San Pedro Mártir, San Lorenzo–Las Ánimas–Salsipuedes, Rasa, and Partida. However, in spite of their tremendous practical and conceptual significance, the establishment of protected areas in the Gulf during the 1990s accounted for less than 4 percent of the Gulf's marine area.

The creation of SEMARNAP brought an additional management tool to ongoing discussions on the Gulf's sustainability in the form of Mexico's territorial-use planning regulations, or Ordenamiento Ecológico. The *ordenamiento* is a process established in the Mexican environmental law to regulate the spatial use of natural resources and land use at different scales. It demands the best available scientific knowledge, participation of the regional stakeholders, and a series of comprehensive hearings and negotiations with local governments, local businesses, and nongovernmental organizations. The planning studies for the Gulf of California started in 1997; however, because of their sheer complexity, the hearings and discussions are still ongoing. Almost a decade after it was started, the territorial-use planning of the Gulf has yet to be accomplished and thus is one of the conservation agenda's most important pending objectives.

Legal Status of Protected Areas in the Gulf of California

Mexico's environmental legislation, the Ley General del Equilibrio Ecológico y la Protección al Ambiente (*Diario Oficial* 1988, 1996b) recognizes seven categories of natural protected areas that can be established by

TABLE 11.2. Protected natural areas in the Sea of Cortez region, including the Mexican northwestern Pacific Ocean.

	Date of Creation	Area (ha)	Location(s)
Biosphere Reserves			
Complejo Lagunar Ojo de Liebre	14 Jan. 1972	60,343	Baja California Sur
El Vizcaíno	30 Nov. 1988	2,546,790	Baja California Sur
Alto Golfo de California y Delta del Río Colorado*	10 Jun. 1993	934,756	Baja California, Sonora
El Pinacate y Gran Desierto de Altar	10 Jun. 1993	714,557	Sonora
Sierra La Laguna	6 Jun. 1994	112,437	Baja California Sur
Archipiélago de Revillagigedo*	6 Jun. 1994	636,685	Baja California Sur
Islas Marías*	27 Nov. 2000	641,285	Nayarit
Isla San Pedro Mártir*	13 Jun. 2002	30,165	Sonora
Isla Guadalupe*	25 Apr. 2005	476,971	Baja California
National Parks			
Sierra de San Pedro Mártir	26 Apr. 1947	72,911	Baja California
Constitución de 1857	27 Apr. 1962	5,009	Baja California
Isla Isabel	8 Dec. 1980	194	Nayarit
Cabo Pulmo*	6 Jun. 1995	7,111	Baja California Sur
Bahía de Loreto*	19 Jul. 1996	206,581	Baja California Sur
Archipiélago de San Lorenzo*	25 Apr. 2005	58,442	Baja California
Islas Marietas*	25 Apr. 2005	1,383	Nayarit
Wildlife Protection Areas			
Cabo San Lucas	29 Nov. 1973 (7 Jun. 2000)	3,996	Baja California Sur
Islas del Golfo de California	2 Aug. 1978 (7 Jun. 2000)	358,000	Baja California, Baja California Sur, Sonora, Sinaloa
Valle de los Cirios	2 Jun. 1980 (7 Jun. 2000)	2,521,776	Baja California
Sierra de Álamos-Río Cuchujaqui	19 Jul. 1996	92,890	Sonora

Notes: Areas marked with an asterisk (*) contain open marine waters decreed as part of the reserve and are hence true marine protected areas. Reserves showing a second date of creation (in parentheses) were first decreed in some other category and and later re-categorized to their current status.

federal authority. These are: (1) biosphere reserves (*reservas de la biosfera*); (2) national parks (*parques nacionales*), including both terrestrial and marine parks; (3) natural monuments (*monumentos naturales*); (4) areas for the protection of natural resources (*áreas de protección de recursos naturales*); (5) areas for the protection of wildlife (*áreas de protección de flora y fauna*); (6) natural sanctuaries (*santuarios*); and (7) unclassified areas.

The term *Biosphere Reserve* is applied both to protected areas as defined by Mexican law and to areas integrated within UNESCO's Man and the Biosphere (MAB) network of protected areas. Some reserves in Mexico are considered Biosphere Reserves under Mexican law but have not yet fulfilled the conditions to be incorporated within MAB's international system; conversely, other Mexican protected areas are formally recognized by MAB as Biosphere Reserves but do not have formal Biosphere Reserve status under Mexican law (Gómez-Pompa and Dirzo 1995; SEDUE 1989). Within the Gulf of California Region, UNESCO's MAB Program has designated three protected areas as part of its international network of Biosphere Reserves: (1) El Pinacate y Gran Desierto de Altar, in the core of the Sonora Desert and designated in 1993; (2) Alto Golfo de California y Delta del Río Colorado in the Upper Gulf, designated in 1995 and also dedicated as a site of global significance within the Ramsar International Convention of Wetlands; and (3) Islas del Golfo de California, which were designated in 1995. However, the Gulf islands do not have a formal recognition as Biosphere Reserves under Mexican federal law. Originally decreed as a reserve zone and refuge for wildlife and migratory birds, the islands were re-categorized in June 2000 as an Area for the Protection of Wildlife (Área de Protección de Flora y Fauna Islas del Golfo de California; see *Diario Oficial* 2000a and table 11.2).

As a wildlife protection area, the islands do not enjoy the same strict restrictions that are imposed on Biosphere Reserves. The reasons to nationally designate the islands within a different category from the one they hold internationally are possibly related to the large size and spatial complexity of the whole archipelago and to the difficulties involved in law enforcement within the larger protected area. In spite of their less-restrictive status under Mexican law, the islands of the Gulf of California are in practice managed as a large reserve, and substantial efforts are devoted to their protection (Breceda et al. 1995; INE 1994). The relevance given by federal authorities to the Gulf islands may be the result of an effort to fulfill the Mexican government's commitment to the UNESCO-MAB network and to the Global Environmental Facility (GEF), which has funded part of the conservation work on the islands. In 1996 the administration of the islands was divided into three regional headquarters: (a) the southern islands are managed from an administrative office at La Paz; (b) Tiburón and San Esteban are managed from an office in Guaymas; and (c) the western Midriff Islands area is managed from headquarters in Ensenada.

Additionally, it is an explicit policy of the Mexican Commission for Protected Natural Areas (CONANP) to re-decree many of the Gulf islands as smaller protected areas with a "donut" ring of marine protected waters around them. This policy, established in the late 1990s, has already yielded a number of islands with protected adjacent marine territories, such as Archipiélago de Revillagigedo (decreed in 1994), Islas Marías (2000), Isla San Pedro Mártir (2002), and Isla Guadalupe (2005), as Biosphere Reserves in addition to Archipiélago de San Lorenzo (2005) and Islas Marietas (2005) as National Parks (table 11.2).

The Case of the Alto Golfo

There are myriad stories of dedicated work and heated debates over each of the protected areas of the Gulf of California. In a previous paper (Ezcurra et al. 2002) we analyzed eight of these case studies, the conflicts behind their creation, and the ongoing discussions about their future land use. In this section we concentrate on one case where discussions and debate have been (and continue to be) especially heated: the Alto Golfo, in the northernmost tip of the Upper Gulf of California. The history and evolution of conservation efforts in the Upper Gulf provide an outstanding case study for reflecting on how the conservation movement has evolved in the region and how it has matured in Mexico as a country. Understanding past and ongoing conflicts in the Upper Gulf is of great importance for understanding the viability of the conservation movement in Mexico. In this section we present some historic facts, discuss the events that led to the creation of this important reserve, and examine the conflicts that ensued.

Background

The Upper Gulf of California and Colorado River Delta Biosphere Reserve is formed by part of the surrounding Sonora Desert, the northern marine waters of the Gulf of California, and the lowermost part of the Colorado River. Its high marine biological productivity is based on the churning of nutrients in Colorado River sediment deposits by one of the biggest tidal fluxes on the planet (Thomson et al. 2000). This productivity makes the Upper Gulf an extremely important area for the reproduction, nursery space, and growth of many resident and migratory species. Currently, the total number of marine species recorded for the reserve is 1,438, of which 11 are in danger of extinction—notably the vaquita (*Phocoena sinus*), or Gulf of California harbor porpoise, and the totoaba (*Totoaba macdonaldi*), the largest member of the worldwide croaker (drum) family (Sciaenidae). Both of these species are endemic to the Upper Gulf.

The Upper Gulf's marine richness is reflected in its highly valuable fisheries, especially shrimp, which make the Alto Golfo one of the most important fishing grounds in Mexico. Historically, the most significant economic activity for the reserve's inhabitants, and for some outsiders as well, has been gillnet and trawler fishing (McGuire and Greenberg 1993). Fishers began to establish camps in the Upper Gulf of California at the beginning of twentieth century, and by the 1940s the totoaba fishery was at its maximum and enjoyed a well-developed and profitable export market.

In the 1950s it became well known that the Upper Gulf and the delta of the Colorado River were important sites for the reproduction and breeding of many species of birds and fish. For that reason, in 1955 the Mexican Fishing Authority first declared the area as protected for reproduction and as a nursery. As the years passed, however, it also became apparent that this productive region was still suffering from increased and unsustainable fishing pressure. By 1975, the totoaba was facing extinction through overfishing. This problem forced the federal government to decree a moratorium on harvesting the totoaba in the Gulf of California. Thus, the area was re-decreed in 1974 as a reserve zone for fisheries resource restocking, but the depletion of natural resources continued in spite of this decree (*Diario Oficial* 1974). In 1975, the Ministry of Fisheries established a permanent ban on totoaba captures, which remains in force today.

However, other problems continued to mount. In the mid-1980s, marine mammalogists became concerned about the vaquita harbor porpoise population in the Upper Gulf. The vaquita is indeed a unique and extremely rare marine mammal. First described in 1958, only a few specimens have been studied to date. The occurrence of vaquita specimens as incidental take in gill nets in the Upper Gulf signaled an alert to Mexican and international conservation groups.

In the early 1990s, the vaquita population was estimated to be fewer than 500 animals. The vaquita was classified as endangered, and the International Whaling Commission labeled it as one of the highest-priority conservation marine mammals in the world. It was then that the Mexican federal government created, through the Secretary of Fisheries, the "Technical Committee for the Protection of the Totoaba and the Vaquita" (Comité Técnico para la Preservación de la Totoaba y la Vaquita) for the purpose of evaluating and studying the issue and then recommending adequate measures for the conservation of both endangered species. After a few sessions, it became evident that serious discrepancies existed between various constituents of the committee. While some members favored immediate action to protect the Upper Gulf of California from the devastating effects of overfishing, others were of the opinion that regulating fisheries in any way would harm the local economy. As a result of these conflicts, it was decided to request two of the most recognized research centers in Sonora—the Centro Ecológico de Sonora (CES) and the Centro de Investigación y Desarrollo de los Recursos Naturales de Sonora (CIDESON)—to develop and elaborate upon a feasibility study for a Biosphere Reserve.

The study was completed near the end of 1992 and argued in favor of establishing a reserve in the Upper Gulf (CTPTV 1993). Different research and conservation groups spend the first months of 1993 discussing the costs and benefits of a protected area with local communities (El Golfo de Santa Clara, Puerto Peñasco, and San Felipe, as well as the ejidos in the delta of the Colorado River). Slowly, the people in the area began to accept and then to support the idea. With the support of local businessmen, scientists, conservationists, social leaders of the small-scale fisheries, and traditional authorities of the indigenous peoples around the Gulf, the project was presented to Secretary of Social Development Luis Donaldo Colosio, a native of northern Sonora who was very interested in the idea. With the support of Colosio, the project moved forward.

On June 10, 1993, President Carlos Salinas de Gortari decreed the establishment of the Biosphere Reserve of the Upper Gulf of California and Delta of the Colorado River (Reserva de la Biosfera del Alto Golfo de California y Delta del Río Colorado; see *Diario Oficial* 1993). At that time the project had strong support from both the local population and conservation groups. Important decision makers attended the ceremony, including many cabinet members from the Mexican federal government; the governors of Sonora, Baja California, and Arizona; U.S. Secretary of the Interior Bruce Babbitt; and the traditional governor of the Tohono O'Odham (Papago) people, whose lands extend on both sides of the Mexico–U.S. border.

This reserve was the first one established including parts of the territory of two states, Baja California and Sonora, as well as federal marine waters (INE 1995). Thus, coordination between these entities was from the very beginning a crucial factor in the successful pursuit of the reserve's objectives.

The Political Juncture

It is interesting at this point to reflect on the particular political juncture that drove the establishment of the reserve. The area had twice previously (in 1952 and 1973) been decreed as a protected area and the decrees had not had the desired effect, so why did this third decree carry so many expectations and have so much support?

The main reasons given by federal authorities for officially decreeing that the area be protected were the uniqueness of its ecological attributes and the catastrophic depletion of natural resources in the Upper Gulf. On the one hand, the problem of overfishing in the Gulf had started to appear in the international arena, harming Mexico's reputation on environmental conservation and appropriate natural resource management.

On the other hand, in 1992 a severe crisis had struck the fishers of the Upper Gulf of California (El Golfo de Santa Clara, Puerto Peñasco, and San Felipe). Their shrimp catches had fallen precipitously (Arvizu 1987), and the fishers blamed the federal authorities in general, and the Secretary of Fisheries in particular, for failing to enforce fishing bans that would allow recovery of the resource. The idea started to grow among the fishers themselves that the sea had to rest and that its fisheries had to recover—that things had to change in some way. A decree establishing a protected area seemed to affirm this imminent change and offer an opportunity to start again (McGuire and Greenberg 1993).

Additionally, by 1992 biodiversity conservation and natural resource sustainability had become high-profile international priorities for most countries in the wake of the Rio Summit (also known as UNCED, the United Nations Conference on Environment and Development), from which the specific commitments of Agenda 21 emerged (Cicin-Sain et al. 2002). It was a priority for the Mexican government to signal its willingness and resolution to comply with Agenda 21, and the Upper Gulf had been identified both internally and internationally as a significant conservation problem and hence a potential stumbling block in international affairs.

Finally, the North American Free Trade Agreement (NAFTA, signed in 1992) among Mexico, Canada, and the United States had raised a number of environmental objections in all three countries, and the Mexican government was eager to give positive signals of environmental commitment. Because of its proximity to the U.S. border, the Upper Gulf Biosphere Reserve seemed an ideal proposition. By federally protecting this area, Mexico improved its international environmental image and attained a more favorable position in the NAFTA negotiations. At the same time, however, international and national expectations were raised.

The Aftermath

In 1993, when the Biosphere Reserve was established, the Upper Gulf was facing a deep historic and socioeconomic crisis. Perhaps for this reason, the fishing communities of the region initially supported the reserve project. In retrospect, however, the fishers' support seems now to have been based on the expectation of a temporary resolution that would ban trawling and harmful gillnet sizes for three or five years and thus allow the recovery of the Upper Gulf. After a few years, as the restrictions imposed by the reserve on natural resource started to become clear, the initial enthusiasm waned.

Additionally, in 1993 there was little actual oversight or administration of protected natural areas in Mexico, and most of the reserves existed only on paper. There was no governmental field experience and very little budget for conservation and management of protected areas. The Environmental Attorney General (PROFEPA, the federal authority in charge of environmental enforcement) had just been created, and it did not have the capacity to enforce regulations in remote protected areas.

In short, it seems now that neither the local communities nor the federal government were prepared for the long-term commitment required by the establishment of this reserve. Yet many things have changed since 1993: the National Commission for Natural Protected Areas (CONANP) was created in 1999, and the reserve now boasts a management plan, core funding for its field operation, dedicated staff, and a director. The Mexican Fisheries authority was briefly moved to SEMARNAP between 1995 and 2000 but was then moved back to the Ministry of Agriculture (SAGARPA), where it now resides. As a result, coordination between the federal administrations of fisheries and the environment has become more difficult and complex.

At present, the biggest obstacles to achieving the objectives of the Upper Gulf reserve are poor intergovernmental coordination, conflicts between sectors, low institutional capacity of conservation organizations, and lack of political will to enforce the law. These factors facilitate increased illegal fishing in the reserve—and with it an increment in vaquita mortality, among many other species that reproduce in this unique and fragile area. The dice are still in the air, and the future of the Upper Gulf remains in question.

Learning from Conflict

The challenge has remained unchanged during the past fifty years: all of the conservation and protection efforts for the Upper Gulf have hinged on the need to design and implement a program that guarantees the preservation of the area's ecological value as well as the sustained use of its natural resources by local inhabitants. The sociopolitical efforts articulated by authorities and conservation organizations have been mainly limited to the three decrees already mentioned, and true achievements in conservation have remained elusive. Three important lessons can be derived from this experience.

1. *The objectives of a new protected area must be balanced with capacity and experience.* It is difficult to imagine a better juncture to establish a protected area in the Upper Gulf than 1993, but the decree revealed an immense lack of awareness regarding the long-term commitments needed to consolidate an operational reserve in an area as conflictive as the Upper Gulf. After eleven years, this is still the reserve's weakest point, and much work is needed on it.
2. *The raised expectations of different actors must be met.* Owing to seventy years of open-access fishing without any type of resource management, the Upper Gulf of California is one of the most difficult areas of Mexico in which to achieve conservation and sustainable management. Establishing the Upper Gulf Biosphere Reserve raised great expectations, but the high social pressure on the area's natural resources, the historical bi-national conflicts over Colorado River water, and the convergence of several authorities making decisions within the area have combined to create a sense of disillusionment in the local people that is now working against the aims of the reserve.
3. *Implementation of alternative solutions is important.* For more than fifty years the conservation and management of the Upper Gulf have been addressed by federal decrees, but these have proved to be insufficient. Alternative solutions that are based on active participation of the local stakeholders are of paramount importance in resolving the regional conflicts. Among these, the fishers' own proposals for a sustainable use of their resource should be taken more seriously by authorities and social organizations.

Despite the current impasse in the Upper Gulf Biosphere Reserve, it is essential to recognize that this was the first marine reserve established in Mexico and thus represented an initiative of immense significance. Despite the opposition of Mexican fisheries authorities, it opened the way for new marine protected areas in the Gulf of California, in the Mexican Pacific Ocean, and on the other coasts of Mexico. Specifically, it led to discussions on the possibility of protecting the waters surrounding each of the islands in the Gulf. The Upper Gulf debate established these discussions as an ongoing process and showed that conservation at sea and the establishment of marine protected areas, although difficult, can be achieved in Mexico.

Toward a Regional Conservation Agenda

In spite of some local successes, the depletion of the Gulf and its marine and coastal resources remains a predominant driving force and has serious economic consequences. The perceptions of national and foreign investigators of this depletion are already driving investment and economic opportunities away. It follows that the maintenance of fundamental ecological processes and ecosystem functions is critically needed to protect long-term economic investments as well as biodiversity in the region, which are necessary for lasting economic development and improvements in quality of life. Today, a growing number of leaders from different sectors are becoming more aware of the need to come together and jointly address these issues.

The region is governed by a complex mix of authority emanating from the federal government, the surrounding five states, and 40 coastal municipalities, and there is often little synergy among them. Poorly planned economies—often the result of poor coordination among authorities—are one of the greatest threats to ecosystems in the Gulf of California. As a consequence, the Gulf faces often strongly opposed priorities for regional development among the three levels of government, which are characterized by poor coordination among authorities and increasing dispersion of resources and actions. Thus, although political change (in the form of open elections) has swiftly occurred around the Gulf of California, the reformed institutions have not yet been able to tackle the urgent issues of resource degradation in a coordinated fashion.

Furthermore, there is a notorious absence of a well-developed, consolidated relationship between environmental scientists and decision makers. Indeed, the region is a large, interconnected set of biological communities, but governance is fragmented and most research done in the Gulf has been based on specific, simple aspects of conservation—small, disjunct parts of a complexly functioning regional ecosystem. Few of the regional research and conservation efforts have concentrated on a multidisciplinary approach, and little effort has been placed on large-scale, ecosystem-based research and management.

On the other hand, regional cooperation among nongovernmental agencies has yielded positive results in the past. We described previously how, through their rich network of cooperation, ALCOSTA (the regional alliance of NGOs) was capable of raising concerns about the Escalera Náutica development project, thereby transforming it into one that was much more open to environmental conservation issues. Other cooperative efforts have also yielded successful results for conservation. In December 1997, a group of scientists and conservationists teamed together in a joint project known as the Coalition for the Sustainability of the Gulf of California. After three years of working together, they produced a comprehensive map defining areas of biological importance in the region (CSGC 2004; Enríquez-Andrade et al. 2005; see color plate 7B), which became a milestone document in regional planning and a tool for assisting in the determination of regional priorities. Among other uses, the documents and the maps produced by the coalition became critical inputs to the governmental plans for use of the Gulf and its surrounding coasts.

Finally, cooperation between regional NGOs and research groups allowed the presentation of a regional agenda at the 2003 "Defying Ocean's End" meeting in Los Cabos (Carvajal et al. 2004; see color plate 8A). This cooperative agenda established seven specific objectives in approaching sustainability within the Gulf of California.

1. *Improve the management of regional marine and coastal protected areas.* Although impressive progress has been attained during the last decade by the Mexican government in the funding and management of its protected natural areas, many of them still subsist as "paper parks," with inadequate funding and little effective management while facing increasingly difficult conflicts among sectors. If the regional protected areas are to be effective in their conservation goals, they must improve in their capacities to address these complex issues and receive enhanced funding, equipment, and staffing.
2. *Enlarge the system of marine and coastal protected areas.* Although some marine protected areas have been created in the Gulf

(namely, the Upper Gulf, Islas Marías, and San Pedro Mártir Biosphere Reserves and the Cabo Pulmo, Islas Marietas, San Lorenzo, and Loreto Bay National Parks), these cover only 4 percent of the Gulf's marine area. Effective conservation in the region cannot be achieved without securing the protection of spawning aggregation areas, critically endangered species, and endangered ecosystems such as seamounts, coastal lagoons, coral reefs, estuaries, and marine mammal habitats (Sala et al. 2002). Therefore, a significant increase in marine protected areas must be obtained, probably reaching 15 percent of the Gulf's marine and coastal areas.

3. *Develop a comprehensive plan to manage and protect priority coastal wetlands.* The degradation of coastal wetlands is one of the Gulf's most serious threats. With little consideration given to the ecological purposes they serve, mangrove forests are being cut for the development of aquaculture (mostly shrimp farms) and tourism projects. Furthermore, coastal wetlands in general are threatened by consumptive water use upstream and by pollution of rivers and waterways. The ecological services provided by estuaries and lagoons are critical for the survival of the Gulf of California fisheries and for the health of the large marine ecosystem as a whole. A comprehensive plan to effectively protect coastal wetlands, stop mangrove deforestation, and maintain the ecological services of coastal lagoons and estuaries must be developed and its actions implemented with urgency.
4. *Reduce the ecological impact of shrimp trawling.* Many of the strongest issues of unsustainability in the Gulf stem from the destructive effect and economic inefficiency of the current shrimp bottom-trawling fleet. The only alternative to solve this growing problem is to reduce the fleet size by at least half through a legal buyout. If effective legal means are put in place to ensure that no new fishing permits will be issued in the future—and hence that the fleet will not grow again to unsustainable levels—then action of this sort would allow negotiation of effective enforcement of the existing no-take zones and introduction of exclusion-efficient fishing gear that will reduce by-catch by 40 percent.

5. *Implement a regional plan that regulates the use of land, coasts, and waters.* The main instrument in Mexican legislation for regulating the use of space within environmental guidelines is the "Ordenamiento Ecológico," or Ecological Planning of the Territory, which demands full and comprehensive hearings and negotiations with local governments, local businesses, and nongovernmental organizations. Because of its complexity, effective territorial planning has been difficult to achieve in the Gulf of California and is now one of the most urgent objectives. For this purpose, the participation of civil society and of local conservation alliances is critical.
6. *Re-orient regional tourism toward low-impact, environmentally sustainable resource use.* The Escalera Náutica has become one of the most debated projects in the region. On the one hand, most environmentalists agree that the primary sector of the Gulf economy has reached its limits and that it would be desirable to shift the economy away from unsustainable fisheries and water-intensive agriculture and toward the services sector (including tourism). On the other hand, experiences in Mexico with unsustainable tourism have made environmentalists wary of the dangers and impacts of development projects. The development of a culture of environmentally sustainable tourism is still one of the biggest challenges in the Gulf of California.
7. *Articulate and implement a common regional development vision.* The last point of the agenda, the development of a regional vision, is possibly the most crucial aspect of the Gulf's conservation agenda. The various sectors involved—including government, private business, communities, conservation organizations, and civil society at large—need to transcend their own agendas and move toward the development of a joint way of seeing the region. A new, proactive paradigm is needed that would allow different sectors and social groups to propose new and sustainable modes of development instead of defending unsustainable alternatives. Regional conservation will be successful if, in collaboration with local business and political leaders, a regional development vision based on the long-term protection of the Gulf and its resources can be pieced together collectively and agreed upon.

Aware of the growing problems that the Gulf is facing, a cluster of socially and environmentally concerned leaders who recognize that conservation is essential to creating long-lasting economic and social prosperity came together and created the Sustainable Northwest Initiative (NOS). NOS is working under the model and inspiration of the Chesapeake Bay program and the Great Barrier Reef, adapting the lessons learned in those places to advance an agenda for sustainability for the Gulf of California. As a first endeavor, NOS is promoting the development of a common societal vision that would allow a sustainable future and a shared journey toward a mode of development that also conserves the region's rich natural heritage.

Conservation and the Search for a Viable Future

It is hoped that the increasing pace of conservation efforts will be able to stall the environmental degradation that the Gulf of California has been suffering and diminish the threats to its long-term sustainability. There seems to be a growing awareness in the region, as never seen before, of the need to take urgent action to protect the environment and develop in a sustainable manner. Conservation groups, research institutions, federal and state governments, conscientious businesspersons, and eco-tourism operators have all been contributing to the growing appreciation of the environment and to the attendant conservation actions.

In the past, conservation in the Gulf has progressed through the support of researchers, nongovernmental organizations, local communities, and local, state, and federal governments. The involvement of local groups as allies in conservation has possibly been the single most important element in successful conservation efforts. Local commitment has been the driving force behind environmental protection and the key to the success of such conservation programs as the Bay of Loreto and the Biosphere Reserve at Bahía de Los Ángeles.

At this moment in time, environmental conservation needs to become part of a larger vision, developed jointly by all sectors, that can drive regional development for years to come with increasing consideration for growing social needs, the environment, and the Gulf's natural resources and their sustainability. The Gulf of California receives what little remains of the discharges of the Colorado River Basin, and the survival of the Upper Gulf is a challenge for both Mexico and the United States. Thus, its larger basin is part of a binational collection of linked ecosystems, where both Mexico and the United States share the responsibility of protecting their joint natural heritage. In order to achieve this, both countries must develop further efforts to promote true cooperative work. The region is one large continuum, with shared watersheds and estuaries, species, and natural resources. The protection of these unique environments is of great importance for the survival and well-being of all of us—now and for countless generations to come.

There are plenty of opportunities for creative solutions to the problems the Gulf is facing today. In the end, however, the solution lies in the hands of the local actors of all sectors and in their ability to come together. If we are to conserve the amazing beauty, remarkable biological productivity, and magnificent biological richness of this unique place on the planet, then we must find new ways of coordinating and cooperating among ourselves. We need to change the way we work, combining forces and using our collective knowledge, creativity, and abilities to achieve our common goals.

Bibliography

The following bibliography on the Gulf of California region incorporates all references cited in the chapters as well as additional key works. With a few exceptions, works dealing specifically with the terrestrial realm are not included.

Abbott, I. A. 1966. Elmer Yale Dawson (1918–1966) [obituary]. *Journal of Phycology* 2: 129–132.

Abbott, I. A., and G. J. Hollenberg. 1976. *Marine Algae of California.* Stanford University Press, Stanford, CA.

Abbott, P. L., A. D. Hanson, C. N. Thomson, D. L. Logue, K. D. Bradshaw, W. J. Pollard, and T. E. Seeliger. 1993. Geology of the Paleocene sepulture formation, Mesa de la Sepultura, Baja California [Geología de la formación sepultura del Paleoceno, en mesa de la Sepultura, Baja California]. *Ciencias Marinas* 19: 75–93.

Abitia-Cárdenas, L. A., J. Rodríguez-Romero, F. Galván-Magaña, J. De la Cruz-Agüero, and H. Chávez-Ramos. 1994. Systematic list of the ichthyofauna of La Paz Bay, Baja California Sur, Mexico. *Ciencias Marinas* 20: 159–181.

Aburto-Oropeza, O., and E. F. Balart. 2001. Community structure of reef fish in several habitats of a rocky reef in the Gulf of California. *P.S.Z.N.I.: Marine Ecology* 22(4): 283–305.

Aburto-Oropeza, O., B. Erisman, V. Valdez-Ornelas, and G. Danemann. 2008. Commercially important serranid fishes from the Gulf of California: Ecology, fisheries, and conservation. *Ciencia Marina y Conservación* 1: 1–44.

Aburto-Oropeza, O., E. Ezcurra, G. Danemann, V. Valdez, J. Murray, and E. Sala. 2008. Mangroves in the Gulf of California increase fishery yields. *Proceedings of the National Academy of Sciences* 105(30): 10456–10459.

Aburto-Oropeza, O., and C. A. Sánchez-Ortiz (eds.). 2000. *Recursos arrecifales del Golfo de California, estrategias de manejo para las especies marinas de ornato* [Reef resources of the Gulf of California, management strategies for the marine ornate species]. Universidad Autónoma de Baja California Sur, La Paz.

Aceves-Medina, G., S. P. A. Jiménez-Rosenberg, A. Hinojosa-Medina, R. Funes-Rodríguez, R. J. Saldierna, D. Lluch-Belda, P. E. Smith, and W. Watson. 2003a. Fish larvae from the Gulf of California. *Scientia Marinea* 67: 1–11.

Aceves-Medina, G., S. P. A. Jiménez-Rosenberg, A. Hinojosa-Medina, R. Funes-Rodríguez, R. J. Saldierna-Martínez, and P. E. Smith. 2003b. Fish larvae assemblages in the Gulf of California. *Journal of Fish Biology* 65: 832–847.

Addicott, W. O. 1970. Tertiary paleoclimatic trends in the San Joaquin Basin, California. U.S. Geolological Survey Professional Paper, 644-D, U.S. GPO, Washington, DC.

Adey, W., and D. McKibben. 1970. Studies on the maerl species *Phymatolithon calcareum*

(Pallas) nov. comb. and *Lithothamnion corallioides* Crouan in the Rio de Vigo. *Botanica Marina* 13:100–106.

Agardy, T. 1997. *Marine Protected Areas and Ocean Conservation.* Academic Press, San Diego, CA.

Agler, W. E. 1913. Green turtles in lower California. U.S. Bureau of Foreign and Domestic Commerce, Daily Consular and Trade Reports, no. 55, March 6, p. 1181.

Aguilar-Rosas, L. E., R. Aguilar-Rosas, L. E. Mateo-Cid, and A. C. Mendoza-Gonzáles. 2002. Marine algae from the Gulf of Santa Clara, Sonora, México. *Hydrobiologia* 477: 231–238.

Aguilar-Rosas, L. E., R. Aguilar-Rosas, A. C. Mendoza-Gonzáles, and L. E. Mateo-Cid. 2000a. Marine algae from the northeast coast of Baja California, México. *Botanica Marina* 43: 127–139.

Aguilar-Rosas, R. P., I. Pacheco-Ruíz, and L. E. Aquilar-Rosas. 1984. New records and some notes about the marine algal flora of the northwest coast of Baja California, México. *Ciencias Marinas* 10(2): 159–166.

Aguilar-Rosas, R. P., T. R. Van Devender, and R. S. Felger. 2000b. *Cactáceas de Sonora, México: Su diversidad, uso y conservación.* Arizona-Sonora Desert Museum, Tucson, AZ.

Aguirre, J., R. Riding, and J. C. Braga. 2000. Diversity of coralline red algae: Origination and extinction patterns from Early Cretaceous to the Pleistocene. *Paleobiology* 26(4): 651–667.

Alcalá, G. 2003. Políticas pesqueras en México (1946–2000). Contradicciones y aciertos en la planificación de la pesca nacional. El Colegio de México, Centro de Investigación Científica y de Educación Superior de Ensenada, El Colegio de Michoacán. México, DF, Ensenada, BC, Morelia, MI.

Allard, D. C. 1999. The origins and early history of the steamer *Albatross,* 1880–1887. *Marine Fisheries Review* 61: 1–21.

Allen, G. R., and D. R. Robertson 1994. *Fishes of the Tropical Eastern Pacific.* Crawford House Press, Bathurst, NSW, Australia, and University of Hawaii Press, Honolulu.

———. 1998. *Peces del Pacífico Oriental Tropical* (traducción de M. I. López). México, D.F.: CONABIO, Agrupación Sierra Madre, y CEMEX.

Altringham, J. 1996. *Bats: Biology and Behaviour.* Oxford University Press, Oxford.

Alvarado, J., and A. Figueroa. 1992. Recapturas post-anidatorias de hembras de tortuga marina negra (*Chelonia agassizi*) marcadas en Michoacán, México. *Biotropica* 24: 560–566.

Alvarado-Díaz, J., C. Delgado-Trejo, and I. Suazo-Ortuño. 2001. Evaluation of black turtle project in Michoacán, Mexico. *Marine Turtle Newsletter* 92: 4–7.

Alvarez-Borrego, S. 1983. Gulf of California. In B. H. Ketchum (ed.), *Ecosystems of the World,* 26: *Estuaries and Enclosed Seas,* pp. 427–449. Elsevier, New York.

———. 1990. Evidence of an ENSO event in the data of the 1889 Albatross cruise to the Gulf of California. *Ciencias Marinas* 16(2): 131–135.

———. 2002. Physical oceanography. In T. Case, M. Cody, and E. Ezcurra (eds.), *A New Island Biogeography of the Sea of Cortés,* pp. 41–60. Oxford University Press, New York.

———. 2003. Physical and biological linkages between the upper and lower Colorado delta. In D. Rapport, W. Lasley, D. Rolston, N. Nielsen, C. Qualset, and A. Damania (eds.), *Managing for Healthy Ecosystems,* pp. 1081–1083. Lewis Publishers, Washington, DC.

Alvarez-Borrego, S., B. P. Flores-Báez, and L. A. Galindo-Bect. 1975a. Hidrología del Alto Golfo de California II. Condiciones durante invierno, primavera y verano. *Ciencias Marinas* 2(1): 21–36.

Alvarez-Borrego, S., L. A. Galindo-Bect, and B. P. Flores-Baez. 1973. Hidrología. In Estudio químico sobre la contaminación por insecticidas en la desembocadura del Río Colorado, Tomo I, Reporte a la Dirección de Acuacultura de la Secretaría de Recursos Hidráulicos, pp. 6–177. [Available from Universidad Autónoma de Baja California, Ensenada, México.]

Alvarez-Borrego, S., D. Guthrie, C. H. Culberson, and P. K. Park. 1975b. Test of Redfield's model for the oxygen-nutrient relationships using regression analysis. *Limnology and Oceanography* 20: 795–805.

Alvarez-Borrego, S., and J. R. Lara-Lara. 1991. The physical environment and primary productivity of the Gulf of California. In B. R. T. Simoneit and J. P. Drophin (eds.), *The Gulf and Peninsular Province of the Californias,* pp. 555–567. Memoir 47, American Association of Petroleum Geologists, Tulsa, OK.

Alvarez-Borrego, S., J. A. Rivera, G. Gaxiola-Castro, M. J. Acosta-Ruíz, and R. A. Schwartzlose. 1978. Nutrientes en el Golfo de California. *Ciencias Marinas* 5: 21–36.

Alvarez-Borrego, S., and R. A. Schwartzlose. 1979. Water masses of the Gulf of California. *Ciencias Marinas* 6: 43–63.

Alvarez-Sánchez, L. G., A. Badan-Dangon, and J. M. Robles. 1984. Lagrangian observations of near-surface currents in Canal de Ballenas. *CalCOFI Reports* 25: 35–42.

American Rivers. 1998. Most Endangered Rivers Report. http://www.amrivers.org.

Ames, P. L. 1966. DDT residues in the eggs of the Osprey in the northeastern USA and their relation to nest success. *Journal of Applied Ecology* 3(suppl.): 87–97.

Ames, P. L., and G. S. Mersereau. 1964. Some factors in the decline of the Osprey in Connecticut. *Auk* 81: 173–185.

Amezcua-Linares, F. 1996. *Peces demersales de la plataforma continental del Pacífico central de México.* Instituto de Ciencias del Mar y Limnología, Universidad Nacional Autónoma de México, México, D.F.

Anderson, D. G., and J. C. Gillam. 2000. Paleoindian colonization of the Americas: Implications from an examination of physiography, demography, and artifact distribution. *American Antiquity* 65: 43–66.

Anderson, D. L. 1971. The San Andreas Fault. *Scientific American* 225: 58–68.

Anderson, D. W. 1983. The seabirds. In T. J. Case and M. L. Cody (eds.), *Island Biogeography in the Sea of Cortez,* pp. 246–264. University of California Press, Berkeley.

———. 2003. Overview: The Colorado River Delta Ecosystem: Ecological issues at the U.S.–Mexico border. In D. J. Rapport, W. L. Lasley, D. E. Rolston, N. O. Nielsen, C. Q. Qualset, and A. B. Damania (eds.), *Managing for Healthy Ecosystems,* pp. 1069–1070. Lewis Publishers, Boca Raton, FL.

Anderson, D. W., S. Alvarez-Borrego, H. R. Carter, F. Gress, C. Harrison, J. O. Keith, B. Keitt, P. R. Kelly, E. Palacios, and E. Velarde. 2005. Pacific Seabird Group seabird conservation policy statement for the Gulf of California and waters off western Baja California, Mexico. *Pacific Seabirds* 32(2): 46–49.

Anderson, D. W., F. Gress, K. F. Mais, and P. R. Kelly. 1980. Brown pelicans as anchovy stock indicators and their relationships to commercial fishing. *CalCOFI Reports* 21: 54–61.

Anderson, D. W., C. J. Henny, C. Godinez-Reyes, F. Gress, E. L. Palacios, K. Santos del Prado, and J. Bredy. 2007. Size of the California brown pelican metapopulation during a non–El Niño year. Open-File Report 2007-1299, U.S. Geological Survey, Reston, VA.

Anderson, D. W., and J. O. Keith. 1980. The human influence on seabird nesting success: Conservation implications. *Biological Conservation* 18: 65–80.

Anderson, D. W., and E. Palacios. 2008. Seabirds and bird islands in the Bahía de los Ángeles region: Status and conservation. In *Fauna and Flora of the Bahía de los Ángeles National Park*. ProNatura Noroeste, Mexico.

Anderson, D. W., E. Palacios, E. Mellink, and C. Valdés-Casillas. 2003. Migratory bird conservation and ecological health in the Colorado River delta region. In D. Rapport, W. Lasley, D. Rolston, N. Nielsen, C. Qualset, and A. Damania (eds.), *Managing for Healthy Ecosystems,* pp. 1091–1109. Lewis Publishers, Washington, DC.

APLIC (Avian Power Line Interaction Committee). 2006. Suggested practices for avian protection on power lines: The state of the art in 2006. Edison Electric Institute, APLIC, and the California Energy Commission, Washington, DC, and Sacramento, CA.

Applegate, S. P., F. Soltelo-Macías, and L. Espinosa-Arrubarrena. 1993. An overview of the Mexican shark fisheries, with suggestions for shark conservation in Mexico. In S. Branstetter (ed.), *Conservation Biology of Elasmobranchs,* pp. 31–37. National Oceanic and Atmospheric Administration, National Marine Fisheries Service Technical Report 115.

Aragón-Noriega, E. A., and L. E. Calderón-Aguilera. 2000. Does damming of the Colorado River affect the nursery area of blue shrimp *Litopenaeus stylirostris* (Decapoda; Penaeidae) in the upper Gulf of California? *Revista de Biología Tropical* 48: 867–871.

Argote, M. L., A. Amador, M. F. Lavín, and J. Hunter. 1995. Tidal dissipation and stratification in the Gulf of California. *Journal of Geophysical Research* 100: 103–118.

Arita, H., and J. Ortega. 1998. The middle American bat fauna: Conservation in the Neotropical-Nearctic border. In T. Kunz and P. Racey (eds.), *Bat Biology and Conservation,* pp. 295–308. Smithsonian Institution Press, Washington, DC.

Arnaud, S. 1970. The Sefton Foundation *Orca* expedition to the Gulf of California, March–April, 1953. General account. *Occasional Papers,* no. 86, California Academy of Sciences, San Francisco.

Arnaud, S., M. Monteforte, N. Galtier, F. Bonhomme, and F. Blanc. 2000. Population structure and genetic variability of pearl oyster *Pinctada mazatlanica* along Pacific coast from Mexico to Panama. *Conservation Genetics* 1: 299–307.

Arnold, B. A. 1957. Late Pleistocene and recent changes in land forms, climate, and archaeology in central Baja California. *University of California Publications in Geography* 10: 201–317.

Arreguín-Sánchez, F., E. Arcos, and E. A. Chávez. 2002. Flows of biomass and structure in an exploited benthic ecosystem in the Gulf of California, Mexico. *Ecological Modeling* 156: 167–183.

Arreola-Robles, J. L., and J. F. Elorduy-Garay. 2002. Reef fish diversity in the region of La Paz, Baja California Sur, Mexico. *Bulletin of Marine Science* 70: 1–18.

Arvizu, M. J. 1987. Fisheries activities in the Gulf of California, Mexico. *CalCOFI Report* 28: 26–32.

Aschmann, H. 1959. *The Central Desert of Baja California: Demography and Ecology.* Iberoamericana, no. 42. University of California Press, Berkeley, CA.

——— (ed.). 1966. The natural and human history of Baja California—From manuscripts by Jesuit missionaries in the decade 1752–1762. Dawson's Book Shop, Los Angeles, CA.

Ashby, J. R. 1989. A resume of the Miocene stratigraphic history of the Rosarito Beach Basin, northwestern Baja California, Mexico. In P. L. Abbott (ed.), *Geologic Studies in Baja California,* pp. 37–46. Pacific Section, SEPM, Book 63, Society for Sedimentary Geology, Los Angeles, CA.

Ashby, J. R., and J. A. Minch. 1988. Miocene tectonosedimentary history of the La Mission Basin, northwestern Baja California, Mexico: Implications for the early tectonic development of the Southern California Continental Borderland. *Bulletin of the American Association of Petroleum Geologists* 72: 373.

Athanasiadis, A., P. A. Lebednik, and W. H. Adey. 2004. The genus *Mesophyllum* (Melobesioideae, Corallinales, Rhodophyta) on the northern Pacific coast of North America. *Phycologia* 43(2): 126–165.

ATRT (Advanced Tuna Ranching Technologies, Inc.). 2005. Special Update. Tuna ranching intelligence unit report, November 25, 2005, Madrid.

Aubert, H., and D. V. Lightner. 2000. Identification of genetic populations of the Pacific blue shrimp *Penaeus stylirostris* of the Gulf of California, Mexico. *Marine Biology* 137: 875–885.

Aurioles, D., B. J. Le Boeuf, and L. T. Findley. 1993. Registros de pinnípedos poco comunes para el Golfo de California. *Revista de Investigación Científica, Universidad Autónoma de Baja California Sur (SOMEMMA)* 1: 13–19.

Aurioles-Gamboa, D. 1993. Biodiversidad y estado actual de los mamíferos marinos en México. *Revista Sociedad Mexicana Historia Natural* 44(vol. esp.): 397–412.

Aurioles-Gamboa, D., and G. A. Zavala. 1999. Algunos factores ecológicos que determinan la distribución y abundancia del lobo marino *Zalophus californianus,* en el Golfo de California. *Ciencias Marinas* 20(4): 535–553.

Averett, W. E. 1920. Lower California green turtle fishery. *Pacific Fishermen* 18: 24–25.

Avila-Serrano, G. E., K. W. Flessa, M. A. Téllez-Duarte, and C. E. Cintra-Buenrostro. 2006. Distribution of the intertidal macrofauna of the Colorado River delta, northern Gulf of California, Mexico. *Ciencias Marinas* 32(4): 649–661.

Badan-Dangon, A., C. E. Dorman, M. A. Merrifield, and C. D. Winant. 1991a. The lower atmosphere over the Gulf of California. *Journal of Geophysical Research* 96: 16877–16896.

Badan-Dangon, A., M. C. Hendershott, and M. F. Lavín. 1991b. Underway Doppler current profiles in the Gulf of California. *EOS, Transactions, American Geophysical Union* 72: 209–214.

Badan-Dangon, A., C. J. Koblinsky, and T. Baumgartner. 1985. Spring and summer in the Gulf of California: Observations of surface thermal patterns. *Oceanologica Acta* 8: 13–22.

Bahre, C. J., and L. Bourillón. 2002. Human impact in the Midriff Islands. In T. Case, M. Cody, and E. Ezcurra (eds.), *A New Island Biogeography of the Sea of Cortés*, pp. 383–406. Oxford University Press, New York.

Bahre, C. J., L. Bourillón, and J. Torre. 2000. The Seri and commercial totoaba fishing (1930–1965). *Journal of the Southwest* 42: 559–575.

Baird, S. F., T. M. Brewer, and R. Ridgway. 1874. *A History of North American Birds: Land Birds*. Little, Brown, Boston.

Baker, C. S., L. Medrano-González, J. Calambokidis, A. Perry, F. Pichler, H. Rosenbaum, J. M. Straley, J. Urbán R., M. Yamaguchi, and O. von Ziegesar. 1998. Population structure of nuclear and mitochondrial DNA variation among humpback whales in the North Pacific. *Molecular Ecology* 7(6): 695–708.

Balart, E. F., J. L. Castro-Aguirre, D. Aurioles-Gamboa, F. García-Rodríguez, and C. Villavicencio-Garayzar. 1995. Adiciones a la ictiofauna de Bahía de La Paz, Baja California Sur, México. *Hidrobiológica* 5(1–2): 79–85.

Balart, E. F., J. L. Castro-Aguirre, and F. De Lachica-Bonilla. 1997. Análisis comparativo de las comunidades ícticas de fondos blandos y someros de la Bahía de La Paz. In J. Urbán-Ramírez and M. Ramírez-Rodríguez (eds.), *La Bahía de La Paz, investigación y conservación*, pp. 163–176. Universidad Autónoma de Baja California Sur, La Paz.

Balart, E. F., J. L. Castro-Aguirre, and R. Torres-Orozco. 1993. Ictiofauna de las bahías de Ohuira, Topolobampo, y Santa Mária, Sinaloa, México. *Investigaciones Marinas CICIMAR (La Paz)* [for 1992] 7(2): 91–103.

Balcomb, K. C., III. 1989. Baird's beaked whale *Berardius bairdii* Stejneger, 1883; Arnoux's beaked whale *Berardius arnuxii* Durvenoy, 1851. In S. H. Ridway and R. Harrison (eds.), *Handbook of Marine Mammals*, vol. 4, pp. 261–288. Academic Press, London.

Ballantine, D. L., A. Bowden-Kerby, and N. E. Aponte. 2000. *Cruoriella* rhodoliths from shallow-water back reef environments in La Parguera, Puerto Rico (Caribbean Sea). *Coral Reefs* 19: 75–81.

Ballesteros, E. 1988. Composición y estructura de los fondos de *maerl* de Tossa de Mar (Gerona, Espana). *Collectana Botanica* (Barcelona) 17: 161–182.

Bancroft, G. 1927. Notes on the breeding coastal and insular birds of central Lower California. *Condor* 29: 188–195.

Banks, R. C. 1962a. A history of exploration for vertebrates on Cerralvo Island, Baja California. *Proceedings of the California Academy of Sciences* 30(6): 117–125.

———. 1962b. Birds of the Belvedere expedition to the Gulf of California. *Transactions, San Diego Society of Natural History* 13: 51–60.

Barham, E. 1970. Retracing of the Ricketts-Steinbeck voyage, aboard the RV Saluda. Naval Ocean Systems Center (NOSC) Technical Publication 27, San Diego, CA. [Available from Public Affairs officer, NOSC, Rosecrans Bldg., San Diego, CA 92152.]

Barlow, G. W. 1961. Intra- and interspecific differences in rate of oxygen consumption in gobiid fishes of the genus *Gillichthys*. *Biological Bulletin* 121 (2): 209–229.

Barlow, J., K. Forney, A. von Saunder, and J. Urbán R. 1997. A report of cetacean acoustic detection and dive interval studies (CADDIS) conducted in the southern Gulf of California, 1995. Technical Memorandum NOAA-NMFS-SWFSC-250.

Barlow, J., M. C. Furguson, W. F. Perrin, L. Balance, T. Gerrodette, G. Joyce, C. D. Macleod, K. Mullin, D. L. Palka, and G. Waring. 2006. Abundance and densities of beaked and bottlenose whales (family Ziphiidae). *Journal of Cetacean Research and Management* 7 (3): 263–270.

Barlow, J., T. Gerrodette, and G. Silber. 1997. First estimates of vaquita abundance. *Marine Mammal Science* 13: 44–58.

Barnes, L. G., and E. D. Mitchell. 1984. *Kentriodon obscurus* (Kellogg 1931), a fossil dolphin (Mammalia: Kentridontidae) from the Miocene Sharktooth Hill Bonebed in California. *Los Angeles County Museum Contributions in Science* 353: 1–23.

Barrera-Guevara, J. C. 1990. The conservation of *Totoaba macdonaldi* (Gilbert) (Pisces: Sciaenidae) in the Gulf of California, Mexico. *Journal of Fish Biology* 37: 201–202.

Bartholomew, G. A., and C. L. Hubbs. 1952. Winter population of pinnipeds about Guadalupe, San Benitos, and Cedros Islands, Baja California. *Journal of Mammalogy* 33: 160–171.

———. 1960. Population growth and seasonal movements of the northern elephant seal, *Mirounga angustirostris*. *Mammalia* 24: 313–324.

Basurto, X. 2007. Commercial diving and the callo de hacha fishery in Seri Territory. *Journal of the Southwest* 48(2): 189–209.

Baumgartner, T. R., V. Ferreira-Bartrina, H. Schrader, and A. Soutar. 1985. A 20-year varve of siliceous phytoplankton variability in the central Gulf of California. *Marine Geology* 64: 113–129.

Beal, C. H. 1948. *Reconnaissance of the Geology and Oil Possibilities of Baja California, Mexico.* Memoir 31, Geological Society of America, Boulder, CO.

Behrens, D. W., and A. Hermosillo. 2005. *Eastern Pacific Nudibranchs. A Guide to the Opisthobranchs from Alaska to Central America.* Sea Challengers, Monterey, CA.

Beier, E. 1997. A numerical investigation of the annual variability in the Gulf of California. *Journal of Physical Oceanography* 27: 615–632.

Bellan-Santini, D., and S. Ruffo. 2003. Biogeography of benthic marine amphipods in Mediterranean Sea. *Biogeographia* 24: 273–292.

Bello, G. 2003. The biogeography of Mediterranean cephalopods. *Biogeographia* 24: 210–226.

Berdegué, A. J. 1956a. La foca fina, el elefante marino y la ballena gris en Baja California y el problema de su conservación. Instituto Mexicano de Recursos Renovables, México, D.F., México.

———. 1956b. Peces de importancia comercial en la Costa Noroccidental de México. Dirección General de Pesca e Industrias Conexas, Secretaría de Marina, México, D.F.

Bergman, C. 2002. *Red Delta. Fighting for Life at the End of the Colorado River.* Defenders of Wildlife and Fulcrum Press, Golden, CO.

Bermingham, E., S. S. McCafferty, and A. P. Martin. 1997. Fish biogeography and molecular clocks: Perspectives from the Panamanian isthmus. In T. D. Kocher and C. A. Stepien (eds.), *Molecular Systematics of Fishes*, pp. 113–128. Academic Press, San Diego, CA.

Bernardi, G., L. Findley, and A. Rocha-Olivares. 2003. Vicariance and dispersal across Baja California in disjunct marine fish populations. *Evolution* 57(7): 1599–1609.

Bernardi, G., and J. Lape. 2005. Tempo and mode of speciation in the Baja California disjunct fish species *Anisotremus davidsonii*. *Molecular Ecology* 14: 4085–4096.

Berón-Vera, F. J., and P. Ripa. 2002. Seasonal salinity balance in the Gulf of California. *Journal of Geophysical Research* 107: 1–15.

Berry, R. W., and J. Ledesma-Vázquez. 1998. Clay and zeolite mineralogy of Miocene and Pliocene (proposed) volcaniclastic deposits, Mulegé, Baja California Sur, Mexico. Paper presented at the 11th International Clay Conference, Ontario, Canada.

Bérubé, M., J. Urbán R., A. Dizon, R. L. Brownell, and P. J. Palsbøll. 2002. Genetic identification of a small and highly isolated population of fin whales (*Balaenopetra physalus*) in the Sea of Cortez, Mexico. *Conservation Genetics* 3: 183–190.

Best, P. B. 2007. *Whales and Dolphins of the Southern African Subregion.* Cambridge University Press, Cape Town.

Beveridge, C., G. Kocurek, R. C. Ewing, N. Lancaster, P. Morthekai, A. K. Singhvi, and S. A. Mahan. 2006. Development of spatially diverse and complex dune-field patterns: Gran Desierto Dune Field, Sonora, Mexico. *Sedimentology* 53: 1391–1409.

Bird, C. J., and J. L. Mc Lachlan. 1992. *Seaweed Flora of the Maritimes. 1. Rhodophyta–Red Algae.* Biopress, Bristol, U.K.

Bird, K., and W. J. Nichols. 2002. Community-based research and its application to sea turtle conservation in Bahía Magdalena, Baja California Sur, Mexico. In A. Mosier, A. Foley, and B. Brost (compilers), *Proceedings of the Twentieth Annual Symposium on Sea Turtle Biology and Conservation,* pp. 339–340. NOAA Technical Memorandum NMFS-SEFSC-477.

Birkett, D., C. Maggs, and M. Dring. 1998. *Maerl (volume V). An Overview of Dynamic and Sensitivity Characteristics for Conservation Management of Marine SACs.* Scottish Association for Marine Science (UK Marine SACs Project).

Bjorndal, K. A. 1997. Foraging ecology and nutrition of sea turtles. In P. L. Lutz and J. A. Musick (eds.), *The Biology of Sea Turtles,* pp. 199–232. CRC Press, Boca Raton, FL.

Bjorndal, K. A., and J. B. C. Jackson. 2003. Roles of sea turtles in marine ecosystems: Reconstructing the past. In P. L. Lutz, J. A. Musick, and J. A. Wyneken (eds.), *The Biology of Sea Turtles,* vol. II, pp. 259–273. CRC Marine Biology Series, CRC Press, Boca Raton, FL.

Blake, E. R. 1953. *Birds of Mexico: A Guide for Field Identification.* University of Chicago Press, Chicago.

Blunden, G., W. Farnham, N. Jephson, C. Barwell, R. Fenn, and B. Plunkett. 1981. The composition of maerl beds of economic interest in Northern Brittany, Cornwall and Ireland. *International Seaweed Symposium* 10: 651–656.

Blunden, G., W. Farnham, N. Jephson, R. Fenn, and B. Plunkett. 1977. The composition of maerl from the Glenan Islands of Southern Brittany. *Botanica Marina* 20: 121–125.

Boehm, M. C. 1984. An overview of the lithostratigraphy, biostratigraphy, and paleoenvironments of the late Neogene San Felipe marine sequence, Baja California, Mexico. In V. A. Frizzell (ed.), *Geology of the Baja California,* pp. 219–236. Pacific Section, SEPM, Book 39, Society for Sedimentary Geology, Los Angeles, CA.

Bogan, M. 1999. Family Vespertilionidae. In S. Alvarez-Casteñada and J. Patton (eds.), *Mamíferos del noroeste de México,* pp. 139–181. Centro de Investigaciones Biológicas del Noroeste, S.C., La Paz, Baja California Sur, México.

Bohannon, R. G., and T. Parsons. 1995. Tectonic implications of post–30 Ma Pacific and North American relative plate motions. *Geological Society of America, Bulletin* 107: 937–959.

Bold, H. C., and M. J. Wynne. 1985. *Introduction to the Algae: Structure and Reproduction.* Prentice-Hall, Englewood Cliffs, NJ.

Bolton, H. E. 1936. *The Rim of Christendom.* Macmillan, New York.

———. 1948. *Kino's Historical Memoir of Pimería Alta.* University of California Press, Berkeley, CA.

Bond, J. 1947. *Field Guide to Birds of the West Indies.* Macmillan, New York.

Bonnichsen, R., and D. Gentry Steele (eds.). 1994. *Method and Theory for Investigating the Peopling of the Americas.* Center for the Study of the First Americans, Oregon State University, Corvallis, OR.

Bordehore, C., A. A. Ramos-Esplá, and R. Riosmena-Rodríguez. 2003. Comparative study of two maerl beds with different otter trawling history, southeast Iberian Peninsula. *Aquatic Conservation Marine Freshwater Ecosystems* 13: S43–S54.

Bordehore, C., R. Riosmena-Rodríguez, and A. A. Ramos-Esplá. 2002. Maerl-forming species in Alicante province (SE Spain): Taxonomic analysis. *Proceedings of the First Mediterranean Symposium on Marine Vegetation* UNEP MAP RAC/SPA Mednature 1: 101–104.

Bosence, D. W. J. 1976. Ecological studies on two unattached coralline algae from western Ireland. *Palaeontology* 19: 365–395.

———. 1979. Live and dead faunas from coralline algal gravels, Co. Galway. *Palaeontology* 22: 449–478.

———. 1983. The occurrence and ecology of recent rhodoliths—A review. In T. M. Peryt (ed.), *Coated Grains,* pp. 225–242. Springer-Verlag, Berlin.

Boscence, D., and J. Wilson. 2003. Maerl growth, carbonate production rates and accumulation rates in the northeast Atlantic. *Aquatic Conservation Marine Freshwater Ecosystems* 13: S21–S31.

Bossellini, A., and R. N. Ginsburg. 1971. Form and internal structure of recent algal nodules (rhodolites) from Bermuda. *Journal of Geology* 79: 669–682.

Boswall, J., and M. Barrett. 1978. Notes on the breeding birds of Isla Rasa, Baja California. *Western Birds* 9: 93–108.

Bouchard, S. S., and K. A. Bjorndal. 2000. Sea turtles as biological transporters of nutrients and energy from marine to terrestrial ecosystems. *Ecology* 81: 2305–2313.

Bourillón, L., A. Cantú, F. Eccardi, E. Lira, J. Ramírez, E. Velarde, and A. Zavala. 1988. *Islas del Golfo de California.* Secretaría de Gobernación–Universidad Nacional Autónoma de México, México, D.F.

Bowen, B. W., F. A. Abreu Grobois, G. H. Balazs, N. Kamezaki, C. J. Limpus, and R. J. Ferl. 1995. Trans-Pacific migrations of the loggerhead sea turtle demonstrated with mitochondrial DNA markers. *Proceedings of the National Academy of Sciences* 92: 3731–3734.

Bowen, B. W., W. S. Nelson, and J. C. Avise. 1993. A molecular phylogeny for marine turtles: Trait mapping, rate assessment, and conservation relevance. *Proceedings of the National Academy of Sciences* 90: 5574–5577.

Bowen, T. 2000. *Unknown Island. Seri Indians, Europeans, and San Esteban Island in the Gulf of California.* University of New Mexico Press, Albuquerque, NM.

———. 2004. Archaeology, biology and conservation on islands in the Gulf of California. *Environmental Conservation* 31: 199–206.

———. 2009. *The Record of Native People on Gulf of California Islands.* Arizona State Museum Archaeological Series, no. 201, Tucson, AZ.

Bracken, M. E. S. 2004. Invertebrate-mediated nutrient loading increases growth of an intertidal macroalga. *Journal of Phycology* 40: 1032–1041.

Braniff, B., and R. S. Felger (eds.). 1976. *Sonora, antropología del desierto.* Colección Cientifica Diversa, 27. Instituto Nacional de Antropología e Historia, Mexico City. [Reprinted 1994, Centro INAH Sonora, Hermosillo, Mexico.]

Bray, N. A. 1988. Thermohaline circulation in the Gulf of California. *Journal of Geophysical Research* 93: 4993–5020.

Bray, N. A., and J. M. Robles. 1991. Physical oceanography of the Gulf of California. In J. P. Dauphin and B. R. T. Simoneit (eds.), *The Gulf and Peninsular Province of the Californias,* pp. 511–553. Memoir 47, American Association of Petroleum Geologists, Tulsa, OK.

Breceda, A., A. Castellanos, L. Arriaga, and A. Ortega. 1995. Nature conservation in Baja California Sur Mexico. *Protected Natural Areas Journal* 15(3): 267–273.

Breese, D., and B. R. Tershy. 1993. Relative abundance of cetaceans in the Canal de Ballenas, Gulf of California. *Marine Mammal Science* 9: 319–324.

Brewer, G. D. 1973. Midwater fishes from the Gulf of California and the adjacent eastern tropical Pacific. *Los Angeles County Museum Contributions in Science* 242: 1–47.

Briand, X. 1991. Seaweed harvesting in Europe. In M. Guiry and G. Blunden (eds.), *Seaweed Resources in Europe: Uses and Potential,* pp. 293–308. Wiley, London.

Briggs, J. C. 1951. A review of the clingfishes (Gobiesocidae) of the eastern Pacific, with descriptions of new species. *Proceedings of the California Zoological Club* 1(11): 57–108.

———. 1955. A monograph of the clingfishes (Order Xenopterygii). *Stanford Ichthyological Bulletin* 6: 1–224.

———. 1960. Fishes of worldwide (circumtropical) distribution. *Copeia* 1960(3): 171–180.

———. 1961. The East Pacific barrier and the distribution of marine shore fishes. *Evolution* 15: 545–554.

———. 1974. *Marine Zoogeography.* McGraw-Hill, New York.

Brinton, E., A. Fleminger, and D. C. Siegel. 1986. The temperate planktonic biotas of the Gulf of California. *CalCOFI Reports* 27: 228–266.

Brittan, M. R. 1997. The Stanford school of ichthyology: Eighty years (1891–1970) from Jordan (1851–1933) to Myers (1905–1985). In T. W. Pietsch and W. D. Anderson Jr. (eds.), *Collection Building in Ichthyology and Herpetology. American Society of Ichthyologists and Herpetologists,* Special Publication 3, pp. 233–263. Lawrence, KS.

Brogan, M. W. 1994. Distribution and retention of larval fishes near reefs in the Gulf of California. *Marine Ecology Progress Series* 115: 1–13.

———. 1996. Larvae of the eastern Pacific snapper *Hoplopagrus guntheri* (Teleostei: Lutjanidae). *Bulletin of Marine Science* 58(2): 329–343.

Brusca, R. C. 1980. *Common Intertidal Invertebrates of the Gulf of California,* 2nd ed. University of Arizona Press, Tucson, AZ. [1st ed. printed 1975.]

———. 1981. A monograph on the Isopoda Cymothoidae (Crustacea) of the Eastern Pacific. *Zoological Journal Linnean Society (London)* 73(2): 117–199.

———. 1983. Two new idoteid isopods from Baja California and the Gulf of California

(Mexico) and an analysis of the evolutionary history of the genus *Colidotea* (Crustacea: Isopoda: Idoteidae). *Transactions San Diego Society of Natural History* 20(4): 69–79.

———. 1984. Phylogeny, evolution and biogeography of the marine isopod subfamily Idoteinae (Crustacea: Isopoda: Idoteidae). *Transactions San Diego Society of Natural History* 20(7): 99–134.

———. 1991. The genus *Rocinela* (Crustacea: Isopoda: Aegidae) in the tropical eastern Pacific. *Zoological Journal of the Linnean Society (London)* 106: 231–275.

———. 1993. The Arizona/Sea of Cortez years of J. Laurens Barnard. *Journal Natural History* 27: 727–730.

———. 2000. Unraveling the history of arthropod biodiversification. *Annals Missouri Botanical Garden* 87: 13–25.

———. 2002. On the Vermilion Sea. *Sonorensis,* Winter 2002, pp. 2–9.

———. 2004a. The Gulf of California—An overview. In R. C. Brusca, E. Kimrey, and W. Moore, *Seashore Guide to the Northern Gulf of California,* pp. 1–8. Arizona-Sonora Desert Museum, Tucson, AZ.

———. 2004b. A history of discovery in the northern Gulf of California. In R. C. Brusca, E. Kimrey, and W. Moore, *Seashore Guide to the Northern Gulf of California,* pp. 9–24. Arizona-Sonora Desert Museum, Tucson, AZ.

———. 2005. Where the desert meets the sea. In C. Conte (ed.), "Conservation without Borders," *Sonorensis,* Winter 2005, pp. 20–24.

———. 2007a. Father Eusebio Kino, the Baja California peninsula, and the Sea of Cortez. *Sonorensis,* Winter 2007, pp. 6–13.

———. 2007b. Harvesting the Sea of Cortez: From Kino's era to modern times. *Sonorensis,* Winter 2007, pp. 42–47.

———. 2007c. Invertebrate biodiversity in the northern Gulf of California. In R. S. Felger and W. Broyles (eds.), *Dry Borders. Great Natural Reserves of the Sonoran Desert,* pp. 418–504. University of Utah Press, Salt Lake City, UT.

———. 2009. Seafood and the Sea of Cortez: An ecological bouillabaisse. In C. Conte (ed.), *Sonorensis*, pp. 20–29. Arizona-Sonora Desert Museum.

Brusca, R. C., and G. J. Brusca. 2002. *Invertebrates,* 2nd ed. Sinauer Associates, Sunderland, MA. [Spanish edition published 2005 as *Invertebrados,* McGraw-Hill/Interamericana de Espana, S.A.U., Madrid.]

Brusca, R. C., and G. C. Bryner. 2004. A case study of two Mexican biosphere reserves: The Upper Gulf of California/Colorado River Delta and Pinacate/Gran Desierto de Altar Biosphere Reserves. In N. E. Harrison and G. C. Bryner (eds.), *Science and Politics in the International Environment,* pp. 21–52. Rowman & Littlefield, New York.

Brusca, R. C., and L. T. Findley. 2005. The Sea of Cortez [El Mar de Cortés]. In M. E. Hendrickx, R. C. Brusca, and L. T. Findley (eds.), *A Distributional Checklist of the Macrofauna of the Gulf of California, Mexico, Part I. Invertebrates* [Listado y distribución de la macrofauna del Golfo de California, México, Parte I. Invertebrados]. Arizona-Sonora Desert Museum, Tucson, AZ, and Conservation International, Washington, DC.

Brusca, R. C., L. T. Findley, P. A. Hastings, M. E. Hendrickx, J. Torre Cosio, and A. M. van der Heiden. 2005. Macrofaunal biodiversity in the Gulf of California. In J.-L. E. Cartron, G. Ceballos, and R. Felger (eds.), *Biodiversity, Ecosystems, and Conservation in Northern Mexico,* pp. 179–203. Oxford University Press, New York.

Brusca, R. C., and M. R. Gilligan. 1983. Tongue replacement in a marine fish (*Lutjanus guttatus*) by a parasitic isopod (Crustacea: Isopoda). *Copeia* 1980(3): 813–816.

Brusca, R. C., and M. E. Hendrickx (eds.). 1992. *Benthic Macro-Crustaceans of the Eastern Tropical Pacific*. San Diego Natural History Museum and Universidad Nacional Autónoma de México.

———. 2008. The Gulf of California Invertebrate Database: The invertebrate portion of the *Macrofauna Golfo* Database. www.desertmuseum.org/center/seaofcortez/database.php.

Brusca, R. C., E. Kimrey, and W. Moore. 2004. *Seashore Guide to the Northern Gulf of California*. Arizona-Sonora Desert Museum, Tucson, AZ.

Brusca, R. C., and D. A. Thomson. 1977. The Pulmo Reefs of Baja California—True coral reef formation in the Gulf of California. *Ciencias Marinas* 1(3): 37–53.

Brusca, R. C., and B. R. Wallerstein. 1979a. Zoogeographic patterns of idoteid isopods in the northeast Pacific, with a review of shallow-water zoogeography for the region. *Bulletin Biological Society Washington* 3: 67–105.

———. 1979b. The marine isopod crustaceans of the Gulf of California. II. Idoteidae. New genus, new species, new records, and comments on the morphology, taxonomy and evolution within the family. *Proceedings of the Biological Society Washington* 92(2): 253–271.

Brusca, R. C., R. Wetzer, and S. France. 1995. Cirolanidae (Crustacea; Isopoda; Flabellifera) of the tropical eastern Pacific. *Proceedings of the San Diego Society of Natural History*, no. 30.

Burckhalter, D. 1999. *Among Turtle Hunters and Basket Makers. Adventures with the Seri Indians*. Treasure Chest Books, Tucson, AZ.

Burrus, E. J. 1971. *Kino and Manje: Explorers of Sonora and Arizona*. Rome: Jesuit Historical Institute.

Bussing, W. A., and R. J. Lavenberg. 2003. Four new species of eastern tropical Pacific jawfishes (*Opistognathus:* Opistognathidae). *Revista Biologia Tropical* 51: 529–550.

Cabioch, J. 1969. Les fonds de maerl de la Baie de Morlaix et leur peuplement vegetal. *Cahiers de Biologie Marine* 9: 131–169.

Cabioch, J., J. Y. Floc'h, A. L. Toquin, C. F. Boudouresque, A. Meinesz, and M. Verlaque. 1992. *Guía de las algas de los mares de Europa: Atlantico y Mediterraneo*. Omega, España.

Calambokidis, J., G. H. Steiger, J. M. Straley, L. M. Herman, S. Cerchio, D. R. Salden, J. Urbán R., J. K. Jacobsen, O. von Ziegesar, K. C. Balcomb, C. M. Gabriele, M. E. Dahlheim, S. Uchida, G. Ellis, Y. Miyamura, P. Ladron de Guevara P., M. Yamaguchi, F. Sato, S. A. Mizroch, L. Schlender, K. Rasmussen, J. Barlow, and T. J. Quinn II. 2001. Movements and population structure of humpback whales in the North Pacific. *Marine Mammal Science* 17(4): 769–794.

Calderon-Aguilera, L. E., S. G. Marinone, and E. A. Aragon-Noriega. 2003. Influence of oceanographic processes on the early life stages of the blue shrimp (*Litopenaeus stylirostris*) in the Upper Gulf of California. *Journal of Marine Systems* 39: 117–128.

Caldwell, D. K. 1962. Sea turtles in Baja California waters (with special reference to those of the Gulf of California), and the description of a new subspecies of north-eastern Pacific green turtle. *Los Angeles County Museum Contributions in Science* 61: 1–31.

———. 1963. The sea turtle fishery of Baja California, Mexico. *California Fish and Game* 49: 140–151.

Caldwell, D. K., and M. C. Caldwell. 1962. The black "steer" of the Gulf of California. *Los Angeles County Museum of Science and History Quarterly* 1(1): 1–15.

———. 1989. Pygmy sperm whale, *Kogia breviceps* (de Blainville, 1838). Dwarf sperm whale, *Kogia simus* Owen, 1866. In S. H. Ridgway and R. J. Harrison (eds.), *Handbook of Marine Mammals*, vol. 4: *River Dolphins and the Larger Toothed Whales*, pp. 235–260. Academic Press, London.

Calmus, T., C. Pallares, R. C. Maury, H. Bellon, E. Pérez-Segura, A. Aguillón-Robles, A. L. Carreno, J. Bourgois, J. Cotten, and M. Benoit. 2008. Petrologic diversity of Plio-Quaternary post-subduction volcanism in Baja California: An example from Isla San Esteban (Gulf of California, Mexico). *Bulletin de la Société Géologique de France* 179(5): 465–481.

Calvert, S. E. 1964. Factors affecting distribution of laminated diatomaceous sediments in the Gulf of California. In T. H. Van Andel and G. G. Shor Jr. (eds.), *Marine Geology of the Gulf of California: A Symposium*, pp. 311–330. Memoir 3, American Association of Petroleum Geologists, Tulsa, OK.

———. 1966. Accumulation of diatomaceous silica in the sediments of the Gulf of California. *Geological Society of America Bulletin* 77: 569–596.

Campos-Dávila, L., V. H. Cruz-Escalona, F. Galván-Magaña, A. Abitia-Cárdenas, F. J. Gutiérrez-Sánchez, and E. F. Balart. 2005. Fish assemblages in a Gulf of California marine reserve. *Bulletin of Marine Science* 77(3): 347–362.

Cantú, J. C., and M. E. Sánchez. 2000. *Tráfico ilegal de tortugas marinas en México, situación histórica y actual.* Investigación de TEYELIZ, México City.

Carabias-Lillo, J., J. de la Maza, D. Gutiérrez-Carbonell, M. Gómez-Cruz, G. Anaya-Reina, A. Zavala-González, A. L. Figueroa, and B. Bermúdez-Almada. 2000. Programa de Manejo Área de Protección de Flora y Fauna Islas del Golfo de California, México. Comisión Nacional de Áreas Naturales protegidas, Secretaría de Medio Ambiente, Recursos Naturales y Pesca, México.

Carballo, J. L., J. A. Cruz-Barraza, and P. Gómez. 2004. Taxonomy and description of clionaid sponges (Hadromerida, Clionaidae) from the Pacific Ocean of Mexico. *Zoological Journal of the Linnean Society* 141: 353–397.

Carballo, J. L., P. Gómez, J. A. Cruz-Barraza, and D. M. Flores-Sánchez. 2003. Sponges of the family Chondrillidae (Porifera: Demospongiae) from the Pacific Coast of Mexico, with the description of three new species. *Proceedings of the Biological Society of Washington* 116(2): 515–527.

Carballo, J. L., C. Olabarria, and T. Garza Osuna. 2002. Analysis of four macroalgal assemblages along the Pacific Mexican coast during and after the 1997–1998 El Niño. *Ecosystems* 5: 749–760.

Cárdenas, H. G. 2008. Distribución y habitat de zífidos en la costa sudoccidental del Golfo de California. Cetacea: Ziphiidae. Master's thesis, Posgrado en Ciencias Marinas y Costera, Universidad Autónoma de Baja California Sur.

Cariño, M. 1995. *Historia de las relaciones hombre/naturaleza en Baja California Sur 1500–1940.* UABCS, SEP-Fomes, México.

———. 1998. *El porvenir de la Baja California está en sus mares. Vida y legado de Don Gastón J. Vives, primer maricultor de América.* H. Congreso del Estado de Baja California Sur, La Paz.

Cariño, M., and M. Monteforte. 1999. *El primer emporio perlero sustentable del mundo.* UABCS-SEP, México.

Carlton, J. T. (ed.). 2007. *The Light and Smith Manual. Intertidal Invertebrates from Central California to Oregon,* 4th ed. University of California Press, Berkeley, CA.

Carmona, R., and G. D. Danemann. 1994. Nesting waterbirds of Santa Maria Bay, Sinaloa, Mexico, April 1988. *Western Birds* 25: 158–162.

Carmona, R., J. Guzmán, S. Ramírez, and G. Fernández. 1994. Breeding waterbirds of La Paz Bay, Baja California Sur, Mexico. *Western Birds* 25: 151–157.

Carmona, R., S. Ramírez, B. Zárate, and F. Becerril. 1996. Some nesting waterbirds from southern San Jose Island and adjacent islands, Gulf of California, Mexico. *Western Birds* 27: 81–85.

Carmona, R., G. Ruíz Campos, J. A. Castillo Guerrero, and G. Brabata. 2005. Patterns of occurrence and abundance of land birds on Espíritu Santo Island, Gulf of California, Mexico. *Southwestern Naturalist* 50: 440–447.

Carr, A. 1961. Pacific turtle problem. *Natural History* 70: 64–71.

———. 1987. New perspectives on the pelagic stage of sea turtle development. *Conservation Biology* 1: 103.

Carr, A., and A. B. Meylan. 1980. Evidence of passive migration of green turtle hatchlings in Sargassum. *Copeia* 1980: 366–368.

Carreño, A. L. 1983. Ostrácodos y foraminíferos planctónicos de la localidad Loma del Tirabuzón, Santa Rosalía, Baja California Sur e implicaciones bioestratigráficas y paleoecológicas. Universidad Nacional Autónoma de México, Instituto de Geología. *Revista* 5(1): 55–64.

———. 1985. Biostratigraphy of the late Miocene to Pliocene on the Pacific island María Madre, Mexico. *Micropaleontology* 31: 139–166.

———. 1992a. Neogene microfossils from the Santiago Diatomite, Baja California Sur, Mexico. In M. Alcayde-Orraca and A. Gómez-Caballero (eds.), "Calcareous Neogene Microfossils of Baja California Sur, Mexico." Instituto de Geología, Universidad Nacional Autónoma de México. *Paleontología Mexicana* 59(1): 1–3.

———. 1992b. Early Neogene foraminifera and associated microfossils of the Cerro Tierra Blanca member (El Cien Formation), Baja California Sur, México. In M. Alcayde-Orraca and A. Gómez-Caballero (eds.), "Calcareous Neogene Microfossils of Baja California Sur, Mexico." Instituto de Geología, Universidad Nacional Autónoma de México. *Paleontología Mexicana* 59(2): 39–93.

Carreño, A. L., and J. Helenes. 2002. Geology and ages of the islands. In T. J. Case, M. L. Cody, and E. Ezcurra (eds.), *A New Island Biogeography of the Sea of Cortés,* pp. 14–40. Oxford University Press, New York.

Carreño, A. L., and J. T. Smith. 2007. Stratigraphy and correlation for the ancient Gulf of California and Baja California Peninsula, Mexico. *Bulletin of American Paleontology* 371: 1–146.

Carrillo, L. E., M. F. Lavín, and E. Palacios-Hernández. 2002. Seasonal evolution of the geostrophic circulation in the northern Gulf of California. *Estuarine, Coastal and Shelf Science* 54: 157–173.

Carriquiry, J. D., and A. Sánchez. 1999. Sedimentation in the Colorado River Delta and Upper Gulf of California after nearly a century of discharge loss. *Marine Geology* 158: 125–145.

Carriquiry, J. D., A. Sánchez, and V. F. Camacho-Ibar. 2001. Sedimentation in the northern Gulf of California after cessation of the Colorado River discharge. *Sedimentary Geology* 144: 37–62.

Cartron, J.-L. E. 2000. Status and productivity of Ospreys along the eastern coast of the Gulf of California: 1992–1997. *Journal of Field Ornithology* 71: 298–309.

Cartron, J.-L. E., G. Ceballos, and R. S. Felger (eds.). 2005. *Biodiversity, Ecosystems, and Conservation in Northern Mexico.* Oxford University Press, New York.

Cartron, J.-L. E., G. L. Garber, C. Finley, C. Rustay, R. P. Kellermueller, M. P. Day, P. Manzano Fisher, and S. H. Stoleson. 2000. Power pole casualties among raptors and ravens in northwestern Chihuahua, Mexico. *Western Birds* 31: 255–257.

Cartron, J.-L. E., R. Harness, R. Rogers, and P. Manzano. 2005. Impact of concrete power poles on raptors and ravens in northwestern Chihuahua, Mexico. In J.-L. E. Cartron, G. Ceballos, and R. S. Felger (eds.), *Biodiversity, Ecosystems, and Conservation in Northern Mexico,* pp. 357–369. Oxford University Press, New York.

Cartron, J.-L. E., and M. C. Molles Jr. 2002. Osprey diet along the eastern side of the Gulf of California. *Western North American Naturalist* 62: 249–252.

Cartron, J.-L. E., R. Rodríguez-Estrella, R. C. Rogers, L. B. Rivera, and B. Granados. 2006a. Raptor electrocutions in northwestern Mexico: A preliminary regional assessment of the impact of concrete power poles. In R. Rodríguez Estrella (ed.), *Current Raptor Studies in Mexico,* pp. 202–230. CIBNOR, La Paz, B.C.S.

Cartron, J.-L. E., S. A. Sommer, D. E. Kilpatrick, and T. A. Pfister. 2006b. Ecology and population status of Ospreys in coastal Sonora. In R. Rodríguez Estrella (ed.), *Current Raptor Studies in Mexico,* pp. 120–148. CIBNOR, La Paz, B.C.S.

Carvajal, M. A., E. Ezcurra, and A. Robles. 2004. The Gulf of California: Natural resource concerns and the pursuit of a vision. In L. K. Glover and S. A. Earle (eds.), *Defying Ocean's End: An Agenda for Action,* pp. 105–123. Island Press, Washington, DC.

Carwardine, M., and M. Camm. 1995. *Whales, Dolphins and Porpoises.* Smithsonian Handbooks. Dorling Kindersley, London.

Case, T. J., and M. L. Cody. 1983. *Island Biogeography in the Sea of Cortez.* University of California Press, Berkeley, CA.

Case, T. J., M. L. Cody, and E. Ezcurra (eds.). 2002. *A New Island Biogeography of the Sea of Cortés.* Oxford University Press, New York.

Caso, M., C. González-Abraham, and E. Ezcurra. 2007. Divergent ecological effects of oceanographic anomalies on terrestrial ecosystems of the Mexican Pacific coast. *Proceedings of the National Academy of Sciences* 104(25): 10530–10535.

Castillo Geniz, J. L., O. Sosa Nishizaki, and J. C. Perez Jiménez. 2007. Morphological variation and sexual dimorphism in the California skate, *Raja inornata* Jordan and Gilbert, 1881 from the Gulf of California, Mexico. *Zootaxa* 1545: 1–16.

Castro, R., M. F. Lavín, and P. Ripa. 1994. Seasonal heat balance in the Gulf of California. *Journal of Geophysical Research* 99: 3249–3261.

Castro-Aguirre, J. L. 1980. Posible impacto sobre la fauna del Pacífico oriental tropical, con especial referencia al Golfo de California, por la apertura de un nuevo canal interoceánico a nivel del mar. In Memorias del 1er. Simposium sobre Biología Marina, 6–8 Dic. 1978, pp. 111–128. Universidad Autónoma de Baja California Sur, Area de Ciencias del Mar, La Paz, B.C.S.

Castro-Aguirre, J. L., A. Antuna-Mendiola, A. F. González-Acosta, and J. De la

Cruz-Agüero. 2005. *Mustelus albipinnis* sp. nov. (Chondrichthyes: Carcharhiniformes: Triakidae) de la costa suroccidental de Baja California Sur, México. *Hidrobiológica* 15(2 esp.): 123–130.

Castro-Aguirre, J. L., J. Arvizu-Martínez, and J. Paez. 1970. Contribución al conocimiento de los peces del Golfo de California. *Revista de la Sociedad Mexicana de História Natural* 31: 107–181.

Castro-Aguirre, J. L., E. F. Balart, and J. Arvizu-Martínez. 1995. Contribución al conocimiento del origen y distribución de la ictiofauna del Golfo de California, México. *Hidrobiológica* 5(1–2): 57–78.

Castro-Aguirre, J. L., H. S. Espinosa-Pérez, and J. J. Schmitter-Soto. 1999. *Ictiofauna estuarino-lagunar y vicaria de México*. Editorial Limusa, México, D.F.

Castro-Aguirre, J. L., J. C. Ramírez-Cruz, and M. A. Martínez-Muñoz. 1992. Nuevos datos sobre la distribución de lenguados (Pisces: Pleuronectiformes) en la costa oeste de Baja California, México, con aspectos biológicos y zoogeográficos. *Anales de la Escuela Nacional de Ciencias Biológicas de México* 37: 97–119.

Castro-Aguirre, J. L., G. Ruíz-Campos, and E. F. Balart. 2002. A new species of the genus *Lile* (Clupeiformes: Clupeidae) of the eastern tropical Pacific. *Bulletin of the Southern California Academy of Sciences* 101(1): 1–12.

Castro-Aguirre, J. L., J. J. Schmitter-Soto, E. F. Balart, and R. Torres-Orozco. 1993. Sobre la distribución geográfica de algunos peces bentónicos de la costa oeste de Baja California Sur, México, con consideraciones ecológicas y evolutivas. *Anales de la Escuela Nacional de Ciencias Biológicas de México* 38: 75–102.

Castro-Aguirre, J. L., and R. Torres-Orozco. 1993. Consideraciones acerca del origen de la ictiofauna de Bahía Magdalena-Almejas, un sistema lagunar de la costa occidental de Baja California Sur, México. *Anales de la Escuela Nacional de Ciencias Bioligicas de México* 38: 67–73.

Cavallero Carranco, J. 1966 [1668]. *The Pearl Hunters in the Gulf of California 1668* (transcribed, translated, and annotated by W. Michael Mathes). Baja California Travel Series, no. 4, Dawson's Book Shop, Los Angeles, CA.

Ceballos, G. 1975. The rockfishes, genus *Sebastes* (Scorpaenidae), of the Gulf of California, including three new species with a discussion of their origin. *Proceedings of the California Academy of Sciences* 40: 109–141.

Ceballos, G., J. Arroyo-Cabrales, and R. A. Medellín. 2002. The mammals of Mexico: Composition, distribution, and conservation status. *Occasional Papers of the Museum,* Texas Tech University, Lubbock, TX.

CEC. 1997. *Ecological Regions of North America. Toward a Common Perspective.* Commission for Environmental Cooperation, Montréal, Québec.

Chávez, H. 1985. Bibliografía sobre los peces de la Bahía de La Paz, Baja California Sur, México. *Investigaciones Marinas CICIMAR (La Paz)* 2(num. esp. II): 1–75.

———. 1986. Bibliografía sobre peces del Golfo de California. *Investigaciones Marinas CICIMAR (La Paz)* 3(num. esp. I): 1–267.

———. 1989. Presencia de tortuga carey, *Eretmochelys imbricata,* en Playa Platanitos, Nayarit, México. In *Memorias del VI Encuentro Interuniversitario Mexicano sobre Tortugas Marinas,* 7–10 July 1989, pp. 28–29. Universidad Nacional Autónoma de México.

Chen, L.-C. 1975. The rockfishes, genus *Sebastes* (Scorpaenidae), of the Gulf of California, including three new species with a discussion of their origin. *Proceedings of the California Academy of Sciences* 40: 109–141.

Chimenz Gusso, C., and L. Lattanzi. 2003. Mediterranean Pycnogonida: Faunistic, taxonomical and zoogeographical considerations. *Biogeographia* 24: 251–262.

Cho, T. O., R. Riosmena-Rodríguez, and M. S. Boo. 2001. The developmental morphology of *Ceramium procumbens* (Ceramiaceae, Rhodophyta) from the Gulf of California, Mexico. *Algae* 16: 45–52.

———. 2002. Developmental morphology of a poorly documented alga, *Ceramium recticorticum* (Ceramiaceae, Rhodophyta) from the Gulf of California, Mexico. *Cryptogamie Algologie* 23(4): 277–289.

Cicin-Sain, B., P. Bernal, V. Vandeweerd, S. Belfiore, and K. Goldstein. 2002. *A Guide to Oceans, Coasts and Islands at the World Summit on Sustainable Development.* Center for the Study of Marine Policy, Newark, DE.

Cintra-Buenrostro, C. E. 2001. Los asteroideos (Echinodermata: Asteroidea) de agues someras del Golfo de California, México. *Oceánides* 16(1): 49–90.

Cintra-Buenrostro, C. E., and K. W. Flessa. 2004. Cavidades, mordiscos y peladas: Herramientas para determinar la importancia trófica de una especie en desvanecimiento dentro del delta del Río Colorado, México [Holes and scars on the shells of *Mulinia coloradoensis* are used to determine the trophic importance of this vanishing species in the Colorado River Delta, Mexico]. *Ciencia y Mar* 8(24): 3–19.

Cintra-Buenrostro, C. E., K. W. Flessa, and G. Avila-Serrano. 2004. Who cares about a vanishing clam? Trophic importance of *Mulinia coloradoensis* inferred from predatory damage. *Palaios* 20: 295–301.

Cintra-Buenrostro, C. E., M. S. Foster, and K. H. Meldahl. 2002. Response of nearshore marine assemblages to global change: A comparison of molluscan assemblages in Pleistocene and modern rhodolith beds in the southwestern Gulf of California, Mexico. *Palaeogeography, Palaeoclimatology, Palaeoecology* 183: 299–320.

Cisneros-Mata, M. A. 2004. Sustainability in complexity: From fisheries management to conservation of species, communities and spaces in the Sea of Cortez. In R. C. Brusca (ed.), *Proceedings of the Gulf of California Conference 2004,* 13–16 June, pp. 20–23. Arizona-Sonora Desert Museum, Tucson, AZ.

Cisneros-Mata, M. A., L. W. Botsford, and J. F. Quinn. 1997. Projecting viability of *Totoaba macdonaldi,* a population with unknown age-dependent variability. *Ecological Applications* 7(3): 968–980.

Cisneros-Mata, M. A., G. Montemayor-López, and M. O. Nevárez-Martínez. 1996. Modeling deterministic effects of age structure, density dependence, environmental forcing and fishing in the population dynamics of the Pacific sardine (*Sardinops sagax caeruleus*) stock of the Gulf of California. *CalCOFI Report* 37: 201–208.

Cisneros-Mata, M. A., G. Montemayor-López, and M. J. Román-Rodríguez. 1995a. Life history and conservation of *Totoaba macdonaldi. Conservation Biology* 94: 806–814.

Cisneros-Mata, M. A., M. O. Nevárez-Martínez, and M. G. Hammann. 1995b. The rise and fall of the Pacific sardine, *Sardinops sagax caeruleus* Giard, in the Gulf of California, Mexico. *CalCOFI Reports* 36: 136–143.

Clapham, P. 2002. Humpback whale. In W. F. Perrin, B. Würsig, and J. G. M. Thewissen (eds.), *Encyclopedia of Marine Mammals*, pp. 589–592. Academic Press, San Diego, CA.

Clark, R. N. 2000. The chiton fauna of the Gulf of California rhodolith beds (with descriptions of four new species). *Nemouria* 43: 1–20.

Clavigero, F. J. 1937 [1789]. *The History of [Lower] California* (translated by Sara E. Lake and edited by A. A. Gray). Stanford University Press, Stanford, CA.

Cliffton, K., D. O. Cornejo, and R. S. Felger. 1982. Sea turtles of the Pacific coast of Mexico. In K. A. Bjorndal (ed.), *Biology and Conservation of Sea Turtles*, pp. 199–209. Smithsonian Institution Press, Washington, DC.

Coates, A. G., and J. A. Obando. 1996. The geologic evolution of the Central American isthmus. In J. B. C. Jackson, A. F. Budd, and A. G. Coates (eds.), *Evolution and Environments in Tropical America*, pp. 21–56. University of Chicago Press, Chicago.

Cohen, M., and C. Henges-Jeck. 2001. Missing water: The uses and flows of water in the Colorado River Delta region. Report of the Pacific Institute, Oakland, CA.

Cohen, M. J., C. Henges-Jeck, and G. Castillo-Moreno. 2001. A preliminary water balance for the Colorado River delta, 1992–1998. *Journal of Arid Environments* 49(1): 35–48.

Collins, C. A., N. Garfield, A. S. Mascarenhas, M. G. Spearman, and T. A. Rago. 1997. Ocean currents across the entrance to the Gulf of California. *Journal of Geophysical Research* 102: 927–936.

Collins, T. 1996. Molecular comparisons of transisthmian species pairs: Rates and patterns of evolution. In J. B. C. Jackson, A. F. Budd, and A. G. Coates (eds.), *Evolution and Environments in Tropical America*, pp. 303–334. University of Chicago Press, Chicago.

CONABIO. 1999. *Ecorregiones de México*. México, D.F., http://www.conabio.gob.mx/ [digital map scale 1:1,000,000].

CONANP. 2004. Program de Conservación y Manejo de la Reserva de la Biósphera Alto Golfo de California y Delta del Río Colorado, http://conanp.gob.mx/anp/consulta.php.

Cornelius, S. E. 1982. Status of sea turtles along the Pacific coast of Middle America. In K. A. Bjorndal (ed.), *Biology and Conservation of Sea Turtles*, pp. 211–219. Smithsonian Institution Press, Washington, DC.

Correll, D. S., and H. B. Correll. 1972. Aquatic and wetland plants of the southwestern United States. Environmental Protection Agency, Washington, DC.

Cortés-Lara, M. C., S. Alvarez-Borrego, and A. D. Giles-Guzmán. 1999. Efecto de la mezcla vertical sobre la distribución de nutrients y fitoplancton en dos regiones del Golfo de California, en verano. *Revista de la Sociedad Mexicana de Historia Natural* 49: 193–206.

Cortés-Lara, A., A. López-López, L. L. Moctezuma-Torres, F. J. Mosqueda-Martínez, E. Paredes Arellano, and F. A. Sandova. 2002. Revestimiento del Canal Todo Americano y sus efectos en Baja California. Gobierno del Estado de Baja California, Mexicali, Baja California, México.

Council Directive 92/43/EEC. 1992. Conservation of natural habitats and of wild flora and fauna. *International Journal of the European Communities* L206: 7–49.

Coyne, M. S., and R. D. Clark (compilers). *Proceedings of the 21st Annual Symposium on Sea Turtle Biology and Conservation*. NOAA Technical Memorandum NMFS-SEFSC-528, Philadelphia, PA.

Crabtree, C. B. 1989. A new silverside of the genus *Colpichthys* (Atheriniformes: Atherinidae) from the Gulf of California, Mexico. *Copeia* 1989(3): 558–568.

Craig, J. A. 1926. A new fishery in Mexico. *California Fish and Game* 12: 166–169.

Craig, M. T., P. A. Hastings, D. J. Pondella II, D. R. Robertson, and J. A. Rosales-Casián. 2006. Phylogeography of the flag cabrilla *Epinephelus labriformis* (Serranidae): Implications for the biogeography of the tropical eastern Pacific and the early stages of speciation in a marine shore fish. *Journal of Biogeography* 33: 969–979.

CSGC. 2001. *Prioridades de conservación para la Región del Golfo de California.* Coalición para la Sustentabilidad del Golfo de California, with support from Comisión Nacional para el Conocimiento y Uso de la Biodiversidad (CONABIO), Cementos Mexicanos (CEMEX), Conservation International (CI), Fondo Mexicano para la Conservación de la Naturaleza (FMCN), Instituto Nacional de Ecología (INE), Instituto Tecnológico de Monterrey (ITESM), and World Wildlife Fund (WWF), México, D.F. [map scale 1:1,500,000].

———. 2004. *Prioridades de conservación para el Golfo de California.* Coalición para la Sustentabilidad del Golfo de California. WWF Mexico and Gulf of California Program, México, D.F.

CTPTV. 1993. Propuesta para la Declaración de Reserva de la Biosfera Alto Golfo de California y Delta del Río Colorado. Internal report, Comité Ténico para la Preservación de la Totoaba y la Vaquita, Hermosillo, Sonora.

Cudney-Bueno, R., and X. Basurto. 2009. Lack of cross-scale linkages reduces robustness of community-based fisheries management. *PLoS ONE* 4(7) e6253: 1–8.

Cudney-Bueno, R., L. Bourillón, A. Sáenz-Arroyo, J. Torre-Cosío, P. Turk-Boyer, and W. W. Shaw. 2008. Governance and effects of marine reserves in the Gulf of California, Mexico. *Ocean and Coastal Management* 52: 207–218.

Cudney-Bueno, R., M. F. Lavín, S. G. Marinone, P. T. Raimondi, and W. W. Shaw. 2009. Rapid Effects of Marine Reserves via Larval Dispersal. *PLoS ONE* 4(1) e4140: 1–7.

Cudney-Bueno, R., R. Prescott, and O. Hinojosa-Huerta. 2008. The black murex snail, *Hexaplex nigritus* (Mollusca, Muricidae), in the Gulf of California, Mexico: I. Reproductive ecology and breeding aggregations. *Bulletin of Marine Science* 83(2): 285–298.

Cudney-Bueno, R., and K. Rowell. 2008a. Establishing a baseline for management of the rock scallop, *Spondylus calcifer* (Carpenter, 1857): Growth and reproduction in the upper Gulf of California, Mexico. *Journal of Shellfish Research* 27(4): 625–632.

———. 2008b. The black murex snail, *Hexaplex nigritus* (Mollusca, Muricidae) in the Gulf of California, Mexico. II. Growth, longevity and morphological variations with implications for management of a rapidly declining fishery. *Bulletin of Marine Science* 83(2): 299–313.

Cudney-Bueno, R., and P. J. Turk Boyer. 1998. Pescando entre mareas del alto Golfo de California. Una guía sobre la pesca artesanal, su gente y sus propuestas de manejo. CEDO Technical Series (Puerto Peñasco), no. 1.

D'Agrosa, C., C. Lennert-Cody, and O. Vidal. 2000. Vaquita bycatch in Mexico's artisanal gillnet fisheries: Driving a small population to extinction. *Conservation Biology* 14(4): 1110–1119.

Dahlheim, M. E., S. Leatherwood, and W. F. Perrin. 1982. Distribution of killer whales in the warm temperate and tropical eastern Pacific. *Reports of the International Whaling Commission* 36: 647–653.

Danemann, G. D., and E. Ezcurra (eds.). 2008. *Bahía de Los Ángeles: Recursos naturales y comunidad. Línea base 2007*. Pronatura Noroeste, Instituto Nacional de Ecología, México, D.F., and San Diego Natural History Museum.

Dauphin, J. P., and G. E. Ness. 1991. *Bathymetry of the Gulf and Peninsular Province of the Californias*. Memoir 47, American Association of Petroleum Geologists, Tulsa, OK.

Dauphin, J. P., and B. R. T. Simoneit. 1991. *The Gulf and Peninsular Province of the Californias*. Memoir 47, American Association of Petroleum Geologists, Tulsa, OK.

Dávila, G., and F. García-Badillo. 1986. Análisis de la biología y condiciones del stock del calamar gigante *Dosidicus gigas* en el Golfo de California, México, durante 1980. *Ciencia Pesquera* 5: 63–76.

Davis, R. W., N. Jaquet, D. Gendron, U. Markaida, G. Bazzino, and W. Gilly. 2007. Diving behavior of sperm whales in relation to behavior of a major prey species, the jumbo squid, in the Gulf of California, Mexico. *Marine Ecology Progress Series* 333: 291–302.

Dawson, C. 1974. Studies on eastern Pacific sand stargazers (Pisces: Dactyloscopidae) 1. *Platygillelus* new genus, with descriptions of new species. *Copeia* 1974(1): 39–55.

———. 1975. Studies on eastern Pacific sand stargazers (Pisces: Dactyloscopidae). 2. Genus *Dactyloscopus*, with descriptions of new species and subspecies. *Natural History Museum of Los Angeles County, Scientific Bulletin* 22: 1–61.

———. 1976. Studies on eastern Pacific sand stargazers. 3. *Dactylagnus* and *Myxodagnus*, with description of a new species and subspecies. *Copeia* 1976(1): 13–43.

———. 1985. *Indo-Pacific Pipefishes (Red Sea to the Americas)*. Gulf Coast Research Laboratory, Ocean Springs, MS.

Dawson, E. Y. 1944. The marine algae of the Gulf of California. *Allan Hancock Pacific Expeditions* 3: 189–453.

———. 1945. Some new and unreported sublittoral algae from Cedros Island, Mexico. *Bulletin Southern California Academy Sciences* 43: 102–122.

———. 1949. Resultados preliminarios de un reconocimiento de las algas marinas de la costa Pacifica de Mexico. *Revista Sociedad Mexicana Historia Naturalia* 9: 215–255.

———. 1950. A review of *Ceramium* along the Pacific coast of North America with special reference to its Mexican representatives. *Farlowia* 4: 113–138.

———. 1953. Marine red algae of Pacific Mexico. Part 1, Bangiales to Corallinaceae Subf. Corallinoideae. *Allan Hancock Pacific Expeditions* 17: 1–238.

———. 1954a. Resumen de las investigaciones recientes sobre algas marinas de la costa Pacifica de Mexico, con una synopsis de la literatura, sinonimia y distribución de las especies descritas. *Revista Sociedad Mexicana Historia Natural* 13: 97–197.

———. 1954b. Marine red algae of Pacific Mexico. Part 2, Cryptomeniales (cont.): Dermocorynidaceae to Choreccolacaeae. *Allan Hancock Pacific Expeditions* 1: 240–396.

———. 1959. Marine algae from the 1958 cruise of the Stella Polaris in the Gulf of California. *Los Angeles County Museum Contributions in Science* 27: 1–39.

———. 1960a. A review of the ecology, distribution, and affinities of the benthic flora. In "The Biogeography of Baja California and Adjacent Seas. Part II: Marine Biotas." *Systematic Zoology* 9: 93–100.

———. 1960b. Marine red algae of Pacific Mexico, Part 3: Cryptomeniales; Corallinaceae subf. Melobesioideae. *Pacific Naturalist* 2: 1–125.

———. 1960c. New records of marine algae from Pacific Mexico and Central America. *Pacific Naturalist* 1(20): 31–52.

———. 1961a. Marine red algae of Pacific Mexico, Part 4: Gigartinales. *Pacific Naturalist* 2: 191–343.

———. 1961b. Marine red algae of Pacific Mexico. Part 7, Ceramiales, Ceramiaceae, Deles seriaceae. *Allan Hancock Pacific Expeditions* 26: 1–207.

———. 1961c. A guide to the literature and distributions of Pacific benthic algae from Alaska to the Galapagos Islands. *Pacific Science* 15: 370–461.

———. 1963a. Marine red algae of Pacific Mexico. Part 6, Rhodymeniales. *Nova Hedwigia* 5: 437–476.

———. 1963b. Marine red algae of Pacific Mexico. Part 8, Ceramiales: Dasyaceae, Rhodomelaceae. *Nova Hedwigia* 6: 401–481.

———. 1966a. *Marine Algae in the Vicinity of Puerto Peñasco, Sonora, Mexico.* Gulf of California Field Guide Series, no. 1. University of Arizona, Tucson, AZ.

———. 1966b. New records of marine algae from the Gulf of California. *Journal of the Arizona Academy of Sciences* 4: 55–66.

Dawson, E. Y., and M. S. Foster. 1982. *Seashore Plants of California.* University of California Press, Berkeley, CA.

Dayton, P. K. 2004. The importance of the natural sciences to conservation. *American Naturalist* 162: 1–13.

Dayton, P. K., S. Thrush, and F. C. Coleman. 2002. Ecological effects of fishing in marine ecosystems of the United States. Pew Oceans Commission, Arlington, VA.

Dean, M. A. 1996. Neogene Fish Creek gypsum and associated stratigraphy and paleontology, southwestern Salton Trough, California. In P. Abbott and D. Seymour (eds.), *Sturzstroms and Detachment Faults, Anza-Borrego Desert State Park California, South Coast,* pp. 123–148. Annual Field Trip Guide, no. 24, Geological Society of America, Boulder, CO.

de Diego-Forbis, T. A., R. Douglas, D. Gorsline, E. Nava-Sánchez, L. Marrack, and J. Banner. 2004. Late Pleistocene (last interglacial) terrace deposits, Bahía Coyote, Baja California Sur, México; coastal environmental change during sea-level highstands. IGCP 437 Symposium, Barbados. *Quaternary International* 120(1): 29–40.

Dedina, S. 2000. *Saving the Gray Whale. People, Politics, and Conservation in Baja California.* University of Arizona Press, Tucson, AZ.

De Francia, G. 1930 [1596]. Memorial of Gonzalo de Francia. In H. R. Wagner, "Pearl Fishing Enterprises in the Gulf of California." *Hispanic American Historical Review* 10(2): 188–220.

De Grave, S. 1999. The influence of sedimentary heterogeneity on maerl bed differences in infaunal crustacean community. *Estuarine, Coastal and Shelf Science* 49: 153–163.

De Grave, S., H. Fazakerley, L. Kelly, M. D. Guiry, M. Ryan, and J. Walshe. 2000. *A Study of Selected Maërl Beds in Irish Waters and Their Potential for Sustainable Extraction.* Marine Resource Series, no. 10. The Marine Institute, Dublin.

De Grave, S., and A. Whitaker. 1999. Benthic community re-adjustment following dredging of a muddy-maerl matrix. *Marine Pollution Bulletin* 38: 102–108.

De la Cruz-Agüero, J., and F. Galván-Magaña. 1992. Peces mesopelágicos de la costa occidental de Baja California Sur y del Golfo de California. Universidad Nacional Autónoma de México. *Anales del Instituto de Ciencias del Mar y Limnología* 19: 25–31.

De la Cruz-Agüero, J., M. Arellano-Martínez, V. M. Cota-Gómez, and G. De la Cruz-Agüero. 1997. Catálogo de los peces marinos de Baja California Sur. Instituto Politécnico Nacional, Centro Interdisciplinario de Ciencias Marinas, La Paz, Baja California Sur, México.

De la Cruz González, J., L. F. Beltrán-Morales, E. A. Aragón-Noriega, C. Salinas-Zavala, J. I. Urciaga-García, and M. A. Cisneros-Mata. 2005. Socioeconomic analysis of the jumbo squid and shrimp fishery in Guaymas. In CICIMAR/IPN, *Proceedings of the Symposium on Fishery Sciences in Mexico,* May 2–4, 2005, p. 107. La Paz, B.C.S.

Del Barco, Miguel. 1768. *Correcciones y adiciones a la historia o noticia de la California en su primera edición de Madrid, año de 1757.* [Edited by M. León-Portilla under the title *Historia natural y crónica de la antigua California,* Universidad Nacional Autónoma de México, Instituto de Investigaciones Históricas, 1988, México, D.F.]

———. 1980 [ca. 1770–1780]. *The Natural History of Baja California* (translated by Froylán Tiscareno). Baja California Travel Series, no. 43. Dawson's Book Shop, Los Angeles, CA.

Delebout, M. L., G. J. B. Ross, C. S. Baker, R. C. Anderson, P. B. Best, V. G. Cockcroft, H. L. Hinz, V. Peddemors, and R. L. Pitman. 2003. Appearance, distribution and genetic distinctiveness of Longman's beaked whale, *Indopacetus pacificus. Marine Mammal Science* 1(3): 421–461.

Delgadillo-Hinojosa, F., J. V. Macías-Zamora, J. A. Segovia-Zavala, and S. Torres-Valdés. 2001. Cadmium enrichment in the Gulf of California. *Marine Chemistry* 75: 109–122.

Delgado-Argote, L. A., M. López-Martínez, and M. C. Perilliat. 2000. Geologic reconnaissance and Miocene age of volcanism and associated fauna from sediments of Bahía de los Ángeles, Baja California. In H. Delgado-Granados, G. Aguirre-Gutiérrez, and J. Stock (eds.), "Cenozoic Tectonics and Volcanism of Mexico." *Geological Society of America Special Paper* 334: 111–121.

Delgado-Estrella, A., J. G. Ortega-Ortíz, and A. Sánchez-Ríos. 1994. Varamientos de mamíferos marinos durante primavera y otoño y su relación con la actividad pesquera. Universidad Nacional Autónoma de México. *Anales Instituto de Biología, serie Zoología* 65: 287–295.

De Mets, C. 1995. A reappraisal of seafloor spreading lineations in the Gulf of California: Implications for the transfer of Baja California to the Pacific Plate and estimates of Pacific–North America motion. *Geophysical Research Letters* 22: 3545–3548.

Department of Conservation [New Zealand]. 1998. Kapiti Marine Reserve: Conservation Management Plan. Wellington Conservancy Conservation Management Planning Series, no. 4.

Des Lauriers, M. R. 2006. Terminal Pleistocene and Early Holocene occupations of Isla de Cedros, Baja California, Mexico. *Journal of Island and Coastal Archaeology* 1(2): 255–270.

Diario Oficial [*Diario Oficial de la Federación*]. 1963. Decreto por el que se declara zona de reserva natural y refugio para la fauna silvestre, la Isla de Tiburón, situada en el Golfo de California. *Diario Oficial de la Federación,* México, D.F., 15 March 1963.

———. 1964. Decreto que declara zona de reserva natural y refugio de aves a Isla Rasa, Estado de Baja California. *Diario Oficial de la Federación,* México, D.F., 30 May 1964.

———. 1974. Decreto de zona de reserva, cultivo y/o repoblación para todas las especies de pesca, al área del delta del Río Colorado, en el Golfo de California, delimitada ésta por una línea imaginaria trazada de Este a Oeste, tangente al extremo sur de Isla Montague y Gore, desde las costas del Golfo de Santa Clara al litoral oriente de Baja California. *Diario Oficial de la Federación,* México, D.F., 30 May 1974.

———. 1978. Decreto por el que se establece una zona de reserva y refugio de aves migratorias y de la fauna silvestre en las islas que se relacionan, situadas en el Golfo de California. *Diario Oficial de la Federación,* México, D.F., 2 August 1978.

———. 1988. Ley General del Equilibrio Ecológico y la Protección al Ambiente. *Diario Oficial de la Federación,* México, D.F., 28 January 1988.

———. 1990. Acuerdo por el que se establece veda para las especies y subespecies de tortuga marina en aguas de jurisdicción federal del Golfo de México y Mar Caribe, así como en las costas del Océano Pacífico, incluyendo el Golfo de California. *Diario Official de la Federación.* México, D.F., May 28, 1990.

———. 1991. Articulo 254 bis al Codigo Penal. *Diario Official de la Federación,* México, D.F., 30 December 1991.

———. 1993. Decreto por el que se declara área natural protegida con el carácter de Reserva de la Biosfera, la región conocida como Alto Golfo de California y Delta del Río Colorado, ubicada en aguas del Golfo de California y los municipios de Méxicali, B.C., de Puerto Peñasco y San Luis Río Colorado, Sonora. *Diario Oficial de la Federación,* México, D.F., 10 June 1993, pp. 24–28.

———. 1996a. Decreto por el que se declara área natural protegida con el carácter de parque marino nacional, la zona conocida como Bahía de Loreto, ubicada frente a las costas del Municipio de Loreto, Estado de Baja California Sur, con una superficie de 11,987–87–50 hectáreas. *Diario Oficial de la Federación,* México, D.F., 19 July 1996.

———. 1996b. Decreto que reforma, adiciona y deroga diversas disposiciones de la Ley General del Equilibrio Ecológico y la Protección al Ambiente. *Diario Oficial de la Federación,* México, D.F., 13 December 1996.

———. 1999. Modificacion a la Norma Oficial Mexicana 002-PESC-1993. Para ordenar aprovechameinto de las especies de camarón en aguas de juristicción federal de los Estados Unidos Mexicanos, publicada el 31 de diciembre de 1993. *Diario Oficial de la Federación,* México, D.F., 29 December 1999.

———. 2000a. Acuerdo que tiene por objeto dotar con una categoría acorde con la legislación vigente a las superficies que fueron objeto de diversas declaratorias de áreas naturales protegidas emitidas por el Ejecutivo Federal. *Diario Oficial de la Federación,* México, D.F., 7 June 2000.

———. 2000b. Ley General de Vida Silvestre. *Diario Oficial de la Federación,* México, D.F., 4 July 2000.

———. 2000c. Norma Oficial Mexicana NOM-131-ECOL-1998. Que establece lineamientos y especificaciones para el desarrollo de actividades de observación de ballenas, relativas a su protección y conservación de su hábitat. *Diario Oficial de la Federación,* México, D.F., 10 January 2000 (Primera sección), pp. 11–17.

———. 2002a. Norma Oficial Mexicana NOM-059–2001. Protección ambiental de especies nativas de México de flora y fauna silvestres, categorías de riesgo y especificaciones para su inclusion, exclusion o cambio. Secretaría del Medio Ambiente y Recursos Naturales. México, D.F.

———. 2002b. Decreto por el que se reforman diversas disposiciones de la Ley General de Vida Silvestre. *Diario Oficial de la Federación,* México, D.F., 10 January 2002.

———. 2002c. Acuerdo por el que se establece como área de refugio para proteger a las especies de grandes ballenas de los subórdenes Mysticeti y Odontoceti, las zonas marinas que forman parte del territorio nacional y aquellas sobre las que la nación ejerce su soberanía y jurisdicción. *Diario Oficial de la Federación,* México, D.F., 24 May 2002.

———. 2004. Acuerdo mediante el cual se aprueba la actualización de la Carta Nacional Pesquera y su Anexo. Secretaría de Agricultura, Ganadería, Desarrollo Rural, Pesca y Alimentación. Segunda Sección.

Díaz de León-Corral, A., and M. A. Cisneros-Mata. 2000. Buzón del futuro: La Carta Nacional Pesquera. Instituto Nacional de Ecología, SEMARNAP, México, D.F. *Desarrollo Sustentable* 2(16): 10–12.

Díaz-Gamboa, R. E. 2003. Diferenciación entre tursiones *Tursiops truncatus* costeros y oceánicos en el Golfo de California por medio de isótopos estables de carbono y nitrógeno. Master's thesis, Centro Interdisciplinario de Ciencias Marinas, IPN.

Díaz-Guzmán, C. F. 2006. Abundancia y movimientos del rorcual común, *Balanoptera physalus,* en el Golfo de California. Master's thesis, Posgrado en Ciencias del Mar y Limnología, Universidad Nacional Autónoma de México.

Director of National Parks [Australia]. 2005. Australian Bight Marine Park (Commonwealth Waters) Management Plan 2005–2012. Environment Protection and Biodiversity Conservation.

Dizon, A. E., C. A. Lux, R. G. Leduc, J. Urbán, M. Henshaew, and R. L. Brownell. 1995. An interim phylogenetic analysis of sei and Bryde's whale mitochondrial DNA control region sequences. Paper presented at the Scientific Committee of the International Whaling Commission.

Dodd, N. L., and J. R. Vahle. 1998. Osprey (*Pandion haliaetus*). In R. L. Glinski (ed.), *The Raptors of Arizona,* pp. 37–41. University of Arizona Press, Tucson, AZ.

Dorsey, R. J. 1997. Origin and significance of rhodolith-rich strata in the Punta el Bajo section, southeastern Pliocene Loreto basin; Pliocene carbonates and related facies flanking the Gulf of California, Baja California, Mexico. *Geological Society of America Special Paper* 318: 119–126.

Dozier, C. L. 1963. Mexico's transformed northwest: The Yaqui, Mayo and Fuerte examples. *Geographical Reviews* 53: 548–571.

Drinkwater, K. F., and K. T. Frank. 1994. Effects of river regulation and diversion on marine fish and invertebrates. *Aquatic Conservation: Freshwater and Marine Ecosystems* 4: 135–151.

d'Udekem d'Acoz, C. 1999. Inventaire et distribution des crustacés décapodes de l'Atlantique nord-oriental, de la Méditerranée et des eaux douces continentales adjacentes au nord de 25° N. Muséum National d'Histoire Naturelle. Collection Patrimoines Naturels, 40.

Dungan, M. L. 1985. Competition, and the morphology, ecology, and evolution of acorn barnacles: An experimental test. *Paleobiology* 11: 165–173.

Dunn, J. R. 1997. Charles Henry Gilbert (1859–1928): Pioneer ichthyologist of the American West. In T. W. Pietsch and W. D. Anderson Jr. (eds.), *Collection Building in Ichthyology and Herpetology,* pp. 265–278. Special Publication 3, American Society of Ichthyologists and Herpetologists, Lawrence, KS.

Durham, J. W. 1947. *Corals from the Gulf of California and the North Pacific Coast of America.* Memoir 20, Geological Society of America, Boulder, CO.

———. 1950. *E. W. Scripps Cruise to the Gulf of California, Part II. Megascopic Paleontology and Marine Stratigraphy.* Memoir 43, Geological Society of America, Boulder, CO.

Eckert, S. A., K. L. Eckert, P. Ponganis, and G. L. Kooyman. 1989. Diving and foraging behavior of leatherback sea turtles (*Dermochelys coriacea*). *Canadian Journal of Zoology* 67: 2834–2840.

Eckert, S. A., and L. Sarti. 1997. Distant fisheries implicated in the loss of the world's largest leatherback nesting population. *Marine Turtle Newsletter* 78: 2–7.

Eder, T., and I. Sheldon. 2002. *Whales and Other Marine Mammals of California and Baja.* Lone Pine Publishing, Renton, WA.

Edwards, R. R. C. 1978. The fishery and fisheries biology of penaeid shrimp on the Pacific coast of Mexico. *Annual Reviews, Marine Biology* 16: 145–180.

Ehrhardt, N. M. 1991. Potential impact of a seasonal migratory jumbo squid (*Dosidicus gigas*) stock on a Gulf of California sardine (*Sardinops sagax caeruleus*) population. *Bulletin of Marine Science* 49(1/2): 325–332.

Ehrhardt, N. M., P. S. Jacquemin, F. García, G. González, J. M. López, J. Ortiz, and A. Solis. 1983. On the fishery and biology of the giant squid *Dosidicus gigas* in the Gulf of California, Mexico. In J. F. Caddy (ed.), *Advances in Assessment of World Cephalopod Resources,* pp. 306–340. FAO Fisheries Technical Paper no. 231, Food and Agriculture Organization of the United Nations, Rome.

Emilsson, I., and M. A. Alatorre. 1997. Evidencias de un remolino ciclónico de mesoescala en la parte sur del Golfo de California. In M. F. Lavín (ed.), *Contribuciones a la Oceanografía Física en México,* pp. 173–182. Monografía 3, Unión Geofísica Mexicana, Ensenada.

Engel, J., and R. Kvitek. 1998. Effects of otter trawling on a benthic community in Monterey Bay National Marine Sanctuary. *Conservation Biology* 12(6): 1204–1214.

Engilis, A., Jr., L. W. Oring, E. Carrera, J. W. Nelson, and A. M. Lopez. 1998. Shorebird surveys in Ensenada Pabellones and Bahía Santa Maria, Sinaloa, Mexico: Critical winter habitats for pacific flyway shorebirds. *Wilson Bulletin* 110: 332–341.

Enríquez-Andrade, R., G. Anaya-Reyna, J. C. Barrera-Guevara, M. A. Carvajal-Moreno, M. E. Martínez-Delgado, J. Vaca-Rodriguez, and C. Valdés-Casillas. 2005. An analysis of critical areas for biodiversity conservation in the Gulf of California Region. *Ocean and Coastal Management* 48: 31–50.

Erisman, B., M. Buckhorn, and P. Hastings. 2007. Spawning behavior and periodicity in the leopard grouper, *Mycteroperca rosacea,* from the central Gulf of California, Mexico. *Marine Biology* 151: 1849–1861.

Erisman, B. E., J. A. Rosales Casian, and P. A. Hastings. 2008. Evidence of gonochorism in a grouper, *Mycteroperca rosacea,* from the Gulf of California, Mexico. *Environmental Biology of Fishes* 82: 23–33.

Ernst, C. H., and R. W. Barbour. 1989. *Turtles of the World.* Smithsonian Institution Press, Washington, DC.

Escalante-Pliego, B. P. 1988. *Aves de Nayarit.* Universidad Autónoma de Nayarit, Tepic, Nayarit, México.
Escalona-Alcázar, F. J., and L. A. Delgado-Argote. 1998. Descripción estratigráfica de la zona El Paladar y litología de la isla Ángel de la Guarda, Golfo de California. GEOS, Unión Geofísica Mexicana, A.C., September, pp. 197–205.
Escalona-Alcázar, F. J., L. A. Delgado-Argote, M. López-Martínez, and G. Rendón-Márquez. 2001. Late Miocene volcanism and marine incursions in the San Lorenzo Archipelago, Gulf of California, Mexico. *Revista Mexicana de Ciencias Geológicas* 18(2): 111–128.
Espinosa-Pérez, M. del C., and M. E. Hendrickx. 1999. Isópodos. *Biodiversitas* 25: 1–6.
———. 2001a. A new species of *Exosphaeroma* Stebbing (Crustacea: Isopoda: Sphaeromatidae) from the Pacific coast of Mexico. *Proceedings of the Biological Society of Washington* 114(3): 640–648.
———. 2001b. Checklist of isopods (Crustacea: Peracarida: Isopoda) from the eastern tropical Pacific. *Belgian Journal of Zoology* 131(1): 41–54.
———. 2002. Distribution and ecology of isopods (Crustacea: Peracarida: Isopoda) of the Pacific coast of Mexico. In E. Escobar-Briones and F. Alvarez (eds.), *Modern Approaches to the Study of Crustacea,* pp. 95–104. Kluwer, Dordrecht.
———. 2006. A comparative analysis of biodiversity and distribution of shallow water marine isopods (Crustacea: Isopoda) from polar and temperate waters in the East Pacific. *Belgian Journal of Zoology* 136(2): 219–247.
Estados Unidos Mexicanos. 1993. *Ley General del Equilibrio Ecológico y la Protección al Ambiente.* Editorial Porrúa, México, D.F.
Evermann, B. W., and O. P. Jenkins. 1891. Report upon a collection of fishes made at Guaymas, Sonora, Mexico, with descriptions of new species. *Proceedings of the United States National Museum* 14(846): 121–165.
Evertt, W. T., and D. W. Anderson. 1991. Status and conservation of the breeding seabirds on offshore Pacific islands of Baja California and the Gulf of California. In J. P. Croxall (ed.), *Seabird Status and Conservation: A Supplement,* pp. 115–139. International Council for Bird Preservation, Technical Publication no. 11, Cambridge, U.K.
Ezcurra, E. 2003. Conservation and sustainable use of natural resources in Baja California. In J. G. Nelson, J. C. Day, and L. Sportza (eds.), *Protected Areas and the Regional Planning Imperative in North America,* pp. 279–296. University of Calgary Press, Calgary, Alberta, and Michigan State University Press, East Lansing, MI.
———. 2007. Hornaday, Lumholtz, and the grandeur of nature. In "Science on Desert and Lava: A Pinacate Centennial, Part I," Special Issue of *Journal of the Southwest* 49(2): 135–140.
Ezcurra, E., L. Bourillón, A. Cantú, M. E. Martínez, and A. Robles. 2002. Ecological conservation. In T. Case, M. Cody, and E. Ezcurra (eds.), *A New Island Biogeography of the Sea of Cortés,* pp. 417–444. Oxford University Press, New York.
Ezcurra, E., R. S. Felger, A. Russell, and M. Equihua. 1988. Freshwater islands in a desert sand sea: The hydrology, flora and phytogeography of the Gran Desierto oases of northwestern Mexico. *Desert Plants* 9(2): 35–44, 55–63.
Ezcurra, E., H. Fujita, E. Hambleton, and R. Ogarrio (eds.). 2003. *Isla Espíritu Santo. Evolución, rescate y conservación.* Fundación Mexicana para la Educación Ambiental, México, D.F.

Ezcurra, E., and V. Rodrigues. 1986. Rainfall patterns in the Gran Desierto, Sonora, Mexico. *Journal of Arid Environments* 10(1): 13–28.

FAO (Food and Agriculture Organization of the United Nations). 1995. Code of Conduct for Responsible Fisheries. Rome.

Farmer, W. 1968. *Tidepool Animals from the Gulf of California.* Wesword, San Diego, CA.

Fauchald, K. 1977. *The Polychaete Worms—Definitions and Keys to the Orders, Families and Genera.* Natural History Museum, Los Angeles County, CA.

Faulk, O. B. (ed.). 1969. *Derby's Report on the Opening of the Colorado 1850–1851.* University of New Mexico Press, Albuquerque, NM.

Felger, R. S. 1980. Vegetation and flora of the Gran Desierto, Sonora, Mexico. *Desert Plants* 2: 87–114.

———. 2000. *Flora of the Gran Desierto and Río Colorado of Northwestern Mexico.* University of Arizona Press, Tucson, AZ.

———. 2001. Coastal Wetlands. In P. Robles Gil, E. Ezcurra, and E. Mellink (eds.), *The Gulf of California, a World Apart,* pp. 159–181. Agrupación Sierra Madre, Mexico, D.F.

———. 2007. Living resources at the center of the Sonoran Desert: Native American plant and animal utilization. In R. S. Felger and W. Broyles (eds.), *Dry Borders. Great Natural Reserves of the Sonoran Desert,* pp. 147–192. University of Utah Press, Salt Lake City, UT.

Felger, R. S., and B. Broyles (eds.). 2007. *Dry Borders. Great Natural Reserves of the Sonoran Desert.* University of Utah Press, Salt Lake City, UT.

Felger, R. S., K. Cliffton, and P. J. Regal. 1976. Winter dormancy in sea turtles: Independent discovery and exploitation in the Gulf of California, Mexico, by two local cultures. *Science* 191: 283–285.

Felger, R. S., M. B. Johnson, and M. F. Wilson. 2000. *The Trees of Sonora, Mexico.* Oxford University Press, Oxford.

Felger, R. S., and E. Joyal. 1999. The palms (Arecaceae) of Sonora, Mexico. *Aliso* 18: 1–18.

Felger, R. S., and C. H. Lowe. 1976. The island and coastal vegetation and flora of the northern part of the Gulf of California. *Los Angeles County Museum Contributions in Science* 285.

Felger, R. S., and M. B. Moser. 1973. Eelgrass (*Zostera marina* L.) in the Gulf of California: Discovery of its nutritional value by the Seri Indians. *Science* 181: 355–356.

———. 1985. *People of the Desert and Sea: Ethnobotany of the Seri Indians.* University of Arizona Press, Tucson, AZ. [Reprinted in 1991.]

———. 1987. Sea turtles in Seri Indian culture. *Environment Southwest,* Autumn, pp. 18–21.

Felger, R. S., M. B. Moser, and E. W. Moser. 1980. Seagrasses in Seri Indian culture. In R. C. Phillips and C. P. McRoy (eds.), *Handbook of Seagrass Biology, and Ecosystem Perspective,* pp. 260–276. Garland STPM Press, New York.

Felger, R. S., W. J. Nichols, and J. A. Seminoff. 2004. Sea turtle conservation, diversity and desperation in northwestern Mexico. In J.-L. Cartron, G. Ceballos, and R. Felger (eds.), *Biodiversity, Ecosystems, and Conservation in Northern Mexico.* Oxford University Press, New York.

———. 2005. Sea turtles in northwestern Mexico: Conservation, ethnobiology, and desperation. In J.-L. E. Cartron, G. Ceballos, and R. S. Felger (eds.), *Biodiversity, Ecosys-*

tems, and Conservation in Northwestern Mexico, pp. 405–424. Oxford University Press, New York.

Felger, R. S., P. L. Warren, L. S. Anderson, and G. P. Nabhan. 1992. Vascular plants of a desert oasis: Flora and ethnobotany of Quitobaquito, Organ Pipe Cactus National Monument, Arizona. *Proceedings of the San Diego Society of Natural History* 8.

Ferguson, M. C., J. Barlow, S. B. Reilly, and T. Gerrodette. 2006. Predicting Cuvier's (*Ziphius cavostris*) and mesoplodon beaked whale population density from habitat characteristics in the eastern tropical Pacific Ocean. *Journal of Cetacean Research and Management* 7(3): 287–299.

Fernández, G. J., R. Carmona, and H. de la Cueva. 1998. Abundance and seasonal variation of western sandpipers (*Calidris mauri*) in Baja California Sur, Mexico. *Southwestern Naturalist* 43: 57–61.

Ferrusquía-Villafranca, I., and V. Torres-Roldán. 1980. El registro de mamíferos terrestres del Mesozoico y Cenozoico de Baja California. Universidad Nacional Autónoma de México, Instituto de Geología. *Revista* 4: 56–62.

Figueroa, A., J. Alvarado, F. Hernández, G. Rodríguez, and J. Robles. 1993. The ecological recovery of sea turtles of Michoacán, Mexico. Special attention to the black turtle (*Chelonia agassizi*). Final Report to WWF-USFWS, Albuquerque, NM.

Filloux, J. H. 1973. Tidal patterns and energy balance in the Gulf of California. *Nature* 243: 217–221.

Findley, L. T. 1976. Aspectos ecológicos de los esteros con manglares en Sonora y su relación con la explotación humana [Ecological aspects of mangrove estuaries in Sonora and their relation to human exploitation]. In B. Braniff and R. S. Felger (eds.), *Sonora: Antropología del Desierto,* pp. 95–106. Instituto Nacional de Antropología e Historia, Serie Científica Diversa no. 27, México, D.F.

Findley, L. T. (ed.). In prep. *Listado y distribución de la macrofauna del Golfo de California, México. Parte 2. Vertebrados* [A Distributional checklist of the macrofauna of the Gulf of California, Mexico. Part 2. Vertebrates]. Arizona-Sonora Desert Museum, Tucson, AZ.

Findley, L. T., and R. C. Brusca. 2005. Presentación de datos [Presentation of data]. In M. E. Hendrickx, R. C. Brusca, and L. T. Findley (eds.), *Listado y distribución de la macrofauna del Golfo de California, México. Parte 1. Invertebrados* [A distributional checklist of the Macrofauna of the Gulf of California, Mexico. Part 1. Invertebrates], pp. 25–40. Arizona-Sonora Desert Museum, Tucson, AZ.

Findley, L. T., M. E. Hendrickx, R. C. Brusca, A. M. van der Heiden, P. A. Hastings, and J. Torre. In press. *Macrofauna del Golfo de California* [Macrofauna of the Gulf of California]. CD-ROM Version 1.0, Macrofauna Golfo Project. Center for Applied Biodiversity Science, Conservation International, Washington, DC, and Región Golfo de California, Guaymas, Sonora, México.

Findley, L. T., and O. Vidal. 2002. Gray whale (*Eschrichtius robustus*) at calving sites in the Gulf of California. *Journal of Cetacean Research and Management* 14(1): 27–40.

Fischer, W., F. Krupp, W. Schneider, C. Sommer, K. E. Carpenter, and V. H. Niem (eds.). 1995a. *Guía FAO para la identificación de especies para los fines de la pesca, Pacifico centro-oriental, Volumen I, Plantas e invertebrados.* Organización de las Naciones Unidas para la Agricultura y la Alimentación, Roma.

——— (eds.). 1995b. *Guía FAO para la identificación para los fines de la pesca. Pacifico centro-oriental. Volumen II. Vertebrados—Parte 1:647–1200; Volumen III. Vertebrados.* Organización de las Naciones Unidas para la Agricultura y la Alimentación, Roma.

——— (eds.). 1995c. *Guía FAO para la identificación de especies para los fines de la pesca, Pacifico centro-oriental, Volumen 3, Vertebrados Parte 2.* Organización de las Naciones Unidas para la Agricultura y la Alimentación, Roma.

Flanagan, C. A., and J. R. Hendrickson. 1976. Observations on the commercial fishery and reproductive biology of the totoaba, *Cynoscion macdonaldi,* in the northern Gulf of California. *Fishery Bulletin* 74: 531–554.

Fleischner, T. L., and H. River Gates. 2009. Shorebird use of Estero Santa Cruz, Sonora, Mexico: Abundance, diversity and conservation implications. *Waterbirds* 32(1): 36–43.

Flessa, K. 2001. Panel III, Environmental issues. In *Colorado River Delta Bi-National Symposium Proceedings,* 11–12 September, p. 49. International Boundary and Water Commission, United States and Mexico, Mexicali.

Fletcher, J. M., B. P. Kohn, D. A. Foster, and A. J. W. Gleadow. 2000. Heterogeneous Neogene cooling and exhumation of the Los Cabos block, Southern Baja California: Evidence from fission-track thermochronology. *Geology* 28: 107–110.

Flores-Verdugo, F., F. González-Farias, and U. Zaragoza-Araujo. 1993. Ecological parameters of the mangroves of semi-arid regions of Mexico. Importance for ecosystem management. In H. Lieth and A. Masoom (eds.), *Toward the Rational Use of High Salinity Tolerant Plants,* vol. 1, pp. 123–132. Kluwer, Dordrecht.

Fontana, B. L. 1994. *Entrada. The Legacy of Spain and Mexico in the United States.* Southwest Parks and Monuments Association, Tucson, AZ, and University of New Mexico Press, Albuquerque, NM.

Forrest, M., J. Ledesma-Vázquez, W. Ussler III, J. T. Kulongoski, D. R. Hilton, and H. G. Greene. 2005. Gas geochemistry of a shallow submarine hydrothermal vent associated with the El Requesón fault zone, Bahía Concepción, Baja California Sur, Mexico. *Chemical Geology* 224: 82–95.

Foster, M. S. 2001. Rhodoliths: Between rocks and soft places. *Journal of Phycology* 37: 659–667.

Foster, M. S., L. M. McConnico, L. Lundsten, T. Wadsworth, T. Kimball, L. B. Brooks, M. Medina-Lopez, R. Riosmena-Rodríguez, G. Hernandez-Carmona, R. Vasquez-Elisando, S. Johnson, and D. L. Steller. 2007. The diversity and natural history of a *Lithothamnion muelleri–Sargassum horridum* community in the Gulf of California. *Ciencias Marinas* 33(4): 367–384.

Foster, M. S., R. Riosmena-Rodríguez, D. L. Steller, and W. J. Woelkerling. 1997. Living rhodolith beds in the Gulf of California and their implications for paleoenvironmental interpretation; Pliocene carbonates and related facies flanking the Gulf of California, Baja California, Mexico. *Geological Society of America Special Paper* 318: 127–139.

Fradkin, P. L. 1984. *A River No More: The Colorado River and the West.* University of Arizona Press, Tucson, AZ.

Frantz, B. R., M. Kashgarian, K. H. Coale, and M. S. Foster. 2000. Growth rate and potential climate record from a rhodolith using super(14)C accelerator mass spectrometry. *Limnology and Oceanography* 45(8): 1773–1777.

Frazier, J. 2003. Prehistoric and ancient historic interactions between humans and marine turtles. In P. L. Lutz, J. A. Musick, and J. Wyneken (eds.), *The Biology of Sea Turtles,* vol. 2, pp. 1–38. CRC Press, Boca Raton, FL.

Freiwald, A., R. Henrich, P. Schafer, and H. Willkomm. 1991. The significance of high-boreal to subarctic maerl deposits in northern Norway to reconstruct Holocene climatic changes and sea level oscillations. *Facies* 25: 315–340.

Fritzsche, R. A. 1980. Revision of the eastern Pacific Syngnathidae (Pisces: Syngnathiformes), including both recent and fossil forms. *Proceedings of the California Academy of Sciences* 42(6): 181–227.

Frizzell, V. A. (ed.). 1984. *Geology of Baja California.* Pacific Section, SEPM, Book 39, Society for Sedimentary Geology, Los Angeles, CA.

Fujita, H., and G. Poyatos de Paz. 1998. Settlement patterns on Espíritu Santo Island, Baja California Sur. *Pacific Coast Archaeological Society Quarterly* 34(4): 67–105.

Galil, B. S. 2007. Loss or gain? Invasive aliens and biodiversity in the Mediterranean Sea. *Marine Pollution Bulletin* 55: 314–322.

———. 2008. Alien species in the Mediterranean Sea—Which, when, where, why? *Hydrobiologia* 606: 105–116.

Galindo-Bect, M. S., E. P. Glenn, H. M. Page, K. Fitzsimmons, L. A. Galindo-Bect, J. M. Hernández-Ayon, R. L. Petty, J. García-Hernández, and D. Moore. 2000. Penaeid shrimp landings in the upper Gulf of California in relation to Colorado River freshwater discharge. *Fisheries Bulletin* 98: 222–225.

Gallo, J. P., and D. Aurioles. 1984. Distribución y estado actual de la población de foca común (*Phoca vitulina richardsi* [Gray, 1864]), en la Península de Baja California, México. *Anales del Instituto de Biología, serie Zoología* 55: 323–332.

Gallo-Reynoso, J. P. 1998. La vaquita marina y su hábitat crítico en el alto Golfo de California. Instituto Nacional de Ecología, SEMARNAP, México, D.F. *Gaceta Ecológica* 47: 29–44.

Galván-Magaña, F., L. A. Abitia-Cárdenas, J. Rodríguez-Romero, H. Pérez-España, and H. Chávez-Ramos. 1996. Lista sistemática de los peces de la Isla Cerralvo, Baja California Sur, México [Systematic list of the fishes from Cerralvo Island, Baja California Sur, Mexico]. *Ciencias Marinas* 22(3): 295–311.

Galván-Magaña, F., F. Gutiérrez-Sánchez, L. A. Abitia-Cárdenas, and J. Rodríguez-Romero. 2000. The distribution and affinities of the shore fishes of the Baja California Sur lagoons. In M. Munawar, S. G. Lawrence, I. F. Munawar, and D. F. Malley (eds.), *Aquatic Ecosystems of Mexico: Status and Scope,* pp. 383–398. Backhuys, Leiden, The Netherlands.

Galván-Magaña, F., H. J. Nienhuis, and P. A. Klimley. 1989. Seasonal abundance and feeding habits of sharks of the lower Gulf of California. *California Fish and Game* 75(2): 74–84.

Galván-Piña, V. H., F. Galván-Magaña, L. A. Abitia-Cárdenas, F. J. Gutiérrez-Sánchez, and J. Rodríguez-Romero. 2003. Seasonal structure of fish assemblages in rocky and sandy habitats in Bahía de La Paz, Mexico. *Bulletin of Marine Science* 72: 19–35.

Garcia, A., G. Ceballos, and R. Adaya. 2003. Intensive beach management as an improved sea turtle conservation strategy in Mexico. *Biological Conservation* 111: 253–261.

García-Caudillo, J. M. 1999. El uso de los excluidores de peces en la pesca comercial de camarón: Situación actual y perspectivas. *Pesca y Conservación* 3(7): 5.

García-Caudillo, J. M., M. A. Cisneros-Mata, and A. Balmori-Ramírez. 2000. Performance of a by-catch reduction device in the shrimp fishery of the Gulf of California, Mexico. *Biological Conservation* 92: 199–205.

García-Caudillo, J. M., and J. V. Gómez-Palafox. 2005. *La pesca industrial de camarón en el Golfo de California: Situación económico-financiera e impactos socio-ambientales.* Conservación Internacional–Región Golfo de California. Guaymas, Sonora, México.

García-Hernández, J., K. King, A. Velaso, E. Shumilin, M. Morea, and E. Glenn. 2001. Selenium, selected inorganic elements, and organochlorine pesticides in bottom material and biota from the Colorado River delta. *Journal of Arid Environments* 49(1): 65–89.

Garcillán, P. P., and E. Ezcurra. 2003. Biogeographic regions and beta-diversity of woody dryland legumes in the Baja California peninsula. *Journal of Vegetation Science* 14(6): 859–868.

Gardner, S. C., and W. J. Nichols. 2001. Assessment of sea turtle mortality rates in the Bahía Magdalena region, Baja California Sur, Mexico. *Chelonian Conservation and Biology* 4: 197–199.

Garman, S. 1899. The fishes. In "Reports on an exploration off the west coasts of Mexico, Central and South America, and off the Galapagos Islands by the U.S. Fish Commission steamer 'Albatross,' during 1891, no. XXVI." *Memoirs of the Museum of Comparative Zoology* 24: 1–431.

Gastil, R. G., D. Krummenacher, and J. Minch. 1979. The record of Cenozoic volcanism around the Gulf of California. *Geological Society of America Bulletin* 90: 839–857.

Gastil, R. G., J. Minch, and R. P. Phillips. 1983. The geology and ages of the islands. In T. J. Case and M. L. Cody (eds.), *Island Biogeography in the Sea of Cortez,* pp. 13–25. University of California Press, Berkeley, CA.

Gastil, R. G., J. Neuhaus, M. E. Cassidy, J. T. Smith, J. C. Ingle Jr., and D. Krummenacher. 1999. Geology and paleontology of southwestern Isla Tiburón, Sonora, Mexico. *Revista Mexicana de Ciencias Geológicas* 16(1): 1–34 [geologic map scale 1:25,000].

Gastil, R. G., R. P. Phillips, and E. C. Allison. 1975. *Reconnaissance Geology of the State of Baja California.* Memoir 140, pp. 1–70 [1973], Geological Society of America, Boulder, CO [geologic map scale 1:250,000].

Gaxiola-Castro, G., S. Alvarez-Borrego, and R. A. Schwartzlose. 1978. Sistema del bióxido de carbono en el Golfo de California. *Ciencias Marinas* 2: 25–40.

Gell, F., and C. Roberts. 2003. Benefits beyond boundaries: The fishery effects of marine reserves. *Trends in Ecology and Evolution* 18(9): 448–454.

Gendron, D. 2000. Familia Kogiidae. In S. T. Alvarez-Castañeda and J. L. Patton (eds.), *Mamíferos del Noroeste de México II,* pp. 639–641. Centro de Investigaciones Biológicas del Noroeste, S.C., La Paz, Baja California Sur.

———. 2002. Ecología poblacional de la ballena azul *Balaenoptera musculus* de la Península de Baja California. Ph.D. thesis, CICESE. Ensenada, B.C., México.

Gendron, D., and S. Chavez. 1996. Recent sei whale (*Balaenoptera borealis*) sightings in the Gulf of California, Mexico. *Aquatic Mammals* 22: 127–130.

Gendron, D., and T. Gerrodette. 2003. First abundance estimates of blue whales in Baja

California waters from slip and aerial surveys. Abstracts, 15th Biennial Conference on the Biology of Marine Mammals, Greensboro, NC.

Gendron, D., S. Lanham, and M. Carardine. 1999. North Pacific right whale (Eubalaena glacialis) sighting south of Baja California, Mexico. *Aquatic Mammals* 25: 31–34.

Gerrodette, T., and D. M. Palacios. 1996. Estimates of cetacean abundance in EEZ waters of the eastern tropical Pacific. Southwest Fisheries Science Center Administrative Report no. LJ-96-10, National Marine Fisheries Service, La Jolla, CA.

Gieskes, J. M., T. Shaw, T. Brown, A. Sturz, and A. C. Campbell. 1991. Interstitial water and hydrothermal water chemistry, Guaymas Basin, Gulf of California. In J. P. Dauphin and B. R. Simoneit (eds.), *The Gulf of California and Peninsular Province of the Californias,* pp. 753–779. Memoir 47, American Association of Petroleum Geologists, Tulsa, OK.

Gilbert, C. H. 1890. Scientific results of explorations by the U.S. Fish Commission steamer Albatross. XII. A preliminary report on the Pacific coast of North America during the year 1889, with descriptons of twelve new genera and ninety-two new species. *Proceedings of the United States National Museum* 13: 49–126.

———. 1892. Scientific results of explorations by the U.S. Fish Commission steamer Albatross. XXII. Descriptions of thirty-four new species of fishes collected in 1888 and 1889, principally among the Santa Barbara Islands and in the Gulf of California. *Proceedings of the United States National Museum* 14 [for 1891], pp. 539–566.

Gilbert, J. Y., and W. E. Allen. 1943. The phytoplankton of the Gulf of California obtained by the "E.W. Scripps" in 1939 and 1940. *Journal of Marine Research* 5: 89–110.

Gill, T. H. 1862–1863. Catalogue of the fishes of Lower California in the Smithsonian Institution, collected by Mr. J. Xantus, parts 1–4. *Proceedings of the Academy of Natural Sciences of Philadelphia* 1862: 140–151, 242–246, 249–262; 1863: 80–88.

Gilligan, M. R. 1991. Bergmann ecogeographic trends in triplefin blennies (Teleostei: Tripterygiidae) in the Gulf of California, Mexico. *Environmental Biology of Fishes* 31: 301–305.

Gilmore, R. M. 1960. A census of the California gray whale. U.S. Fish and Wildlife Service, Special Scientific Report: Fisheries 342.

Gilmore, R. M., R. L. Brownell Jr., J. G. Mills, and A. Harrison. 1967. Gray whales near Yavaros, southern Sonora, Golfo de California, México. *Transactions of the San Diego Society of Natural History* 14: 197–204.

Glenn, E. P., R. Felger, A. Burquez, and D. Turner. 1992. Cienega de Santa Clara: Endangered wetland in the Colorado Delta. *Natural Resources Journal* 32: 817–824.

Glenn, E. P., J. Garcia, R. Tanner, C. Congdon, and D. Luecke. 1999. Status of wetlands supported by agricultural drainage water in the Colorado River Delta, Mexico. *HortScience* 34(1): 39–45.

Glenn, E. P., C. Lee, R. Felger, and S. Zengel. 1996. Effects of water management on the wetlands of the Colorado River Delta, Mexico. *Conservation Biology* 10(4): 1175–1186.

Glenn, E. P., P. L. Nagler, R. C. Brusca, and Osvel Hinojosa-Huerta. 2006. Coastal wetlands of the northern Gulf of California: Inventory and conservation status. *Aquatic Conservation: Marine and Freshwater Ecosystems* 16: 5–28.

Glenn, E., R. Tanner, S. Mendez, T. Kehret, D. Moore, J. Garcia, and C. Valdés-Casillas.

1998. Growth rates, salt tolerance and water use characteristics of native and invasive riparian plants from the delta of the Colorado River, Mexico. *Journal of Arid Environments* 40: 281–294.

Glenn, E. P., F. Zamora-Arroyo, P. L. Nagler, M. Briggs, W. Shaw, and K. Flessa. 2001. Ecology and conservation biology of the Colorado River delta, Mexico. *Journal of Arid Environments* 49: 5–15.

Goebel, T., M. R. Waters, and D. H. O'Rourke. 2008. The Late Pleistocene dispersal of modern humans in the Americas. *Science* 319: 1497–1502.

Gómez, P. 1998. First record and new species of *Gastrophanella* (Porifera: Demospongiae: Lithistida) from the central east Pacific. *Proceedings of the Biological Society of Washington* 111(4): 774–780.

Gómez, P., and G. J. Bakus. 1992. *Aplysina gerardogreeni* and *Aplysina aztecus* (Porifera: Demospongiae), new species from the Mexican Pacific. *Anales Instituto Ciencias del Mar y Limnología UNAM* 19(2): 175–180.

Gómez, P., J. L. Carballo, L. E. Vázquez, and J. A. Cruz. 2002. New records for the sponge fauna (Porifera: Demospongiae) of the Pacific coast of Mexico (eastern Pacific Ocean). *Proceedings of the Biological Society of Washington* 115(1): 223–237.

Gómez-Noguera, S. E., and M. E. Hendrickx. 1997. Distribution and abundance of meio fauna in a subtropical coastal lagoon in the southeastern Gulf of California Mexico. *Marine Pollution Bulletin* 34(7): 582–587.

Gómez-Pompa, A., and R. Dirzo. 1995. *Reservas de la biosfera y otras áreas naturales protegidas de México.* INE/CONABIO, México, D.F.

Gómez-Ponce, M. 1971. Sobre la presencia de estratos marinos del Mioceno en el Estado de Sonora, México. Instituto Mexicano del Petróleo Revista, Notas Técnicas, Octubre, 77–78.

González-Acosta, G., J. De la Cruz-Agüero, and G. Ruíz-Campos. 1999. Ictiofauna asociada al manglar del estero El Conchalito, Ensenada de La Paz, Baja California Sur, México. *Oceánides* 14: 121–131.

González-Armas, R., R. Polomares-Garca, and R. De Silva-Dávila. 2002. Copepod and macrozooplankton distribution associated to El Bajo Espíritu Santo Seamount. *Contribuciones al Estudio de los Crustáceos del Pacífico Este* 1: 1–11.

González-Cabello, A. 2003. Variabilidad espacio-temporal de las asociaciones de peces crípticos en áreas arrecifales coralinas y rocosas de la región de La Paz, B.C.S. Master's thesis, Centro de Investigaciones Biológicas del Noroeste, S.C., La Paz, B.C.S., México.

González-Peral, U. A. 2006. Identidad poblacional de las ballenas jorobadas (*Megaptera novaeangliae*) que se congregan en Baja California Sur. Master's thesis, Posgrado en Ciencias Marinas y Costeras, Universidad Autónoma de Baja California Sur.

Gordon, H. R., and A. Morel. 1983. *Remote Assessment of Ocean Color for Interpretation of Satellite Visible Imagery: A Review.* Lecture Notes on Coastal and Estuarine Studies, vol. 4, Springer-Verlag, New York.

Gotshall, D. W. 1994. *Guide to Marine Invertebrates: Alaska to Baja California.* Sea Challengers, Monterey, CA.

———. 1998. *Sea of Cortez Marine Animals.* Sea Challengers, Monterey, CA.

Graham, L. E., and L. W. Wilcox. 2000. *Algae.* Prentice-Hall, Upper Saddle River, NJ.
Grall, J., and M. Glemarec. 1997. Biodiversité des fonds de maerl en Bretagne: Approache fonctionnelle et impacts anthropogeniques. *Vie Milieu* 47: 339–349.
Granados-Gallegos, J. L., and R. A. Schwartzlose. 1974. Corrientes superficiales en el Golfo de California. In F. A. Manrrique (ed.), *Memorias del V Congreso Nacional de Oceanografía,* pp. 271–285. Escuela de Ciencias Marítimas del Instituto Tecnológico de Monterrey, Guaymas.
Grant, P. R., and I. McT. Cowan. 1964. A review of the avifauna of the Tres Marias Islands, Nayarit, Mexico. *Condor* 66: 221–228.
Grayson, A. J. 1871. The physical geography and natural history of the islands of the Tres Marias. *Proceedings of the Boston Natural History Society* 14: 261–302.
Greenberg, J. B. In press. Territorialization, globalization, and dependent capitalism in the political ecology of fisheries in the upper Gulf of California. In A. Biersack and J. B. Greenberg (eds.), *Culture, History, Power, Nature: Ecologies for the New Millennium.* Bureau of Applied Research in Anthropology, University of Arizona, Tucson, AZ.
Greenberg, J. B., and C. Vélez-Ibáñez. 1993. Community dynamics in a time of crisis: An ethnographic overview of the upper Gulf. In T. R. McGuire and J. B. Greenberg (eds.), *Maritime Community and Biosphere Reserve: Crisis and Response in the Upper Gulf of California.* Occasional Paper no. 2, Bureau of Applied Research in Anthropology, University of Arizona, Tucson, AZ.
Grigg, R. W., and R. Hey. 1992. Paleoceanography of the tropical eastern Pacific Ocean. *Science* 255: 172–178.
Grijalva-Chon, J. M., S. Núñez-Quevedo, and R. Castro-Longoria. 1996. Ictiofauna de la laguna costera La Cruz, Sonora, México (Ichthyofauna of La Cruz coastal lagoon, Sonora, Mexico). *Ciencias Marinas* 22(2): 129–150.
Grinnell, J. 1928. A distributional summation of the ornithology of Lower California. *University of California Publications in Zoology* 32(1): 1–300.
Grismer, L. L. 1994. The origin and evolution of the peninsular herpetofauna of Baja California, Mexico. *Herpetological Natural History* 2: 51–106.
———. 2000. Evolutionary biogeography on Mexico's Baja California peninsula: A synthesis of molecules and historical geology. *Proceedings of the National Academy of Sciences* 97: 14017–14018.
———. 2002. *Amphibians and Reptiles of Baja California, Including Its Pacific Islands and the Islands in the Sea of Cortés.* University of California Press, Berkeley, CA.
Groombridge, B., and R. Luxmoore. 1989. *The Green Turtle and Hawksbill (Reptilia: Cheloniidae): World Status, Exploitation and Trade.* Secretariat of the Convention on International Trade in Endangered Species of Wild Fauna and Flora, Lausanne, Switzerland.
Guala, I., M. M. Flagella, N. Andreakis, G. Procaccini, W. H. C. F. Kooistra, and M. C. Buia. 2003. Aliens—algal introductions to European shores. *Biogeographia* 24: 45–52.
Guerrero-Ruíz, M., H. Pérez-Cortés M., M. Salinas Z., and J. Urbán R. 2006a. First mass stranding of killer whales (*Orcinus orca*) in the Gulf of California, Mexico. *Aquatic Mammals* 32(3): 265–272.
Guerrero-Ruíz, M., J. Urbán R., and L. Rojas-Bracho. 2006b. *Las ballenas del Golfo de California.* Instituto Nacional de Ecología, Secretaría del Medio Ambiente y Recursos Naturales.

Guevara-Escamilla, S. R. 1974. Ecología de peces del alto Golfo de California y delta del Río Colorado. Master's thesis, Universidad Autónoma de Baja California, Ensenada.

Günther, A. 1864. On some new species of Central-American fishes. *Proceedings of the Zoological Society, London* 1864(part 1): 23–27.

Gutiérrez, O. Q., S. G. Marinone, and A. Parés-Sierra. 2004. Lagrangian surface circulation in the Gulf of California from a 3D numerical model. *Deep-Sea Research II* 51: 659–672.

Guzmán, J., R. Carmona, E. Palacios, and M. Bojórquez. 1994. Seasonal distribution of aquatic birds in Estero de San José del Cabo, B.C.S., Mexico. *Ciencias Marinas* 20: 93–103.

Halfar, J. 1999. Warm-temperate to subtropical shallow water carbonates of the southern Gulf of California and geochemistry of rhodoliths. Ph.D. dissertation, Stanford University.

Halfar, J., L. Godinez-Orta, and J. C. Ingle Jr. 2000a. Microfacies analysis of recent carbonate environments in the southern Gulf of California, Mexico; a model for warm-temperate to subtropical carbonate formation. *Palaios* 15: 323–342.

Halfar, J., L. Godinez-Orta, M. Mutti, J. Valdez-Holguin, and J. Borges. 2004a. Nutrient and temperature controls on modern carbonate production: An example from the Gulf of California, Mexico. *Geology* 32: 213–216.

Halfar, J., J. C. Ingle Jr., and L. Godinez-Orta. 2004b. Modern non-tropical mixed carbonate-siliciclastic sediments and environments of the southwestern Gulf of California, Mexico. *Sedimentary Geology* 165: 93–115.

Halfar, J., and M. Mutti 2005. Global dominance of coralline red-algal facies: A response to Miocene oceanographic events. *Geology* 33: 481–484.

Halfar, J., T. Zack, A. Kronz, and J. C. Zachos. 2000b. Growth and high-resolution paleoenvironmental signals of rhodoliths (coralline red algae): A new biogenic archive. *Journal of Geophysical Research* 105(C9): 22107–22116.

Hall, H. 1992. *Shadows in a Desert Sea.* Video. Howard Hall Productions, Del Mar, CA.

Hall-Spencer, J. M. 1998. Conservation issues relating to maerl beds as habitats for molluscs. *Journal of Conchology Special Publication* 2: 271–286.

———. 1999. Effects of towed demersal fishing gear on biogenic sediments: A 5-year study. In O. Giovanardi (ed.), *Impact of Trawl Fishing on Benthic Communities,* pp. 9–20. ICRAM, Rome.

Hall-Spencer, J. M., and P. G. Moore. 2000. Impact of scallop dredging on maerl grounds. In M. J. Kaiser and S. J. de Groot (eds.), *Effects of Fishing on Non-Target Species and Habitats: Biological, Conservation and Socio-Economic Issues,* pp. 105–117. Blackwell, Oxford, U.K.

Hamilton, W. 1961. Origin of the Gulf of California. *Geological Society of America Bulletin* 72: 1307–1318.

Hammann, M. G., T. R. Baumgartner, and A. Badan-Dangon. 1988. Coupling of the Pacific sardine (*Sardinops sagax caeruleus*) life cycle with the Gulf of California pelagic environment. *CalCOFI Reports* 29: 102–109.

Hammann, M. G., M. O. Nevárez-Martínez, and Y. Green-Ruíz. 1998. Spawning habit of the Pacific sardine (*Sardinops sagax*) in the Gulf of California: Egg and larval distribution 1956–1957 and 1971–1991. *CalCOFI Reports* 39: 169–179.

Hariot, P. 1895. Algues du Golfe de Californie recueillies par M. Diguet. *Journal Botanic* 9: 167–170.

Harris, M. F. 1969. Effects of tropical cyclones upon Southern California. Master's thesis, San Fernando Valley State College, San Fernando, CA.

Harvey, A. S., S. Broadwater, W. J. Woelkerling, and P. Mitrovski. 2003. *Choreonema* (Corallinales, Rhodophyta): 18S rDNA phylogeny and resurrection of the Hapalidiaceae for the subfamilies Choreonematoidea, Austrolithoideae and Melobesioidae. *Journal of Phycology* 39: 988–998.

Harvey, A., and W. J. Woelkerling. 2007. A guide to nongeniculate coralline red algal (Corallinales, Rhodophyta) rhodolith identification. *Ciencias Marinas* 33(4): 411–426.

Hastings, A., and L. W. Botsford. 1999. Equivalence in yield from marine reserves and traditional fisheries management. *Science* 284: 1537–1538.

Hastings, P. A. 1988a. Correlates of male reproductive success in the browncheek blenny, *Acanthemblemaria crockeri* (Chaenopsidae). *Behavioral Ecology and Sociobiology* 22: 95–102.

———. 1988b. Female choice and male reproductive success in the angel blenny, *Coralliozetus angelica* (Chaenopsidae). *Animal Behaviour* 36: 115–124.

———. 1989. Protogynous hermaphroditism in *Paralabrax maculatofasciatus* (Pisces: Serranidae). *Copeia* 1989(1): 184–188.

———. 1991. Flexible responses to predators in a marine fish. *Ethology, Ecology and Evolution* 3(3): 177–184.

———. 1992. Nest-site size as a short-term constraint on the reproductive success of paternal fishes. *Environmental Biology of Fishes* 20: 213–218.

———. 2000. Biogeography of the tropical eastern Pacific: Distribution and phylogeny of chaenopsid fishes. *Zoological Journal of the Linnean Society* 128: 319–335.

Hastings, P. A., and L. T. Findley. 2006. Marine fishes of the Upper Gulf Biosphere Reserve, Northern Gulf of California. In R. S. Felger and W. Broyles (eds.), *Dry Borders. Great Natural Reserves of the Sonoran Desert,* pp. 364–382. University of Utah Press, Salt Lake City, UT.

Hastings, P. A., and C. W. Petersen. 1986. A novel sexual pattern in serranid fishes: Simultaneous hermaphrodites and secondary males in *Serranus fasciatus. Environmental Biology of Fishes* 15: 59–68.

Hastings, P. A., and D. R. Robertson (eds.). 2001. Systematics of tropical eastern Pacific fishes. *Revista de Biología Tropical* 49(suppl. 1).

Hausback, B. P. 1984. Cenozoic volcanic and tectonic evolution of Baja California Sur, Mexico. In V. A. Frizzell (ed.), *Geology of the Baja California,* pp. 219–236. Pacific Section, SEPM, Book 39, Society for Sedimentary Geology, Los Angeles, CA.

Hayden, J. D. 1956. Notes on the archaeology of the central coast of Sonora, Mexico. *Kiva* 21: 19–22.

Hayden, J. D., and J. W. Dykinga (photographer). 1998. *The Sierra Pinacate.* Southwest Center Series, University of Arizona Press, Tucson, AZ.

Heath, W. G. 1967. Ecological significance of temperature tolerance in Gulf of California shore fishes. *Journal of the Arizona Academy of Sciences* 4: 172–178.

Hedgpeth, J. W. 1945. The United States Fish Commission Steamer Albatross. *Neptune* 5: 5–26.

———. 1978a. *The Outer Shores,* part 1: *Ed Ricketts and John Steinbeck Explore the Pacific Coast.* Mad River Press, Eureka, CA.

———. 1978b. *The Outer Shores,* part 2: *Breaking Through.* Mad River Press, Eureka, CA.

Helenes, J., and A. L. Carreño. 1999. Neogene sedimentary evolution of Baja California in relation to regional tectonics. In C. Lomnitz (ed.), "Earth Sciences in Mexico: Some Recent Perspectives," *Journal of South American Sciences, Special Issue* 12(6): 589–605.

Helenes, J., A. L. Carreño, M. A. Esparza, and R. M. Carrillo. 2005. Neogene paleontology in the Gulf of California and the geologic evolution of Baja California. VII International Meeting on the Geology of the Baja California Peninsula. Ensenada, B.C. *Proceedings,* p. 2.

Helenes-Escamilla, J. 1980. Stratigraphy, depositional environments and Foraminifera of the Miocene Tortugas Formation, Baja California, Mexico. *Boletín de la Sociedad Geológica Mexicana* 41(1–2): 47–67.

Henderson, D. A., 1972. *Men and Whales at Scammon's Lagoon.* Dawson's Book Shop, Los Angeles, CA.

———. 1984. Nineteenth century gray whale whaling: Grounds, catches and kills. Practices and depletion of the whale population. In M. L. Jones, S. L. Swartz, and S. Leatherwood (eds.), *The Gray Whale* Eschrichtius robustus, pp. 159–186. Academic Press, Orlando, FL.

Hendrickson, J. R. 1973. Study of the marine environment in the northern Gulf of California. Final Report, National Technical Information Service Publication N74-16008, Washington, DC.

Hendrickx, M. E. 1984. The species of *Sicyonia* H. Milne Edwards (Crustacea: Penaeoidea) of the Gulf of California, Mexico, with a key for their identification and a note on their zoogeography. *Revista de Biología Tropical* 32(2): 279–298.

———. 1985. Diversidad de los macroinvertebrados bentónicos acompañantes del camarón en el área del Golfo de California y su importancia como recurso potencial. Cap. 3. In A. Yañez-Arancibia (ed.), *Recursos potenciales pesqueros de México: La pesca del camarón,* pp. 95–148. Programa Universitario de Alimentos, Instituto de Ciencias del Mar y Limnología, Instituto Nacional de Pesca, UNAM, México.

———. 1990. The stomatopod and decapod crustaceans collected during the GUAYTEC II Cruise in the Central Gulf of California, Mexico, with the description of a new species of Plesionika Bate (Caridea: Pandalidae). *Revista de Biología Tropical* 38(1): 35–53.

———. 1993. Crustáceos decápodos del Pacífico Mexicano. In S. I. Salazar-Vallejo and N. E. González (eds.), *Biodiversidad Marina y Costera de México,* pp. 271–318. Comisión Nacional para el Estudio de la Biodiversidad y CIQRO, México.

———. 1995a. Checklist of lobster-like decapod crustaceans (Crustacea: Decapoda: Thalassinidea, Astacidea and Palinuridea) from the Eastern Tropical Pacific. *Anales del Instituto de Biología, UNAM* 66(2): 153–165.

———. 1995b. Checklist of brachyuran crabs (Crustacea: Decapoda) from the Eastern Tropical Pacific. *Bulletin de l'Institut Royal des Sciences Naturelles de Belgique* 65: 125–150.

———. 1996a. Habitats and biodiversity of decapod crustaceans in the southeastern Gulf of California. *Revista de Biología Tropical* 44(2): 603–617.

———. 1996b. *Los camarones Penaeoidea bentónicos (Crustacea: Decapoda: Dendrobranchiata) del Pacífico mexicano.* Comisión Nacional para el Conocimiento y Uso de la Biodiversidad e Instituto de Ciencias del Mar y Limnología, UNAM, México.

———. 1997. *Los cangrejos braquiuros (Crustacea: Brachyura: Dromiidae, hasta Leucosiidae) del Pacífico mexicano.* Comisión Nacional para el Conocimiento y Uso de la Biodiversidad e Instituto de Ciencias del Mar y Limnología, UNAM, México.

———. 1999. *Los cangrejos braquiuros (Crustacea: Brachyura: Majoidea y Parthenopoidea) del Pacífico mexicano.* Comisión Nacional para el Conocimiento y Uso de la Biodiversidad e Instituto de Ciencias del Mar y Limnología, UNAM, México.

———. 2001. Occurrence of a continental slope deep-water decapod crustacean community along the edge of the minimum oxygen zone in the southeastern Gulf of California, Mexico. *Belgian Journal of Zoology* 131: 71–86.

——— (ed). 2002. *Contribuciones al estudio de los crustáceos del Pacífico Este,* vol. 1. Instituto de Ciencias del Mar y Limnología, UNAM, México.

Hendrickx, M. E., and R. C. Brusca. 2003. Biodiversidad de los invertebrados marinos de Sinaloa. In J. L. Cifuentes (ed.), *Atlas de los ecosistemas y la biodiversidad de Sinaloa,* pp. 141–163. Centro de Ciencias de Sinaloa, México.

Hendrickx, M. E., R. C. Brusca, M. Cordero, and G. Ramírez R. 2007. Marine and brackish-water molluscan biodiversity in the Gulf of California, Mexico. *Scientia Marina* 71(4): 637–647.

Hendrickx, M. E., R. C. Brusca, and L. T. Findley (eds.). 2005. *A Distributional Checklist of the Macrofauna of the Gulf of California, Mexico,* part I: *Invertebrates.* Arizona-Sonora Desert Museum, Tucson, AZ, and Conservation International, Washington, DC.

Hendrickx, M. E., R. C. Brusca, and G. Ramirez Resendiz. 2002. Biodiversity of macrocrustaceans in the Gulf of California, Mexico. In M. E. Hendrickx (ed.), *Contributions to the Study of East Pacific Crustaceans,* vol. 1, pp. 349–368. Instituto de Ciencias del Mar y Limnología, UNAM, Mexico.

Hendrickx, M. E., and F. D. Estrada Navarrete. 1989. A checklist of the species of pelagic shrimps (Penaeoidea and Caridea) from the eastern Pacific with notes on their geographic and depth distribution. *CalCOFI Reports* 30: 104–121.

———. 1996. *Los camarones pelágicos (Crustacea: Dendrobranchiata y Caridea) del Pacífico mexicano.* Comisión Nacional para el Conocimiento y Uso de la Biodiversidad e Instituto de Ciencias del Mar y Limnología, UNAM, México.

Hendrickx, M. E., and A. W. Harvey. 1999. Checklist of anomuran crabs (Crustacea: Decapoda) from the Eastern Tropical Pacific. *Belgian Journal of Zoology* 129(2): 327–352.

Hendrickx, M. E., and J. Salgado Barragán. 1989. Ecology and fishery of stomatopods in the Gulf of California. In E. A. Ferrero, R. B. Manning, M. L. Reaka, and W. Wales (eds.), *Biology of Stomatopods.* Collana UZI: Selected Symposia and Monographs, Mucchi Editore, Modena, Italy.

———. 1991. Los estomatópodos (Crustacea: Hoplocarida) del Pacífico mexicano. Publicación Especial, *Instituto de Ciencias del Mar y Limnología, UNAM* 10: 1–200.

Henny, C. J. 1988. Osprey. In R. S. Palmer (ed.), *Handbook of North American Birds,* vol. 4, pp. 73–101. Yale University Press, New Haven, CT.

Henny, C. J., and D. W. Anderson. 1979. Osprey distribution, abundance, and status in western North America: III. The Baja California and Gulf of California population. *Bulletin of the Southern California Academy of Science* 78: 89–106.

———. 2004. Status of nesting ospreys in coastal Baja California, Sonora and Sinaloa, Mexico, 1977 and 1992–1993. *Bulletin of the Southern California Academy of Sciences* 103: 95–114.

Henny, C. J., D. W. Anderson, A. Castellanos Vera, and J.-L. E. Cartron. 2008. Region-wide trends of nesting Ospreys in northwestern Mexico: A three decade perspective. *Journal of Raptor Research* 42: 229–242.

Henny, C. J., and R. Anthony. 1989. Bald eagle and osprey. In *Western Raptor Management Symposium and Workshop,* pp. 66–75. Scientific and Technical Series no. 12, National Wildlife Federation, Washington, DC.

Henny, C. J., and H. M. Wight. 1969. An endangered osprey population: Estimates of mortality and production. *Auk* 86: 188–198.

Hernández-Ayón, M., S. Galindo-Bect, B. P. Flores-Báez, and S. Alvarez-Borrego. 1993. Nutrient concentrations are high in the turbid waters of the Colorado River delta. *Estuarine, Coastal and Shelf Science* 37: 593–602.

Hernández-Herrera, A., E. Morales-Bojórquez, M. Cisneros-Mata, M. O. Nevárez-Martínez, and G. I. Rivera-Parra. 1998. Management strategy for the giant squid (*Dosidicus gigas*) fishery in the Gulf of California, Mexico. *CalCOFI Report* 39.

Hertlein, L. G., and E. K. Jordan. 1927. Paleontology of the Miocene of Lower California. *Proceedings of the California Academy of Sciences, ser. 4* 16(19): 605–647.

Hetzinger, J., J. Halfar, B. Riegl, and L. Godinez-Orta. 2006. Sedimentology and acoustic mapping of modern rhodolith beds on a non-tropical carbonate shelf (Gulf of California, Mexico). *Journal of Sedimentary Research* 76: 670–682.

Heyning, J. E., and W. F. Perrin. 1994. Evidence for two species of common dolphins (genus *Delphinus*) from the eastern North Pacific. *Los Angeles County Museum Contributions in Science* 442: 1–35.

Hickey, J. J., and D. W. Anderson. 1968. Chlorinated hydrocarbons and eggshell changes in raptorial and fish-eating birds. *Science* 162: 271–273.

Hidalgo-González, R. M., and S. Alvarez-Borrego. 2004. Total and new production in the Gulf of California estimated from ocean color data from the satellite sensor SeaWIFS. *Deep Sea Research II* 51: 739–752.

Hidalgo-González, R. M., S. Alvarez-Borrego, and A. Zirino. 1997. Mixing in the region of the Midriff Islands of the Gulf of California: Effect on surface pCO_2. *Ciencias Marinas* 23: 317–327.

Hilton-Taylor, C. (compiler). 2000. *2000 IUCN Red List of Threatened Species.* IUCN, Gland, Switzerland.

———. 2004. *IUCN Red List of Threatened Species.* World Conservation Union, Gland, Switzerland.

Hily, C., P. Potin, and J.-Y. Floc'h. 1992. Structure of subtidal algal assemblages on soft-bottom sediments: Fauna/flora interactions and role of disturbances in the Bay of Brest, France. *Marine Ecology Progress Series* 85: 115–130.

Hinojosa-Arango, G., and R. Riosmena-Rodríguez. 2004. Influence of rhodolith-forming species and growth-form on associated fauna of rhodolith beds in the Central-West Gulf of California, Mexico. *PSZN Marine Ecology* 25: 109–127.

Hinojosa-Huerta, O., S. DeStefano, Y. Carillo-Guerrero, W. Shaw, and C. Valdes. 2004. Waterbird communities and associated wetlands of the Colorado River delta, Mexico. *Studies in Avian Biology* 27: 52–60.

Hinojosa-Huerta, O., J. García-Hernández, and W. Shaw. 2002. Andrade Mesa wetlands of the All-American Canal. *Natural Resources Journal* 42: 899–914.

Hinton, S. 1987. *Seashore Life of Southern California.* University of California Press, Berkeley, CA.

Hobson, E. S. 1968. Predatory behavior of some shore fishes in the Gulf of California. *Bureau of Sports Fishing and Wildlife, Research Reports* 73: 1–73.

Hodgson, W. C. 2000. *Food Plants of the Sonoran Desert.* University of Arizona Press, Tucson, AZ.

Hoelzer, G. A. 1990. Male-male competition and female choice in the Cortez damselfish, *Stegastes rectifraenum. Animal Behaviour* 40: 339–349.

———. 1995. Filial cannibalism and male parental care in damselfishes. *Bulletin of Marine Science* 57: 663–671.

Hoese, D. F. 1976. Variation, synonymy and a redescription of the gobiid fish *Aruma histrio,* and a discussion of the related genus *Ophiogobius. Copeia* 1976(2): 295–305.

Hoese, D. F., and H. K. Larson. 1985. Revision of the eastern Pacific species of the genus *Barbulifer* (Pisces: Gobiidae). *Copeia* 1985(2): 333–339.

Hoffman, S. G. 1983. Sex-related foraging behavior in sequentially hermaphroditic hogfishes (*Bodianus* spp.). *Ecology* 64: 798–808.

———. 1985. Effects of size and sex on the social organization of reef-associated hogfishes, *Bodianus* spp. *Environmental Biology of Fishes* 14: 185–197.

Holmgren-Urba, D., and T. R. Baumgartner. 1993. A 250-year history of pelagic fish abundance from the anaerobic sediments of the central Gulf of California. *CalCOFI Reports* 34: 60–68.

Hornaday, W. T. 1908. *Campfires on Desert and Lava.* University of Arizona Press, Tucson, AZ. [Reprinted in 1985.]

Houston, R. S. 2006. *Natural History Guide to the Northwestern Gulf of California and Adjacent Desert.* Xlibris, Philadelphia, PA.

Howell, S. N. G., and S. Webb. 1995. *A Guide to the Birds of Mexico and Northern Central America.* Oxford University Press, New York.

Huang, D., and G. Bernardi. 2001. Disjunct Sea of Cortez–Pacific Ocean *Gillichthys mirabilis* populations and the evolutionary origin of their Sea of Cortez endemic relative, *Gillichthys seta. Marine Biology* 138(2): 421–428.

Hubbs, C. L. 1948. Changes in the fish fauna of western North America correlated with changes in ocean temperature. *Journal of Marine Research* 7: 459–482.

———. 1952. A contribution to the classification of the blennioid fishes of the family Clinidae, with a partial revision of eastern Pacific forms. *Stanford Ichthyological Bulletin* 4(2): 41–165.

———. 1953. Revision of the eastern Pacific fishes of the clinid genus *Labrisomus. Zoologica* 38: 113–136.

———. 1960. The marine vertebrates of the outer coast [of Baja California]. *Systematic Zoology* 9(3–4): 134–147.

Huddleston, R. W., and G. T. Takeuchi. 2007. First fossil record of *Totoaba* Villamar, 1980 (Teleostei; Scianidae) based upon Early Miocene otoliths from California, with comments on the ontogeny of the saccular otolith. *Bulletin Southern California Academy of Science* 106(1): 1–15.

Huey, L. M. 1927. Birds recorded in spring at San Felipe, northeastern Lower California, Mexico, with a description of a new woodpecker from that locality. *Transactions of the San Diego Society of Natural History* 5: 13–40.

Humphrey, R. R. 1974. *The Boojum and Its Home.* University of Arizona Press, Tucson, AZ.

Hupp, B., and M. Malone. 2008. *The Edge of the Sea of Cortez. Tidewalker's Guide to the Upper Gulf of California.* Operculum LLC.

Hurtado, L. A., M. Frey, P. Gaube, E. Pfeiler, and T. A. Markow. 2007. Geographical subdivision, demographic history and gene flow in two sympatric species of intertidal snails, *Nerita scabricosta* and *Nerita funiculata,* from the tropical eastern Pacific. *Marine Biology* 151: 1863–1873.

Hyman, L. H. 1953. The polyclad flatworms of the Pacific coast of North America. *Bulletin of the American Museum of Natural History* 100(2): 265–392.

Iglesias-Prieto, R., H. Reyes-Bonilla, and R. Riosmena-Rodríguez. 2003. Effects of 1997–1998 ENSO on coral reef communities in the Gulf of California, Mexico. *Geofísica Internacional* 42: 467–471.

Instituto Nacional Ecología [INE]. 1994. Programa de manejo de la Reserva Especial de la Biosfera Islas del Golfo de California. Instituto Nacional de Ecología, SEDESOL, México, D.F.

———. 1995. Programa de Manejo para la Reserva de la Biosfera Alto Golfo de California y Delta del Río Colorado. Primera reimpresión, Feb. 1997. Instituto Nacional de Ecología (INE), SEMARNAP, México, D.F.

Instituto Nacional de la Pesca [INP]. 2002. Evaluación de la migración y el reclutamiento de las poblaciones de camarón en aguas protegidas y en el frente costero de Sinaloa y Sonora. Secretaría de Agricultura, Ganadería, Pesca y Alimentación. Guaymas, Sonora, México; December 2002.

International Union for the Conservation of Nature [IUCN]. 2005. Islands and protected areas of the Gulf of California (Mexico), ID no. 1182. IUCN World Heritage Evaluation Report, May.

———. 2008. The 2008 Red List of Threatened Species, http://www.iucnredlist.org/.

International Whaling Commission. 2001. Appendix 3. Classification of the Order Cetacea (whales, dolphins, and porpoises). *Journal of Cetacean Research and Management* 3: v–xii.

Irvine, L. M., and Y. M. Chamberlain. 1994. *Seaweeds of the British Isles,* vol. 1: *Rhodophyta Part 2B Corallinales, Hildenbrandiales.* HMSO, London.

Ives, R. L. 1950. Puerto Peñasco, Sonora. *Journal of Geography* 49: 349–361.

Jackson, J. B. C. 1997. Reefs since Columbus. *Coral Reefs* 16: S23–S32.

Jackson, J. B. C., and J. D'Cruz. 1998. The ocean divided. In A. G. Coates (ed.), *Central America: A Natural and Cultural History,* pp. 38–71. Yale University Press, New Haven, CT.

Jacobs, D. K., T. A. Haney, and K. D. Louie. 2004. Genes, diversity, and geological process on the Pacific coast. *Annual Review of Earth and Planetary Science* 32: 601–652.

Jacobson, L., J. A. A. De Oliveira, M. Barange, M. A. Cisneros-Mata, R. Félix-Urraga, J. R. Hunter, J. Y. Kim, Y. Matsuura, M. Ñiquen, C. Porteiro, B. Rothschild, R. P. Sánchez, R. Serra, A. Uriarte, and T. Wada. 2001. Surplus production, variability, and climate change in the great sardine and anchovy fisheries. *Canadian Journal of Fisheries and Aquatic Sciences* 58: 1981–1903.

James, D. W. 2000. Diet, movement, and covering behavior of the sea urchin *Toxopneustes roseus* in rhodolith beds in the Gulf of California, Mexico. *Marine Biology* 137(5/6): 913–923.

James, D. W., M. S. Foster, and J. O. Sullivan. 2006. Bryoliths (Bryozoa) in the Gulf of California. *Pacific Science* 60(1): 117–124.

James, N. P. 1979. Facies models 9. Introduction to carbonate facies models. *Geoscience Canada* 4(3): 123–125.

Janisse, C., D. Squires, J. A. Seminoff, and P. H. Dutton. 2009. Conservation investments and mitigation: The California drift gillnet fishery and Pacific sea turtles. In R. Hilorn, D. Squires, M. Williams, M. Tait, and Q. Grafton (eds.), *Handbook of Marine Fisheries Conservation and Management,* pp. 231–240. Oxford University Press, New York.

Jaquet, N., and D. Gendron. 2002. Distribution and relative abundance of sperm whales in relation to key environmental features, squid landings and the distribution of other cetacean species in the Gulf of California, Mexico. *Marine Biology* 141: 591–601.

Jaquet, N., D. Gendron, and A. Coakes. 2003. Sperm whales in the Gulf of California: Residency, movements, behavior, and the possible influence of variation in food supply. *Marine Mammal Science* 19(3): 545–562.

Jaramillo-Legorreta, A., L. Rojas-Bracho, R. L. Brownell Jr., A. J. Read, R. R. Reeves, K. Ralls, and B. L. Taylor. 2007. Saving the vaquita: Immediate action, no more data. *Conservation Biology* 21(6): 1653–1658.

Jaramillo-Legorreta, A. M., L. Rojas-Bracho, and T. Gerrodette. 1999. A new abundance estimate for vaquitas: First step for recovery. *Marine Mammal Science* 15(4): 957–973.

Jehl, J. R., Jr., W. S. Boyd, D. S. Paul, and D. W. Anderson. 2003. Massive collapse and rapid rebound: Population dynamics of eared grebes (*Podiceps nigricollis*) during an ENSO event. *Auk* 119: 1162–1166.

Jenkins, O. P., and B. W. Evermann 1889. Description of eighteen new species of fishes from the Gulf of California. *Proceedings of the United States National Museum* 11(698): 137–158.

Jiménez-Gutiérrez, S. V. 1999. Abundancia y estructura de peces de arrecife rocoso en la zona de Isla Cerralvo, B.C.S., México. Master's thesis, CICIMAR-IPN, La Paz, B.C.S., México.

Johnson, J. B. 1950. *The Opata: An Inland Tribe of Sonora.* Publications in Anthropology, no. 6, University of New Mexico Press, Albuquerque, NM.

Johnson, M. E., and M. L. Hayes. 1993. Dichotomous facies on a Late Cretaceous rocky island as related to wind and wave patterns (Baja California, Mexico). *Palaios* 8: 385–395.

Johnson, M. E., and J. Ledesma-Vazquez. 2001. Pliocene-Pleistocene rocky shorelines trace coastal development of Bahía Concepción, gulf coast of Baja California Sur (Mexico). *Palaeogeography, Palaeoclimatology, Palaeoecology* 166: 65–88.

Jones, G., and P. Ward. 1998. Privatizing the commons: Reforming the ejido and urban development in Mexico. *International Journal Urban and Regional Research* 22: 76.

Jones, M. L., and S. L. Swartz. 1984. Demography and phenology of gray whales and evaluation of human activities in Laguna San Ignacio, Baja California Sur, Mexico. In M. L. Jones, S. L. Swartz, and S. Leatherwood (eds.), *The Gray Whale,* Eschrichtius robustus, pp. 309–374. Academic Press, Orlando, FL.

Jordan, D. S. 1895. The fishes of Sinaloa. *Proceedings of the California Academy of Sciences* 5:

377–514. [Reprinted in *Contributions to Biology*, Hopkins Laboratory of Biology, Stanford University Publications, no. 1, pp. 1–71.]

Jordan, D. S., and B. W. Evermann. 1896–1900. The fishes of North and Middle America: A descriptive catalogue of the species of fish-like vertebrates found in the waters of North America, north of the Isthmus of Panama. Parts I–IV. *Bulletin of the United States National Museum* 47: 1–3313.

Jordan, D. S., and C. H. Gilbert. 1882a. Catalogue of the fishes collected by Mr. John Xantus at Cape San Lucas, which are now in the United States National Museum, with descriptions of eight new species. *Proceedings of the United States National Museum* 5(290): 353–371.

———. 1882b. List of fishes collected by Lieut. Henry E. Nichols, U.S.N., in the Gulf of California and on the west coast of Lower California, with descriptions of four new species. *Proceedings of the United States National Museum* 4: 273–279.

Jordán, F. 1951. *El otro México: Biografía de Baja California.* Biografías Gandesa, México, D.F.

Juárez-Arriaga, E., A. L. Carreño, and J. L. Sánchez-Zavala. 2005. Pliocene marine deposits at Punta Maldonado, Guerrero, Mexico. *Journal of South American Earth Sciences* 19: 537–546.

Judge, D. S. 1981. Productivity and provisioning behavior of Ospreys (*Pandion haliaetus*) in the Gulf of California. Master's thesis, University of California, Davis.

———. 1983. Productivity of ospreys in the Gulf of California. *Wilson Bulletin* 95: 243–255.

Kamenos, N. A., P. G. Moore, and J. M. Hall-Spencer. 2004. Nursery-area function of maerl grounds for juvenile queen scallops *Aequipecten opercularis* and other invertebrates. *Marine Ecology Progress Series* 274: 183–189.

Kamezaki, N., Y. Matsuzawa, O. Abe, et al. 2003. Loggerhead turtles nesting in Japan. In A. Bolten and B. Witherington (eds.), *Loggerhead Sea Turtles*, pp. 210–217. Smithsonian Books, Washington, DC.

Karl, S. A., and B. W. Bowen. 1999. Evolutionary significant units versus geopolitical taxonomy: Molecular systematics of an endangered sea turtle Genus. *Conservation Biology* 13: 990–999.

Karol, K. G., R. M. McCourt, M. T. Cimino, and C. F. Delwiche. 2001. The closest living relatives of land plants. *Science* 294: 2351–2353.

Keegan, B. F. 1974. The macrofauna of maerl substrates on the west coast of Ireland. *Cahiers de Biologie Marine* 4: 513–530.

Keen, A. M. 1971. *Sea Shells of Tropical West America. Marine Mollusks from Baja California to Peru*, 2nd ed. Stanford University Press, Stanford, CA.

Kerstitch, A., and H. Bertsch. 2007. *Sea of Cortez Marine Invertebrates*, 2nd ed. Sea Challengers, Monterey, CA.

Kiff, L. F. 1980. Historical changes in resident populations of California islands raptors. In D. M. Power (ed.), *The California Islands: Proceedings of a Multidisciplinary Symposium*, pp. 651–673. Santa Barbara Museum of Natural History, Santa Barbara, CA.

Kim, W. H., and J. A. Barron. 1986. Diatom biostratigraphy of the upper Oligocene and lowermost Miocene San Gregorio Formation, Baja California Sur, Mexico. *Diatom Research* 1(2): 169–187.

Kino, F. E. 1954 [1683]. Kino Reports to Headquarters: Correspondence of Eusebio

Francisco Kino, S.J. from New Spain to Rome (English translation and notes by Ernest J. Burrus). S.J. Institutum Historicum Societatis Jesus, Rome.

Klimley, A. P. 1993. Highly directional swimming by scalloped hammerhead sharks, *Sphyrna lewini,* and subsurface irradiance, temperature, bathymetry, and geomagnetic field. *Marine Biology* 117: 1–22.

Klimley, A. P., I. Cabrera-Mancilla, and J. L. Castro-Geniz. 1993. Descripción de los movimientos horizontales y verticales del tiburón martillo del sur del Golfo de California, México. *Ciencias Marinas* 19(1): 95–115.

Klimley, A. P., S. J. Forgensen, A. Muhlia-Melo, and S. C. Beavers. 2003. The occurrence of yellowfin tuna (*Thunnus albacares*) at Espíritu Santo Seamount in the Gulf of California. *Fishery Bulletin* 101(3): 684–692.

Klimley, A. P., and D. R. Nelson. 1981. Schooling of the scalloped hammerhead shark, *Sphryna lewini,* in the Gulf of California. *Fishery Bulletin* 79(2): 356–360.

———. 1984. Diel movement patterns of the scalloped hammerhead shark (*Sphryna lewini*) in relation to El Bajo Espiritu Santo: A refuging central-position social system. *Behavioral Ecology and Sociobiology* 15: 45–54.

Klimley, A. P., J. E. Richert, and S. J. Jorgensen. 2005. The home of blue water fish. *American Scientist* 93: 42–49.

Kobayashi, D. R. 1986. Social organization of the spotted sharpnose puffer, *Canthigaster punctatissima. Environmental Biology of Fishes* 15(2): 141–145.

Kopitsky, K. L., R. L. Pitman, and P. H. Dutton. 2005. Aspects of olive ridley feeding ecology in the Eastern Tropical Pacific. In M. S. Coyne and R. Clark (compilers), *Proceedings of the 21st Annual Symposium on Sea Turtle Biology and Conservation,* p. 217. NOAA Technical Memorandum NMFS-SEFSC-528.

Kotrschal, K. 1988. A catalogue of skulls and jaws of eastern tropical Pacific blennioid fishes (Blennioidei: Teleostei): A proposed evolution sequence of morphological change. *Zeitschrift für Zoologische Systematik und Evolutionsforschung* 26: 442–466.

Kotrschal, K., and D. G. Lindquist. 1986. The feeding apparatus in four Pacific tube blennies (Teleostei: Chaenopsidae): Lack of ecomorphological divergence in syntopic species. *P.S.Z.N.I.: Marine Ecology* 7: 241–253.

Kotrschal, K., and D. A. Thomson. 1986. Feeding patterns in eastern tropical Pacific blennioid fishes (Teleostei: Tripterygiidae, Labrisomidae, Chaenopsidae, Blenniidae). *Oecologica (Berlin)* 70: 367–378.

Kowalewski, M., G. E. Avila Serrano, K. W. Flessa, and G. A. Goodfriend. 2000. Dead delta's former productivity: Two trillion shells at the mouth of the Colorado River. *Geology* 28: 1059–1062.

Krutch, J. W. 1961. *The Forgotten Peninsula: A Naturalist in Baja California.* W. Sloane Associates, San Francisco. [Reeditado 1986 por A. H. Zwinger, University of Arizona Press, Tucson, AZ.]

Krutch, J. W., and E. Porter. 1957. *Baja California and the Geography of Hope.* Sierra Club, San Francisco.

Lambeck, K., and J. Chappell. 2001. Sea level change through the last glacial cycle. *Science* 292: 679–686.

Lancaster, N., R. Greeley, and P. R. Christensen. 1987. Dunes of the Gran Desierto Sand-Sea, Sonora, Mexico. *Earth Surface Processes and Landforms* 12: 277–288.

Larson, R. L., H. W. Menard, and S. M. Smith. 1968. Gulf of California: A result of ocean-spreading and transform faulting. *Science* 161: 781–784.

Lavenberg, R. J., and J. E. Fitch. 1966. Annotated list of the fishes collected by mid-water trawl in the Gulf of California, March–April 1964. *California Fish and Game* 52: 92–110.

Lavín, M. F., E. Beier, and A. Badan. 1997a. Estructura hidrográfica y circulación del Golfo de California: Escalas estacional e interanual. In M. F. Lavín (ed.), *Contribuciones a la oceanografía física en México,* pp. 141–171. Monografía no. 3, Unión Geofísica Mexicana, Ensenada.

Lavín, M. F., R. Durazo, E. Palacios, M. L. Argote, and L. Carrillo. 1997b. Lagrangian observations of the circulation in the northern Gulf of California. *Journal of Physical Oceanography* 27: 2298–2305.

Lavín, M. F., V. M. Godínez, and L. G. Alvarez. 1998. Inverse-estuarine features of the upper Gulf of California. *Estuarine, Coastal, Shelf Science* 47: 769–795.

Lavín, M. F., and S. Sánchez. 1999. On how the Colorado River affected the hydrography of the upper Gulf of California. *Continental Shelf Research* 19: 1545–1560.

Lawlor, T. E., D. J. Hafner, P. T. Stapp, B. R. Riddle, and S. T. Álvarez-Castañeda. 2002. Mammals. In T. J. Case, M. L. Cody, and E. Ezcurra (eds.), *A New Island Biogeography of the Sea of Cortés,* pp. 271–312. Oxford University Press, New York.

Lawrence, G. N. 1874. The birds of western and northwestern Mexico, based upon the collections made by Col. A. J. Grayson, Capt. J. Xantus, and Ferd. Bischoff, now in the Museum of the Smithsonian Institution at Washington, DC. *Memoirs of the Boston Society of Natural History* 2(9): 265–319.

Laylander, D., and J. D. Moore (eds.). 2006. *The Prehistory of Baja California.* University Press of Florida, Gainesville, FL.

Lea, R. N., and R. H. Rosenblatt. 2000. Observations on fishes associated with the 1997–98 El Niño off California. *CalCOFI Report* 41: 117–129.

Leaché, A. D., S. C. Crews, and M. J. Hickerson. 2007. Two waves of diversification in mammals and reptiles of Baja California revealed by hierarchical Bayesian analysis. *Biology Letters* 3: 646–650.

Leaché, A. D., and D. G. Mulcahy. 2007. Phylogeny, divergence times, and species limits of spiny lizards (*Sceloporus magister* species-group) in western North American deserts and Baja California. *Molecular Ecology* 16(24): 5216–5233.

Leatherwood, S., W. F. Perrin, V. L. Kirby, C. L. Hubbs, and M. Dahlheim. 1980. Distribution and movements of Risso's dolphin, *Grampus griseus,* in the eastern North Pacific. *Fishery Bulletin (U.S.)* 77: 951–963.

Leatherwood, S., and R. R. Reeves. 1983. *The Sierra Club Handbook of Whales and Dolphins.* Sierra Club, Tokyo.

Leatherwood, S., R. R. Reeves, W. F. Perrin, and W. E. Evans. 1988. *Whales, Dolphins, and Porpoises of the Eastern North Pacific and Adjacent Arctic Waters. A Guide to Their Identification.* Dover, New York.

Le Boeuf, B. J., D. Aurioles, R. Condit, C. Fox, R. Gisiner, R. Romero, and F. Sinsel. 1983. Size and distribution of the California sea lion population in Mexico. *Proceedings of the California Academy of Sciences* 43: 77–85.

Le Boeuf, B. J., H. Pérez-Cortés M., J. Urbán R., B. R. Mate, and F. Ollervides U. 2000.

High gray whale mortality and low recruitment in 1999: Potential causes and implications. *Journal of Cetacean Research and Management* 2(2): 85–99.

Lechuga-Deveze, C. H., M. L. Morquecho-Escamilla, A. Reyes-Salinas, and J. R. Hernandez-Alfonso. 2000. Environmental natural disturbances in Bahía Concepción, Gulf of California. In S. G. L. M. Munawar, I. F. Munawar, and D. F. Malley (eds.), *Aquatic Ecosystems of Mexico: Status and Scope*. Ecovision World Monograph Series, Backhuys, Leiden, The Netherlands.

Ledesma-Vázquez, J. 2002. A gap in the Pliocene invasion of seawater, Gulf of California. *Revista Mexicana de Ciencias Geológicas* 19(3): 145–151.

Ledesma-Vázquez, J., and M. E. Johnson. 1993. Neotectónica del área Loreto-Mulegé. In L. Delgado Argote and A. Martín Barajas (eds.), *Contribuciones a la tectónica del Occidente de México*, Monografía 1, pp. 115–122. Unión Geofísica Mexicana, Ensenada.

———. 2001. Miocene-Pleistocene tectono-sedimentary evolution of Bahía Concepción region, Baja California Sur (Mexico). *Sedimentary Geology* 144: 83–96.

Ledesma-Vázquez, J., M. E. Johnson, and F. Romero-Ríos. 1999. Evolución tectónica del Golfo de California: Mioceno-Plioceno de Bahía Concepción, B.C.S. *Gaceta Asociación Mexicana de Geólogos Petroleros* 3(2): 1–5.

Ledesma-Vázquez, J., and Z. J. J. Kasper. 1989. Interpretación geológica y paleoceanográfica de los miembros La Misión y Los Indios de la formación Rosarito Beach (Mioceno Medio), Baja California, México. *Ciencias Marinas* 15(3): 21–44.

Ledesma-Vázquez, J., A. Y. Montiel-Boehringer, D. Backus, M. Johnson, and V. Z. Fernandez-Díaz. 2007. Armored mud balls in tidal environments, Pliocene in the Gulf of California. *Cuadernos del Museo Geominero* 8: 235–238.

Lee, J., M. Miller, R. Crippen, B. Hacker, and J. Ledesma-Vázquez. 1996. Middle Miocene extension in the Gulf extensional province, Baja California: Evidence from the southern Sierra Juárez. *Geological Society of America Bulletin* 108: 505–525.

Lee, R. E. 2008. *Phycology*. Cambridge University Press, Cambridge, U.K.

Lehner, C. E. 1979. A latitudinal gradient analysis of rocky shore fishes of the eastern Pacific. Ph.D. dissertation, University of Arizona, Tucson.

Leis, J. M. 1984. Larval fish dispersal and the East Pacific Barrier. *Studies in Tropical Oceanography* 19(2): 181–192.

León de la Luz, J. L., R. Coria-Benet, and J. Cansino. 1995. *XI Reserva de la Biósfera El Vizcaíno, Baja California Sur.* Serie Listados Florísticos de México, Instituto de Biología, Universidad Nacional Autónoma de México, México, D.F.

León-Portilla, M. 1989. *Loreto, Capital de las Californias. Las cartas fundacionales de Juan María de Salvatierra.* Universidad Autónoma de Baja California, Centro Cultural Tijuana, FONATUR, México, D.F.

Leopold, A. 1966. *A Sand County Almanac with Other Essays on Conservation from Round River.* Oxford University Press, New York.

Lepley, L. K., S. P. von der Haar, J. R. Hendrickson, and G. Calderon-Riverol. 1975. Circulation in the northern Gulf of California from orbital photographs and ship investigations. *Ciencias Marinas* 2(2): 86–93.

Leslie, H. M., M. Schlüter, R. Cudney-Bueno, and S. A. Levin. 2009. Modeling responses of coupled social–ecological systems of the Gulf of California to anthropogenic and natural perturbations. *Ecological Research* [Ecological Society of Japan], in press.

Lewis, L. A., and R. M. McCourt. 2004. Green algae and the origin of land plants. *American Journal of Botany* 91: 1535–1556.

Libbey, L. K., and M. E. Johnson. 1997. Upper Pleistocene rocky shores and intertidal biotas at Playa La Palmita (Baja California Sur, Mexico). *Journal of Coastal Research* 13: 216–225.

Libes, S. M. 1992. *An Introduction to Marine Biogeochemistry.* Wiley, New York.

Lindquist, D. G. 1980. Aspects of the polychromatism in populations of the Gulf of California browncheek blenny, *Acanthemblemaria crockeri* (Blennioidea: Chaenopsidae). *Copeia* 1980: 137–141.

———. 1985. Depth zonation, microhabitat, and morphology of three species of *Acanthemblemaria* (Pisces: Blennioidea) in the Gulf of California, Mexico. *P.S.Z.N.I.: Marine Ecology* 6: 329–344.

Lindsay, G. E. 1962. The Belvedere expedition to the Gulf of California. *Transactions, San Diego Society of Natural History* 13(1): 144.

———. 1964. Sea of Cortez expedition of the California Academy of Sciences, June 20–July 4, 1964. *Proceedings of the California Academy of Sciences* 30(11): 211–242.

———. 1966. The Gulf islands expedition of 1966. *Proceedings of the California Academy of Sciences* 30(16): 309–355.

———. 1970. Some natural values of Baja California. *Pacific Discovery* 23(2): 1–10.

Link, J. S. 2002. What does ecosystem-based fisheries management mean? *Fisheries* 27(4): 21.

Littler, D. S., and M. M. Littler. 2000. *Caribbean Reef Plants.* OffShore Graphics, Washington, DC.

———. 2003. *South Pacific Reef Plants.* OffShore Graphics, Washington, DC.

Lively, C. M., P. T. Raimondi, and L. F. Delph. 1993. Intertidal community structure: Space-time interactions in the northern Gulf of California. *Ecological Society of America* 74(1): 162–173.

Lluch-Belda, D. 1969. El lobo marino de California, *Zalophus californianus.* Observaciones sobre su ecología y explotación. In *Los amíferos marinos de Baja California,* pp. 1–69. Instituto Mexicano de Recursos Naturales Renobables. México.

Lluch-Belda, D., D. B. Lluch-Cota, and S. E. Lluch-Cota. 2003. Baja California's biological transition zones: Refuges for the California sardine. *Journal of Oceanography* 59(4): 503–513.

Lluch-Belda, D., B. F. J. Magallón, and R. A. Schwartzlose. 1986. Large fluctuations in sardine fishery in the Gulf of California: Possible causes. *CalCOFI Report* 37: 201–208.

Lluch-Cota, S. E. 2000. Coastal upwelling in the eastern Gulf of California. *Oceanologica Acta* 23: 731–740.

Lluch-Cota, S. E., E. A. Aragón-Noriega, F. Arreguín-Sánchez, D. Aurioles-Gamboa, J. J. Bautista-Romero, R. C. Brusca, R. Cervantes-Duarte, R. Cortés-Altamirano, P. Del-Monte-Luna, A. Esquivel-Herrera, H. Herrara-Cervantes, M. Kahru, M. Lavín, D. Lluch-Belda, D. B. Lluch-Cota, J. López-Martinez, S. G. Marinone, M. O. Nevárez-Martinez, S. Ortega-Garcia, E. Palacios-Castro, A. Parés-Sierra, G. Ponce-Diaz, M. Ramírez-Rodriguez, C. A. Salinas-Zavala, R. A. Schwartzlose, and P. Sierra-Beltrán. 2007. The Gulf of California: Review of ecosystem status and sustainability challenges. *Progress in Oceanography* 73: 1–26.

Lluch-Cota, S. E., D. B. Lluch-Cota, D. Lluch-Belda, M. O. Nevárez-Martínez, A. Parés-

Sierra, and S. Hernández-Váquez. 1999. Variability of sardine catch as related to enrichment, concentration, and retention processes in the central Gulf of California. *CalCOFI Reports* 40: 184–190.

Longinos Martínez, José. 1792. *Diario.* [Notes and observations of the Royal Botanical Expedition to Antigua and Alta California. Published in Spanish as *Diario de las expediciones a las Californias de José Longinos,* Serie Textos Clásicos, Editorial Doce Calles, 1994, Madrid; published in English as *Journal of José Longinos Martínez: Notes and Observations of the Naturalist of the Botanical Expedition in Old and New California and the South Coast 1791–1792,* John Howell Books, 1961, San Francisco.]

López, M., and J. García. 2003. Moored observations in the northern Gulf of California: A strong bottom current. *Journal of Geophysical Research* 108(C2): 30.1–30.18.

López-Castro, M. C., R. Carmona, and W. J. Nichols. 2004. Nesting characteristics of the olive Ridley turtle (*Lepidochelys olivacea*) in Cabo Pulmo, southern Baja California. *Marine Biology* 145: 811–820.

López-Mendilaharsu, M., S. C. Gardner, J. A. Seminoff, and R. Riosmena-Rodríguez. 2003. Feeding ecology of the East Pacific green turtle (*Chelonia mydas agassizii*) in Bahía Magdalena, B.C.S., Mexico. In J. A. Seminoff (compiler), *Proceedings of the 22nd Annual Symposium on Sea Turtle Biology and Conservation,* pp. 219–220. NOAA Technical Memorandum NMFS-SEFSC-503.

———. 2005. Identifying critical foraging habitats of the green turtle (*Chelonia mydas*) along the Pacific coast of the Baja California peninsula, Mexico. *Aquatic Conservation: Marine and Freshwater Ecosystems* 15: 259–269.

Love, M. S., C. W. Mecklenburg, T. A. Mecklenburg, and L. K. Thorsteinson. 2005. *Resource Inventory of Marine and Estuarine Fishes of the West Coast and Alaska: A Checklist of North Pacific and Arctic Ocean Species from Baja California to the Alaska-Yukon Border.* Biological Resources Division, U.S. Geological Survey, U.S. Department of Interior, Seattle, WA.

Love, M. S., M. Yoklavich, and L. Thorsteinson. 2002. *The Rockfishes of the Northeast Pacific.* University of California Press, Berkeley, CA.

Lumholtz, C. 1912. *New Trails in Mexico.* Unwin, London. [Reprinted in 1971 by Rio Grande Press, Glorieta, NM.]

Lyle, M., and G. E. Ness. 1990. The opening of the southern Gulf of California. In J. P. Dauphin and B. R. Simoneit (eds.), *The Gulf and Peninsula Province of the Californias,* pp. 403–423. Memoir 47, American Association of Petroleum Geologists, Tulsa, OK.

Lynch, D. J., T. E. Musselman, J. T. Gutman, and P. J. Patchett. 1993. Isotopic evidence for the origin of Cenozoic volcanic rocks in the Pinacate volcanic field, northwestern Mexico. *Lithos* 29: 295–302.

Macleod, C. D., W. F. Perrin, R. Pitman, J. Barlow, L. Balance, A. Dámico, T. Gerrodette, G. Joyce, K. D. Mullin, D. L. Palka, and G. T. Waring. 2006. Known and inferred distributions of beaked whale species (Cetacea: Ziphiidae). *Journal of Cetacean Research and Management* 7(3): 271–286.

Macpherson, E. 2002. Large-scale species-richness gradients in the Atlantic Ocean. *Proceedings of the Royal Society of London, Series B,* 269: 1715–1720.

Macpherson, E., P. A. Hastings, and D. R. Robertson. In press. Macroecology of marine fishes. In J. D. Witman and K. Roy (eds.), *Marine Macroecology.* University of Chicago Press, Chicago.

Magallón-Barajas, F. J. 1987. The Pacific shrimp fishery of Mexico. *CalCOFI Report* 28: 43–52.

Maggs, C. A. 1983. A phenological study of two maerl beds in Galway Bay, Ireland. Ph.D. thesis, National University of Ireland, Galway.

Mailliard, J. 1923. Expedition of the California Academy of Sciences to the Gulf of California in 1921. The birds. *Proceedings of the California Academy of Sciences, 4th series* 12: 443–456.

Maluf, L. Y. 1983. Physical oceanography. In T. J. Case and M. L. Cody (eds.), *Island Biogeography in the Sea of Cortez,* pp. 26–45. University of California Press, Berkeley, CA.

———. 1988. Composition and distribution of the central eastern Pacific echinoderms. Natural History Museum of Los Angeles County, Technical Report no. 2.

Mandryk, C. A. S., H. Josenhans, D. W. Fredje, and R. W. Matthewes. 2001. Late Quaternary paleoenvironments of northwestern North America: Implications for inland versus coastal migration routes. *Quaternary Science Reviews* 20: 301–314.

Mangels, K. F., and T. Gerrodette. 1994. Report of cetacean sightings during a marine mammal survey in the eastern Tropical Pacific Ocean and the Gulf of California aboard the NOAA ships McArthur and David Starr Jordan July 28–November 6, 1993. Technical Memorandum NOAA-TM-NMFS-SWFSC-211.

Manzano-Fischer, P., G. Ceballos, R. List, and J.-L. E. Cartron. 2006. Avian diversity in a priority area for conservation in North America: The Janos–Casas Grandes Prairie Dog Complex and adjacent habitats in northwestern Mexico. *Biodiversity and Conservation* 15: 3801–3825.

Maravilla, M. O., and J. P. Gallo R. 2000. Familia Otariidae. In S. T. Alvarez-Castañeda and J. L. Patton (eds.), *Mamíferos del Noroeste de México II,* pp. 775–780. Centro de Investigaciones Biológicas del Noroeste, S.C, La Paz, Baja California Sur.

Maravilla, M. O., and M. Lowry. 1999. Incipient breeding colony of Guadalupe fur seals at Isla San Benito del Este, Baja California, Mexico. *Marine Mammal Science* 15: 239–241.

Marine Mammal Commission. 2007. Annual Report to the Congress 2006. 4340 East-West Highway, Room 905, Bethesda, MD 20814.

Marinone, S. G. 2003. A three dimensional model of the mean and seasonal circulation of the Gulf of California. *Journal of Geophysical Research* 108(C10): 3325.

Marinone, S. G., and M. F. Lavín. 2003. Residual flow and mixing in the large islands region of the central Gulf of California. In O. U. Velasco Fuentes, J. Sheinbaum, and J. L. Ochoa de la Torre (eds.), *Nonlinear Processes in Geophysical Fluid Dynamics,* pp. 213–236. Kluwer, Dordrecht.

Marinone, S. G., A. Parés-Sierra, R. Castro, and A. Mascarenhas. 2004. Correction to "Temporal and spatial variation of the surface winds in the Gulf of California." *Geophysical Research Letters* 31: L10305.

Marinone, S. G., M. J. Ulloa, A. Parés-Sierra, M. F. Lavín, and R. Cudney-Bueno. 2008. Connectivity in the northern Gulf of California from particle tracking in a three-dimensional numerical model. *Journal of Marine Systems* 71: 149–158.

Markaida, U., J. J. Rosenthal, and W. F. Gilly. 2005. Tagging studies on the jumbo squid (*Dosidicus gigas*) in the Gulf of California, Mexico. *Fishery Bulletin* 103: 219–226.

Márquez, M. R. 1984. Opinión sobre cuotas y franquicias de tortugas marinas, temporada de tortugas marinas. Instituto Nacional de Pesca, México.

Márquez, M. R., M. Carrasco-A., and M. C. Jiménez. 2002. The marine turtles of Mexico: An update. In I. Kinan (ed.), *Proceedings of the Western Pacific Sea Turtle Cooperative Research and Management Workshop,* pp. 281–285. WESTPAC, Honolulu.

Márquez, R., and T. Doi. 1973. Ensayo teórico sobre el análisis de la población de tortuga prieta, *Chelonia mydas carrinegra* Caldwell, en aguas del Golfo de California, México. *Bulletin of the Tokai Regional Fisheries Laboratory* 73: 1–22.

Márquez, R., C. Peñaflores, and J. Vasconcelos. 1996. Olive Ridley turtles (*Lepidochelys olivacea*) show signs of recovery at La Escobilla, Oaxaca. *Marine Turtle Newsletter* 73: 5–7.

Marrack, E. C. 1999. The relationship between water motion and living rhodolith beds in the southwestern Gulf of California, Mexico. *Palaios* 14(2): 159–171.

Martell, M. S., C. J. Henny, P. E. Nye, and M. J. Solensky. 2001. Fall migration routes, timing, and wintering sites of North American Osprey as determined by satellite telemetry. *Condor* 103: 715–724.

Martínez, Maximino. 1947. *Baja California: Reseña histórica del territorio y su flora.* Ediciones Botas, México, D.F.

Massey, B. W., and E. Palacios. 1994. Avifauna of the wetlands of Baja California, Mexico: Current status. *Studies in Avian Biology* 15: 45–57.

Mateo-Cid, L. E., I. Sánchez-Rodríguez, Y. E. Rodríguez-Montesinos, and M. M. Casas-Valdez. 1993. Estudio florístico de las algas marinas bentónicas de Bahía Concepción, B.C.S., México. *Ciencias Marinas* 19: 41–60.

Mathes, W. M. (ed). 1969. *First from the Gulf to the Pacific: The diary of the Kino-Atondo Peninsular Expedition.* Baja California Travel Series, no. 16. Dawson's Book Shop, Los Angeles, CA.

Mayer, L., and K. R. Vincent. 1999. Active tectonics of the Loreto area, Baja California Sur, Mexico. *Geomorphology* 27: 243–255.

McCarthy, J. J. 1980. Nitrogen. Pp. 191–233 in I. Morris (ed.), *The Physiological Ecology of Phytoplankton,* University of California Press, Berkeley and Los Angeles, CA.

McCloy, C. 1984. Stratigraphy and depositional history of the San Jose del Cabo trough, Baja California Sur, Mexico. In V. A. Frizzell Jr. (ed.), *Geology of the Baja California Peninsula,* pp. 267–273. Pacific Section, SEPM, Book 39, Society for Sedimentary Geology, Los Angeles, CA.

McCloy, C., J. C. Ingle Jr., and J. A. Barron. 1988. Neogene stratigraphy, foraminifera, diatoms, and depositional history of María Madre Island, Mexico: Evidence of early Neogene marine conditions in the southern Gulf of California. *Marine Micropaleontology* 13(3): 193–212.

McCosker, J. E., and C. E. Dawson. 1975. Biotic passage through the Panama Canal, with particular reference to fishes. *Marine Biology* 30: 343–351.

McCourt, R. M. 1984a. Niche differences between sympatric Sargassum species in the northern Gulf of California. *Marine Ecology Progress Series* 18: 139–148.

———. 1984b. Seasonal patterns of abundance, distributions, and phenology in relation to growth strategies of three Sargassum species. *Journal of Experimental Marine Biology and Ecology* 74: 141–156.

———. 1985. Biomass allocation patterns in three species of intertidal *Sargassum. Oecologia* 67: 113–117.

McCourt, R. M., K. G. Karol, and C. F. Delwiche. 2004. Charophyte algae and land plant origins. *Trends in Ecology and Evolution* 19(12): 661–666.

McDougall, K., R. Z. Poore, and J. C. Matti. 1999. Age and paleoenvironment of the Imperial Formation near San Gorgonio Pass, southern California. *Journal of Foraminiferal Research* 29(1): 4–25.

McGee, W. J. 1900. The Gulf of California as an evidence of marine erosion. *Science* 11(272): 429–430.

McGuire, T., and J. Greenberg (eds.). 1993. *Maritime Community and Biosphere Reserve: Crisis and Response in the Upper Gulf of California*. University of Arizona, Tucson, AZ.

McKibbin, N. 1989. *The Sea in the Desert. Explorer's Guide to the Gulf of California Seashore*. Golden Puffer Press, Tucson, AZ.

McLean, H. 1988. Reconnaissance geologic map of the Loreto and part of the San Javier quadrangles, Baja California Sur, Mexico. U.S. Geological Survey Map MF-2000 [map scale 1:50,000].

Mead, J. 2003. Crocodilian remains from the late Pleistocene of northeastern Sonora, Mexico. IUCN–World Conservation Union, Species Survival Commission. *Crocodile Specialist Group Newsletter* 22(1): 19–21.

Mead, J. I., A. Baez, S. L. Swift, M. C. Carpenter, M. Hollenshead, N. J. Czaplewski, D. W. Steadman, J. Bright, and J. Arroyo-Cabrales. Tropical marsh and savanna of the late Pleistocene in northeastern Sonora, Mexico. *Southwestern Naturalist* 51(2): 226–239.

Medina-López, M. A. 1999. Estructura de la criptofauna asociada a mantos de rodolitos en el suroeste del Golfo de California, México. Thesis, Universidad Autónoma de Baja California Sur, La Paz.

Meldahl, K. H., K. W. Flessa, and A. H. Cutler. 1997. Time-averaging and postmortem skeletal survival in benthic fossil assemblages: Quantitative comparisons among Holocene environments. *Paleobiology* 23(2): 207–229.

Melin, S. R., and R. L. DeLong. 1999. Guadalupe fur seal (*Arctocephalus townsendi*) female and pup at San Miguel Island, California. *Marine Mammal Science* 15: 995.

Meling-López, A. E., and S. E. Ibarra-Obando. 1999. Annual life cycles of two *Zostera marina* L. populations in the Gulf of California: Contrasts in seasonality and reproductive effort. *Aquatic Botany* 65: 59–69.

Mellink, E. 1995. Status of the muskrat in the Valle de Mexicali and delta del Río Colorado, Mexico. *California Fish and Game* 81: 33–38.

———. 2001. History and status of colonies of Heermann's Gull in Mexico. *Waterbirds* 24: 188–194.

Mellink, E., and V. Ferreira-Bartrina. 2000. On the wildlife of wetlands of the Mexican portion of the Río Colorado delta. *Bulletin of the Southern California Academy of Sciences* 99: 115–127.

Mellink, E., and E. Palacios. 1993. Notes on breeding coastal waterbirds in northwestern Sonora. *Western Birds* 24: 29–37.

Mellink, E., E. Palacios, and S. González. 1996. Notes on nesting birds of the Ciénega de Santa Clara salt flat, northwestern Sonora, Mexico. *Western Birds* 27: 202–203.

———. 1997. Non-breeding waterbirds of the delta of the Río Colorado, México. *Journal of Field Ornithology* 68: 113–123.

Méndez, M. 1981. Claves de identificación y distribución e los longostinos y camorones (Crustacea: Decapoda) del mar y ríos de la costa del Perú. Boletin no. 5. Instituto del Mar del Perú.

Merrifield, M. A., C. D. Winant, J. M. Robles, R. T. Guza, N. A. Bray, J. García, A. Badan-Dangon, and N. Christensen Jr. 1986. Observations of currents, temperature, pressure, and sea level in the Gulf of California 1982–1986. A data report. SIO Ref. 86-11, Scripps Institution of Oceanography, La Jolla, CA.

Michler, Lieut. M. 1857. From the 111th meridian of longitude to the Pacific Ocean. In W. H. Emory, *Report on the United States and Mexican Boundary Survey,* vol. 1, pp. 101–136.

Miller, A. H., H. Friedman, L. Griscom, and R. T. Moore. 1957. Distributional checklist of the birds of Mexico, part 2. *Pacific Coast Avifauna* 33: 1–436.

Miller, M. W. 1998. Coral/seaweed competition and the control of reef community structure within and between latitudes. *Oceanography and Marine Biology: An Annual Review* 36: 65–96.

Miller, R. R., with W. L. Minckley and S. M. Norris. 2005. *Freshwater Fishes of Mexico.* University of Chicago Press, Chicago.

Miller, W. A. 1980. The late Pliocene Las Tunas local fauna from southernmost Baja California. *Journal of Paleontology* 54: 762–805.

Minch, J. A. 1967. Stratigraphy and structure of the Tijuana-Rosarito Beach area, northwestern Baja California, Mexico. In *Ecological Society of America Bulletin* 78(9): 1115–1178.

Minch, J. A., J. R. Ashby, T. A. Deméré, and H. T. Kuper. 1984. Correlation and depositional environments of the middle Miocene Rosarito Beach Formation of northwestern Baja California, Mexico. In J. A. Minch and J. R. Ashby Jr. (eds.), *Miocene and Cretaceous Depositional Environments, Northwestern Baja California, Mexico,* pp. 33–46. Pacific Section, AAPG, Book 54, American Association of Petroleum Geologists, Tulsa, OK.

Minckley, W. L. 2002. Fishes of the lowermost Colorado River, its delta and estuary: A commentary on biotic change. In M. L. Lozano-Vilano (ed.), *Libro Jubilar en Honor al Dr. Salvador Contreras Balderas,* pp. 63–78. Ediciones Universidad Autónoma de Nuevo León, Monterrey, N.L., México.

Minoura, K., and T. Nakamori. 1982. Depositional environment of algal balls in the Ryukyu group, Ryukyu Islands, southwestern Japan. *Journal of Geology* 90: 602–609.

Mitchell, A. J., and K. J. Collins. 2004. Understanding the distribution of maerl, a calcareous seaweed, off Dorset, UK. In *Proceedings of the Second International Symposium on GIS/Spatial Analyses in Fishery and Aquatic Sciences,* pp. 65–82. University of Sussex, Brighton, U.K.

Mittermeier, R. A., C. F. Kormos, C. G. Mittermeier, T. Sandwith, and C. Besançon (eds.). 2005. *Transboundary Conservation: A New Vision for Protected Areas.* CEMEX (Monterrey), Conservation Internacional (Washington, DC), and Agrupación Sierra Madre (México, D.F.).

Mittermeier, R. A., C. G. Mittermeier, P. Robles-Gil, J. Pilgrim, G. A. B. da Fonseca, T. Brooks, and W. R. Konstant (eds.). 2002. *Wilderness. Earth's Last Wild Places.* CEMEX (Monterrey), Conservation Internacional (Washington, DC), and Agrupación Sierra Madre (México, D.F.).

Mittermeier, R. A., P. Robles Gil, and C. G. Mittermeier. 1997. *Megadiversity*. Cemex, Mexico City, Mexico.

Moffat, N. A., and D. A. Thomson. 1975. Taxonomic status of the Gulf grunion (*Leuresthes sardina*) and its relationship to the California grunion (*L. tenuis*). *Transactions of the San Diego Society of Natural History* 18: 75–84.

Møller, P. R., W. Schwarzhans, and J. G. Nielsen. 2005. Review of the American Dinematichthyini (Teleostei: Bythitidae). Part II. *Ogilbia. Aqua (Journal of Ichthyology and Aquatic Biology)* 10(4): 133–205.

Molles, M. C. 1978. Fish species diversity on model and natural reef patches: Experimental insular biogeography. *Ecological Monographs* 48: 289–305.

Mondragon, J., and J. Mondragon. 2003. *Seaweeds of the Pacific Coast. Common Marine Algae from Alaska to Baja California*. Sea Challengers, Monterey, CA.

Montanucci, R. R. 2004. Geographic variation in *Phrynosoma coronatum:* Further evidence for a peninsular archipelago. *Herpetologica* 60(1): 117–139.

Monteforte, M., and M. Cariño. *Del saqueo a la conservación. Historia ambiental contemporánea de Baja California Sur, 1940–2003*. Instituto Nacional de Ecología, México, D.F.

Montgomery, W. L. 1975. Interspecific associations of sea basses (Serranidae) in the Gulf of California. *Copeia* 1975(4): 785–787.

———. 1980. Comparative feeding ecology of two herbivorous damselfishes (Pomacentridae: Teleostei) from the Gulf of California. *Journal of Experimental Marine Biology and Ecology* 47: 9–24.

———. 1981. Mixed-species schools and the significance of vertical territories of damselfishes. *Copeia* 1981: 477–481.

———. 2002. Boyd W. Walker, 1917–2001 [obituary]. *Copeia* 2002(4): 1179–1182.

Montgomery, W. L., T. Gerrodette, and L. D. Marshall. 1980. Effects of grazing by the yellowtail surgeonfish, *Prionurus punctatus,* on algal communities in the Gulf of California, Mexico. *Bulletin of Marine Science* 30: 477–481.

Mora, C., and D. R. Robertson. 2005a. Causes of latitudinal gradients in species richness: A test with the endemic shorefishes of the tropical eastern Pacific. *Ecology* 86: 1771–1782.

———. 2005b. Factors shaping the range-size frequency distribution of the endemic fish fauna of the tropical eastern Pacific. *Journal of Biogeography* 32: 277–286.

Mora, M. A., and D. W. Anderson. 1991. Seasonal and geographical variation of organochlorine residues in birds from northwest Mexico. *Archives of Environmental Contamination and Toxicology* 21: 541–548.

———. 1995. Selenium, boron, and heavy metals in birds from the Mexicali Valley, Baja California, Mexico. *Bulletin of Environmental Contamination and Toxicology* 54: 198–206.

Mora, M. A., D. W. Anderson, and M. E. Mount. 1987. Seasonal variation of body condition and organochlorines in wild ducks from California and Mexico. *Journal of Wildlife Management* 51: 132–141.

Mora, M. A., J. García, M. Carpio-Obeso, and K. A. King. 2003. Contaminants without borders: A regional assessment of the Colorado River Delta ecosystem. In D. J. Rapport, W. L. Lasley, D. E. Rolston, N. Ole Nielsen, C. O. Qualset, and A. B. Damania (eds.), *Managing for Healthy Ecosystems,* pp. 1125–1134. Lewis Publishers, Boca Raton, FL.

Morales-Abril, G. 1994. Reserva de la Biosfera alto Golfo de California y Delta del Río Colorado. *Ecologica* 3: 26–27.

Moser, E. W. 1963. Seri bands. *Kiva* 28: 14–27.

Moser, H. G., E. H. Ahlstorm, D. Kramer, and E. G. Stevens. 1974. Distribution and abundance of fish eggs and larvae in the Gulf of California. *CalCOFI Report* 17: 112–128.

Moyer, J. T., R. E. Thresher, and P. L. Colin. 1983. Courtship, spawning and inferred social organization of American angelfishes (Genera *Pomacanthus, Holacanthus* and *Centropyge;* Pomacanthidae). *Environmental Biology of Fishes* 9: 25–39.

Mueller, G. A., and P. C. Marsh. 2002. *Lost, a Desert River and Its Native Fishes: A Historical Perspective of the Lower Colorado River.* Information and Technology Report USGS/BRD/ITR-2002-0010, Biological Resources Division, U.S. Geological Survey, U.S. Department of the Interior.

Muhlia-Melo, A., P. Klimley, R. González-Armas, S. Jorgensen, A. Trasviña-Castro, J. Rodríguez-Romero, and A. Amador-Buenrostro. 2003. Pelagic fish assemblages at the Espíritu Santo Seamount in the Gulf of California during El Niño 1997–1998 and non El Niño conditions. *Geofísica Internacional* 42(3): 473–481.

Munro Palacio, G. 1994. Historia de la región de Puerto Peñasco. *CEDO News* 6: 20–25.

Murphy, R. W., and G. Aguirre-León. 2002. The nonavian reptiles origins and evolution. In T. J. Case, M. L. Cody, and E. Ezcurra (eds.), *A New Island Biogeography of the Sea of Cortés,* pp. 181–220. Oxford University Press, New York.

Musick, J. A., M. M. Harbin, S. A. Berkeley, G. H. Burgess, A. M. Eklund, L. Findley, R. G. Gilmore, J. T. Golden, D. S. Ha, G. R. Huntsman, J. C. McGovern, S. J. Parker, S. G. Poss, E. Sala, T. W. Schmidt, G. R. Sedberry, H. Weeks, and S. G. Wright. 2000. Marine, estuarine, and diadromous fish stocks at risk of extinction in North America (exclusive of Pacific salmonids). *Fisheries* 25(11): 6–30.

Musick, J. A., and C. J. Limpus. 1997. Habitat utilization and migration in juvenile sea turtles. In P. L. Lutz and J. A. Musick (eds.), *The Biology of Sea Turtles,* pp. 137–163. CRC Press, Boca Raton, FL.

Nabhan, G. P. 2002. Cultural dispersal of plants and reptiles. In T. Case, M. Cody, and E. Ezcurra (eds.), *A New Island Biogeography of the Sea of Cortés,* pp. 407–416. Oxford University Press, New York.

———. 2003. *Singing the Turtles to Sea.* University of California Press, Berkeley, CA.

Nagy, E. A. 2000. Paleomagnetism and extensional deformation at the western margin of the Gulf Extensional Province, Puertecitos Volcanic Province, Baja California, Mexico. *Geological Society of America Bulletin* 112(6): 857–870.

Nagy, E. A., and J. M. Stock. 2000. Structural controls on the continent-ocean transition in the northern Gulf of California. *Journal of Geophysical Research* 105(B7): 251–269.

Nakashima, E. 1916. Notes on the totuava (*Cynoscion macdonaldi* Gilbert). *Copeia* 37: 85–86.

Nava, J. M., and L. T. Findley. 1994. Impact of the shrimp fishery on faunal diversity and stability in the upper Gulf of California, with special emphasis on the vaquita and totoaba. Project final report to Conservation International–Mexico, Gulf of California Program, Guaymas, Sonora.

Nava-Romo, J. M. 1994. Impactos a corto y largo plazo en la diversidad y otras características ecológicas de la comunidad béntico-demersal capturada por la pesquería de camarón en el norte del alto Golfo de California, México. Master's thesis, Instituto Tecnológico y de Estudios Superiores de Monterrey–Campus Guaymas, Sonora, México.

Navarro, S. C. 2003a. *Crocodylus acutus* in Sonora, Mexico. IUCN–World Conservation Union, Species Survival Commission. *Crocodile Specialist Group Newsletter* 22(1): 21.

———. 2003b. Abundance, habitat use and conservation of the American crocodile in Sinaloa. IUCN–World Conservation Union, Species Survival Commission. *Crocodile Specialist Group Newsletter* 22(2): 22–23.

Navarro-Olache, L. F., M. F. Lavín, L. G. Alvarez-Sánchez, and A. Zirino. 2004. Internal structure of SST features in the central Gulf of California. *Deep-Sea Research II* 51: 673–687.

Neat, F. C. 2001. Male parasitic spawning in two species of triplefin blenny (Tripterygiidae): Contrasts in demography, behaviour and gonadal characteristics. *Environmental Biology of Fishes* 61: 57–64.

Nelson, C. S. 1988. An introductory perspective on non-tropical shelf carbonates. *Sedimentary Geology* 60: 3–12.

Nelson, E. W. 1899. Natural history of the Tres Marias Islands, Mexico. Birds of the Tres Marias Islands. *North American Fauna* 14: 21–62.

Neuhaus, J. R. 1989. Volcanic and nonmarine stratigraphy of southwest Isla Tiburón, Gulf of California, Mexico. Master's thesis, San Diego State University, CA.

Nevárez-Martínez, M. O., A. Hernández-Herrera, E. Morales-Bojórquez, A. Balmori-Ramírez, M. A. Cisneros-Mata, and R. Morales-Azpeitia. 2000. Biomass and distribution of jumbo squid (*Dosidicus gigas* d'Orgigny, 1835) in the Gulf of California, Mexico. *Fisheries Research* 49: 129–140.

Nichols, W. J. 2003a. Biology and conservation of the sea turtles of Baja California. Ph.D. dissertation, University of Arizona, Tucson, AZ.

———. 2003b. Sinks, sewers, and speed bumps: The impact of marine development on sea turtles in Baja California, Mexico. In J. A. Seminoff (compiler), *Proceedings of the 22nd Annual Symposium on Sea Turtle Biology and Conservation*, pp. 17–18. NOAA Technical Memorandum NMFS-SEFSC-503.

Nichols, W. J., H. Aridjis, A. Hernandez, B. Machovina, and J. Villavicencio. 2002a. Black market sea turtle trade in the Californias. Unpublished Wildcoast Report, Davenport, CA.

Nichols, W. J., K. E. Bird, and S. Garcia. 2002b. Community-based research and its application to sea turtle conservation in Bahía Magdalena, B.C.S, Mexico. *Marine Turtle Newsletter* 89: 4–7.

Nichols, W. J., L. Brooks, M. Lopez, and J. A. Seminoff. 2001. Record of pelagic east Pacific green turtles associated with *Macrocystis* mats near Baja California Sur, Mexico. *Marine Turtle Newsletter* 93: 10–11.

Nichols, W. J., A. Resendiz, J. A. Seminoff, and B. Resendiz. 2000. Transpacific loggerhead turtle migration monitored with satellite telemetry. *Bulletin of Marine Science* 67: 937–947.

Nichols, W. J., and C. Safina. 2004. Case study: Lunch with a turtle poacher. *Conservation in Practice* 5(4): 30–36.

Nichols, W. J., and J. A. Seminoff. 1994. Sea turtles in the Gulf of California: Where have they gone? *Noticias de Centro de Estudios de Desiertos y Oceanos* 6: 21–33.

Nichols, W. J., J. A. Seminoff, A. Resendiz, P. Dutton, and F. A. Abreu-Grobois. 1999. Using molecular genetics and biotelemetry to study life history and long distance movement: A tale of two turtles. In F. A. Abreu-Grobois, R. Briseño-Dueñas, R. Márquez-Millan, and L. Sarti-Martínez (compilers), *Proceedings of the 18th Annual Symposium on Sea Turtle Biology and Conservation,* pp. 102–103. NOAA Technical Memorandum NMFS-SEFC-436.

Nicholson, C., C. C. Sorlien, T. Atwater, J. C. Crowell, and B. P. Luyendyk. 1994. Microplate capture, rotation of the western Transverse Ranges, and initiation of the San Andreas transform as a low-angle fault system. *Geology* 22: 491–495.

Nieto-García, E. 1998. Nutrientes en el norte del Golfo de California durante condiciones estuarinas y antiestuarinas. Master's thesis. [Available from CICESE, Ensenada, B.C., Mexico.]

Norris, J. 1981. *Articulated Coralline Algae of the Gulf of California.* Smithsonian Press, Washington, DC.

Norris, J. N., and I. A. Abbott. 1972. Some new records of marine algae from the *R/V Proteus* cruise to British Columbia. *Syesis* 5: 87–94.

Norris, J. N., and K. E. Bucher. 1976. New records of marine algae from the 1974 R/V Dolphin Cruise. *Botany* 34: 1–22.

Norris, K. S., and W. N. McFarland. 1958. A new harbor porpoise of the genus *Phocoena* from the Gulf of California. *Journal of Mammalogy* 42: 22–39.

Norris, K. S., and J. H. Prescott. 1961. Observations on Pacific cetaceans of Californian and Mexican waters. *University of California Publications in Zoology* 63: 291–402.

Norris, K. S., R. Villa, G. Nichols, B. Wursig, and K. Miller. 1983. Lagoon entrance and other aggregations of gray whales, *Eschrichtius robustus.* In R. Payne (ed.), *Behavior and Communication of Whales,* pp. 259–293. Westview Press, Boulder, CO.

Novacek, M. J., I. Ferrusquía-Villafranca, J. J. Flynn, A. R. Wyss, and M. Norell. 1991. Wasatchian (early Eocene) mammals and other vertebrates from Baja California, Mexico: The Lomas Las Tetas de Cabra Formation. *Bulletin of the American Museum of Natural History* 208: 1–88.

Odell, D. K., and K. M. McClune. 1999. False killer whale–*Pseudorca crassidens* (Owen, 1846). In S. H. Ridway and R. Harrison (eds.), *Handbook of Marine Mammals,* vol. 6, pp. 213–243. Academic Press, London.

O'Donnell, J. 1974. Green turtle fishery in Baja California waters: History and prospect. Master's thesis, California State University, Northridge.

Olguin-Mena, M. 1990. Las tortugas marinas en la costa oriental de Baja California y costa occidental de Baja California Sur, México. Master's thesis, Universidad Autónoma Baja California Sur, La Paz, México.

Oliverio, M. 2003. The Mediterranean mollusks: the best known malacofauna of the world . . . so far. *Biogeographia* 24: 195–208.

Olson, H. C. 1990. Early and middle Miocene foraminiferal paleoenvironments, southeastern San Joaquin Basin, California. *Journal of Foraminiferal Research* 20(4): 289–311.

Orr, R. T. 1972. *Marine Mammals of California.* University of California Press, Berkeley, CA.

Ortíz de Montellano, G. P. 1987. Impacto de la pesca de arrastre sobre la población juvenil de *Totoaba macdonaldi.* Reporte Técnico no. 7, Centro de Investigación y Desarrrollo de Sonora (CIDESON) [now IMADES], Gobierno de Sonora, Hermosillo.

Oskin, M., and J. Stock. 2003. Marine incursion synchronous with plate boundary localization in the Gulf of California. *Geology* 31(1): 23–26.

Oskin, M., J. Stock, and A. Martín-Barajas. 2001. Rapid localization of Pacific–North America plate boundary motion in the Gulf of California. *Geology* 29(5): 459–462.

Pacheco, M., A. Martín-Barajas, W. Eldres, J. M. Espinosa-Cardeña, J. Belenes, and A. Segura. 2006. Stratigraphy and structure of the Altar basin of NW Sonora: Implications for the history of the Colorado River delta and the Salton Trough. *Revista Mexicana de Ciencias Geológicas* 23(1): 1–22.

Pacheco-Ruíz, I., and J. A. Zertuche-González. 1996a. Green algae (Chlorophyta) from Bahía de Los Angeles, Gulf of California, Mexico. *Botanica Marina* 39: 431–433.

———. 1996b. The commercially valuable seaweeds of the Gulf of California. *Botanica Marina* 1: 201–206.

———. 1996c. Brown algae (Phaeophyta) from Bahía de los Angeles, Gulf of California, Mexico. *Hydrobiologica* 326/327: 169–172.

———. 1999. Population structure and reproduction of the carrageenophyte *Chondracanthus pectinatus* in the Gulf of California. *Hydrobiologia* 398/399: 159–166.

———. 2002. Red algae (Rhodophyta) from Bahía de Los Angeles, Gulf of California, Mexico. *Botanica Marina* 45(5): 465–470.

Pacheco-Ruíz, I., J. A. Zertuche-González, F. Arellano-Carbajal, A. Chee-Barragán, and F. Correa-Díaz. 1999. *Gracilariopsis lemaneiformis* beds along the west coast of the Gulf of California, Mexico. *Hydrobiologia* 398/399: 509–514.

Pacheco-Ruíz, I., J. A. Zertuche-González, A. Cabello-Pasini, and B. H. Brinhuis. 1992. Growth response and seasonal biomass variation of *Gigartina pectinata* Dawson (Rhodophyta) in the Gulf of California. *Journal of Experimental Marine Biology and Ecology* 157(2): 263–274.

Pacheco-Ruíz, I., J. A. Zertuche-González, A. Chee-Barragán, and E. Arroyo-Ortega. 2002. Biomass and potential commercial utilization of *Ulva lactuca* (Chlorophyta, Ulvaceae) beds along the northwest coast of the Gulf of California. *Phycologia* 41(2): 199–201.

Pacheco-Ruíz, I., J. A. Zertuche-González, A. Chee-Barragán, and R. Blanco-Betancourt. 1998. Distribution and quantification of *Sargassum* beds along the west coast of the Gulf of California, Mexico. *Botanica Marina* 41: 203–208.

Pacheco-Ruíz, I., J. A. Zertuche-González, F. Correa-Diaz, F. Arellano-Carbajal, and A. Chee-Barragán. 1999. *Gracilariopsis lemaneiformis* beds along the west coast of the Gulf of Californa, Mexico. *Hydrobiologia* 398/399: 509–514.

Páez-Osuna, F., A. Garcia, F. Flores-Verdugo, L. P. Lyle-Fritch, R. Alonso-Rodríguez, A. Roque, and A. C. Ruíz-Fernández. 2003. Shrimp aquaculture development and the environment in the Gulf of California ecoregion. *Marine Pollution Bulletin* 46: 806–816.

Páez-Osuna, F., S. R. Guerrero-Galván, A. C. Ruíz-Fernández, and R. Espinoza-Angulo. 1997. Fluxes and mass balances of nutrients in a semi-intensive shrimp farm in northwestern Mexico. *Marine Pollution Bulletin* 34(5): 290–297.

Páez-Osuna, F., and A. C. Ruíz-Fernández. 2005. Environmental load of nitrogen and phosphorus from extensive, semi-intensive and intensive shrimp farms in the Gulf of California ecoregion. *Bulletin of Environmental Contamination and Toxicology* 74: 681–688.

Palacios, E., and E. Mellink. 1992. Breeding bird records from Montague Island, northern Gulf of California. *Western Birds* 23: 41–44.

———. 1996. Status of the least tern in the Gulf of California. *Journal of Field Ornithology* 67: 48–58.

———. 2000. Nesting water birds on Islas San Martín and Todos Santos, Baja California. *Western Birds* 31: 184–189.

Palacios, E., E. Beier, M. F. Lavín, and P. Ripa. 2002. The effect of the seasonal variation of stratification on the circulation of the northern Gulf of California. *Journal of Physical Oceanography* 32: 705–728.

Paladino, F. V., M. P. O'Connor, and J. R. Spotila. 1990. Metabolism of leatherback turtles: Gigantothermy and thermoregulation of dinosaurs. *Nature* 344: 858–860.

Pansisni, M., and C. Longo. 2003. A review of the Mediterranean Sea sponge biogeography with, in appendix, a list of the demosponges hitherto recorded from this sea. *Biogeographia* 24: 59–90.

Parker, R. H. 1964. Zoogeography and ecology of some macroinvertebrates, particularly mollusks, in the Gulf of California and the continental slope off Mexico. *Videnskabelige Meddelelser fra den Naturhistoriske Forening i Kjobenhavn* 126: 1–178.

Parsons, J. J. 1962. *The Green Turtle and Man.* University of Florida Press, Gainesville, FL.

Patten, D. R., and L. T. Findley. 1970. Observations and records of *Myotis* (*Pizonyx*) *vivesi* Menegeaux (Chiroptera: Vespertiliontidae). *Los Angeles County Museum Contributions in Science* 183: 1–9.

Paul-Chavez, L., and R. Riosmena-Rodríguez. 2000. Floristic and biogeographical trends in seaweed assemblages from a subtropical insular island complex in the Gulf of California. *Pacific Science* 54: 137–147.

Payri, C. E., and G. Cabioch. 2004. The systematics and significance of coralline red algae in the rhodolith sequence of the Amédée 4 drill core (Southwest New Caledonia). *Palaeo* 204: 187–204.

Peckham, S. H., and W. J. Nichols. 2003. Why did the turtle cross the ocean? Pelagic red crabs and loggerhead turtles along the Baja California coast. In J. A. Seminoff (compiler), *Proceedings of the 22nd Annual Symposium on Sea Turtle Biology and Conservation,* pp. 47–48. NOAA Technical Memorandum NMFS-SEFSC-503.

Pegau, W. S., E. Boss, and A. Martínez. 2002. Ocean color observations of eddies during the summer in the Gulf of California. *Geophysical Research Letters* 29(9): 1295.

Peltier, W. R. 2002. On eustatic sea level history: Last glacial maximum to Holocene. *Quaternary Science Reviews* 21: 377–396.

Peñaflores, C., J. Vasconcelos, E. Albavera, and R. Marquez. 2000. Twenty five years nesting of olive ridley sea turtle *Lepidochelys olivacea* in Escobilla Beach, Oaxaca, Mexico. In F. A. Abreu-Grobois, Márquez R. Briseño, and L. Sarti (compilers), *Proceedings of the 18th International Sea Turtle Symposium,* pp. 27–29. NOAA Technical Memorandum NMFS-SEFSC-436.

Peresbarbosa, E., and E. Mellink. 1994. More records of breeding birds from Montague Isand, northern Gulf of California. *Western Birds* 25: 201–202.

———. 2001. Nesting waterbirds of Isla Montague, northern Gulf of California, Mexico. Loss of eggs due to predation and flooding, 1993–1994. *Waterbirds* 24: 265–271.

Pérez-España, H., F. Galván-Magaña, and L. A. Abitia-Cárdenas. 1996. Variaciones temporales y espaciales en la estructura de la comunidad de peces de arrecifes rocosos del suroeste del Golfo de California, México [Temporal and spatial variations in the structure of the rocky reef fish community of the southwest Gulf of California, Mexico]. *Ciencias Marinas* 22(3): 273–294.

Pérez-Jiménez, J. C., O. Sosa-Nishizaki, and J. L. Castillo-Geniz. 2005. A new eastern North Pacific smoothhound shark (genus *Mustelus,* family Triakidae) from the Gulf of California. *Copeia* 2005(4): 834–845.

Pérez-Mellado, J., and L. T. Findley. 1985. Evaluación de la ictiofauna acompañante del camarón capturado en las costas de Sonora y norte de Sinaloa, México. In A. Yañéz-Arancibia (ed.), *Recursos potentiales de México: La pesca acompañante del camarón,* pp. 201–254. Programa Universitario de Alimentos, Instituto de Ciencias del Mar y Limnología, Instituto Nacional de la Pesca. Universidad Nacional Autónoma de México, México, D.F.

Perrin, W. F. 1998. *Stenella longirostris. Mammal Species* 599: 1–7.

———. 2002. Pantropical spotted dolphin. In W. F. Perrin, B. Wursig, and J. G. M. Thewissen (eds.), *Encyclopedia of Marine Mammals,* pp. 865–867. Academic Press, San Diego, CA.

Perrin, W. F., C. E. Wilson, and F. I. Archer. 1994. Striped dolphin, *Stenella coeruleoalba* (Meyen, 1833). In S. H. Ridway and R. Harrison (eds.), *Handbook of Marine Mammals,* vol. 5, pp. 261–288. Academic Press, London.

PESCA. 1990. XXV Años de Investigación, Conservación, y Protección de la Tortuga Marina. Manufactura Lusag, México, D.F.

Pesenti, C., and J. W. Nichols. 2002. Signs of success: Fourth Annual Meeting of the Sea Turtle Conservation Network of the Californias (Grupo Tortuguero de las Californias). *Marine Turtle Newsletter* 97: 14–16.

Petersen, C. W. 1987. Reproductive behaviour and gender allocation in *Serranus fasciatus,* a hermaphroditic reef fish. *Animal Behaviour* 35: 1601–1614.

———. 1988. Male mating success, sexual size dimorphism, and site fidelity in two species of *Malacoctenus* (Labrisomidae). *Environmental Biology of Fishes* 21: 173–183.

———. 1989. Females prefer mating males in the carmine triplefin, *Axoclinus carminalis,* a paternal brood guarder. *Environmental Biology of Fishes* 26: 213–221.

Petersen, C. W., and K. Marchetti. 1989. Filial cannibalism in the Cortez damselfish *Stegastes rectifraenum. Evolution* 43: 158–168.

Pfeiler, E. 1986. Towards an explanation of the development strategy in leptocephalous larvae of marine fishes. *Environmental Biology of Fishes* 15: 3–13.

Pfeiler, E., L. A. Hurtado, L. L. Knowles, J. Torre-Cosío, L. Bourillón-Moreno, J. F. Márquez-Farías, and G. Montemayor-López. 2005. Population genetics of the swimming crab *Callinectes bellicosus* (Brachyura: Portunidae) from the eastern Pacific Ocean. *Marine Biology* 146: 559–569.

Pfeiler, E., and A. Luna. 1984. Changes in biochemical composition and energy utilization during metamorphosis of leptocephalous larvae of the bonefish (*Albula*). *Environmental Biology of Fishes* 10: 243–251.

Phillips, R. P. 1964. Marine geology in the Gulf of California. Ph.D. dissertation, University of California, San Diego. [Available from University Microfilms Inc., no. 64-9950.]

Phillips, S. J., and P. W. Comus (eds.). 2000. *A Natural History of the Sonoran Desert.* University of California Press, Berkeley, CA.

Pitman, R. L. 1990. Pelagic distribution and biology of sea turtles in the Eastern Tropical

Pacific. In T. H. Richardson, J. I. Richardson, and M. Donnelly (compilers), *Proceedings of the 10th Annual Symposium on Sea Turtle Biology and Conservation,* pp. 143–148. NOAA Technical Memorandum NMFS-SEFC-278.

Pitman, R. L., A. Aguayo L., and J. Urbán R. 1987. Observations of an unidentified beaked whale in the Eastern Tropical Pacific. *Marine Mammal Science* 3(4): 345–352.

Pitman, R. L., and M. S. Lynn. 2001. Biological observations of an unidentified mesoplodont whale in the eastern tropical Pacific and probable identity: *Mesoplodon peruvianus. Marine Mammal Science* 17: 648–657.

Pitt, J., D. Luecke, M. Cohen, E. Glenn, and C. Valdés-Casillas. 2000. Two nations, one river: Managing ecosystem conservation in the Colorado River Delta. *Natural Resources Journal* 40: 819–864.

Plascencia-González, H. G. 1993. Contribución al conocimiento de las comunidades de peces asociadas a los fondos blandos de la plataforma continental del sur de Sinaloa (Proyecto SIPCO). Tesis profesional [Bachelor's thesis], Facultad de Ciencias, Universidad Nacional Autónoma de México.

Polis, G. A., S. D. Hurd, C. T. Jackson, and F. Sánchez-Piñero. 1997. El Niño effects on the dynamics and control of an island ecosystem in the Gulf of California. *Ecology* 78(6): 1884–1897.

Pomeroy, R. S., P. McConney, and R. Mahon. 2004. Comparative analysis of coastal resource co-management in the Caribbean. *Ocean and Coastal Management* 47: 429–447.

Poole, A. F., R. O. Bierregaard, and M. S. Martell. 2002. Osprey (*Pandion haliaetus*). In A. F. Poole and F. Gill (eds.), *The Birds of North America,* no. 683. The Birds of North America, Philadelphia, PA.

Poore, G. C. B. 1984. *Colanthura, Califanthura, Cruranthura* and *Cruregens,* related genera of the Paranthuridae (Crustacea: Isopoda). *Journal of Natural History* 18: 697–715.

Potin, P., J.-Y. Floc'h, C. Augris, and J. Cabioch. 1990. Annual growth rates of the calcareous red alga *Lithothamnion coralliodes* (Corallinales, Rhodophyta) in the Bay of Brest, France. *Hydrobiologia* 204/205: 263–267.

Prescott, R., and R. Cudney-Bueno. 2008. Mobile "reefs" in the northeastern Gulf of California: Aggregations of black murex snails *Hexaplex nigritus* as habitat for invertebrates. *Marine Ecology Progress Series* 367: 185–192.

Prescott, R., V. Koch, C. Yingling, and C. Ruiz Verdugo. 2007. Settlement of Pacific calico scallop larvae (*Argopecten ventricosus,* Sowerby II, 1842) on their predator, the black murex snail (*Hexaplex nigritus,* Philippi, 1845). *Journal of Shellfish Research* 26(4): 1065–1070.

Present, T. M. 1987. Genetic differentiation of disjunct Gulf of California and Pacific coast populations of *Hypsoblennius jenkinsi. Copeia* 1987(4): 1010–1024.

Presti, S., A. Resendiz, A. Sollod, and J. A. Seminoff. 1999. Mercury presence in the scutes of black sea turtles, *Chelonia mydas agassizii,* in the Gulf of California. *Chelonian Conservation and Biology* 3: 531–533.

Pritchard, P. C. H. 1982. Nesting of the leatherback turtle, *Dermochelys coriacea,* in Pacific Mexico, with a new estimate of the world population status. *Copeia* 1982: 741–747.

———. 1999. Status of the black turtle. *Conservation Biology* 13: 1000–1003.

Pritchard, P. C. H., and P. Trebbau. 1984. *Turtles of Venezuela.* Society for the Study of Amphibians and Reptiles, Salt Lake City, UT.

PROFEPA (Procuraduría Federal de Protección al Ambiente)–SEMARNAP (Secretaría de

Medio Ambiente, Recursos Naturales y Pesca). 1995. Mortandad de mamíferos y aves marinas en el alto Golfo de California. Informe final.

Pronzato, R. 2003. Mediterranean sponge fauna: A biological, historical and cultural heritage. *Biogeographia* 24: 91–99.

Raimondi, P. T. 1988a. Rock type affects settlement, recruitment, and zonation of the barnacle *Chthamalus anisopoma* Pilsbury. *Journal of Experimental Marine Biology and Ecology* 123: 253–267.

———. 1988b. Settlement cues and determination of the vertical limit of an intertidal barnacle. *Ecology* 69(2): 400–407.

———. 1990. Patterns, mechanisms, consequences of variability in settlement and recruitment of an intertidal barnacle. *Ecological Monographs* 60(3): 283–309.

———. 1992. Adult plasticity and rapid larval evolution in a recently isolated barnacle population. *Biological Bulletin* 182: 210–220.

Raimondi, P. T., S. Forde, L. Delph, and C Lively. 1993. Processes structuring communities: Evidence for trait-mediated indirect effects through induced polymorphisms. *Oikos* 91(2): 353–361.

Randall, J. E. 1976. The endemic shore fishes of the the Hawaiian Islands, Lord Howe Island and Easter Island. *Travaux et Documents O.R.S.T.O.M.* 47: 49–73.

Readdie, M. D., M. Ranelletti, and R. M. McCourt. 2006. *Common Seaweeds of the Gulf of California.* Sea Challengers, Monterey, CA.

Rebón, G. F. 1997. Análisis de la avifauna presente en el Archipiélago de las Islas Marietas y sus aguas adyacentes, Nayarit, México. Master's thesis, Facultad de Ciencias, Universidad Nacional Autónoma de México, México, D.F.

Reeves, R. R., B. S. Stewart, P. J. Clapham, and J. A. Powell. 2002. *Guide to Marine Mammals of the World.* Knopf, New York.

Reichle, M. S. 1975. A seismological study of the Gulf of California: Sonobuoy and teleseismic observations, and tectonic implications. Ph.D. thesis, University of California, San Diego, La Jolla.

Reish, D. 1972. *Marine Life of Southern California—Emphasizing Marine Life of Los Angeles and Orange Counties.* California State College, Long Beach, CA.

Resendiz, A., B. Resendiz, W. J. Nichols, J. A. Seminoff, and N. Kamezaki. 1998. First confirmation of a trans-Pacific migration of a tagged loggerhead sea turtle (*Caretta caretta*), released in Baja California. *Pacific Science* 52: 151–153.

Reyes, S., and D. L. Cadet. 1986. Atmospheric water vapor and surface flow patterns over the tropical Americas during May–August 1979. *Monographs and Weather Reviews* 114: 582–593.

Reyes-Bonilla, H., R. Riosmena-Rodríguez, and M. S. Foster. 1997. Hermatypic corals associated with rhodolith beds in the Gulf of California, Mexico. *Pacific Science* 51: 328–337.

Reynolds, W. W., and D. A. Thomson. 1974. Responses of young Gulf grunion, *Leuresthes sardina,* to gradients of temperature, light, turbulence and oxygen. *Copeia* 1974: 747–758.

Rice, D. W. 1974. Whales and whale research in the eastern North Pacific. In W. E. Shevill (ed.), *The Whale Problem. A Status Report,* pp. 170–195. Harvard University Press, Cambridge, MA.

———. 1989. Sperm whales (*Physeter macrocephalus*). In S. H. Ridgway and R. Harrison (eds.), *Handbook of Marine Mammals,* vol. 4, pp. 177–233. Academic Press, San Diego, CA.

———. 1998. *Marine Mammals of the World: Systematics and Distribution.* Society of Marine Mammalogy Special Publication 4. University of Central Florida, Orlando, FL.

Rice, D. W., A. A. Wolman, D. E. Withrow, and L. A. Fleischer. 1981. Gray whales on the winter grounds in Baja California. *Report of the International Whaling Commission* 31: 477–489.

Ricketts, E., E. J. Calvin, J. W. Hedgpeth, and D. Phillips. 1985. *Between Pacific Tides,* 5th ed. Stanford University Press, Stanford, CA.

Riddle, B. R., D. J. Hafner, and L. F. Alexander. 2000. Comparative phylogeography of Bailey's pocket mouse (*Chaetodipus baileyi*) and the *Peromyscus eremicus* species group: Historical vicariance of the Baja California peninsular desert. *Molecular Phylogenetics and Evolution* 17: 161–172.

Riddle, B. R., D. J. Hafner, L. F. Alexander, and J. R. Jaeger. 2000a. Phylogeography and systematics of the *Peromyscus eremicus* species group and the historical biogeography of North American warm regional deserts. *Molecular Phylogenetics and Evolution* 17: 145–160.

———. 2000b. Cryptic vicariance in the historical assembly of a Baja California peninsular desert biota. *Proceedings of the National Academy of Sciences* 97: 14438–14443.

Riemann, H., and E. Ezcurra. 2005. Plant endemism and natural protected areas in the peninsula of Baja California, Mexico. *Biological Conservation* 122: 141–150.

———. 2007. Endemic regions of the vascular flora of the peninsula of Baja California, Mexico. *Journal of Vegetation Science* 18: 327–336.

Riginos, C. 2006. Cryptic vicariance in Gulf of California fishes parallels vicariant patterns found in Baja, California mammals and reptiles. *Evolution* 59: 2678–2690.

Riginos, C., and M. W. Nachman. 2001. Population subdivision in marine environments: The contributions of biogeography, geographical distance and discontinuous habitat to genetic differentiation in a blennioid fish, *Axoclinus nigricaudus. Molecular Ecology* 10: 1439–1453.

Riginos, C., and B. C. Victor. 2001. Larval spatial distributions and other early life-history characteristics predict genetic differentiation in eastern Pacific blennioid fishes. *Proceedings of the Royal Society, London B* 268: 1931–1936.

Riosmena-Rodríguez, R. 2002. Taxonomy of the Order Corallinales (Rhodophyta) in the Gulf of California, Mexico. Ph.D. dissertation, La Trobe University, Melbourne, Australia.

Riosmena-Rodríguez, R., and D. A. Siqueiros-Beltrones. 1996. Taxonomy of the genus *Amphiroa* (Corallinales, Rhodophyta) in the southern Baja California Peninsula, Mexico. *Phycologia* 35(2): 135–147.

Riosmena-Rodríguez, R., and W. J. Woelkerling. 2000. Taxonomic biodiversity of Corallinales (Rhodophyta) in the Gulf of California, Mexico: Towards an initial assessment. *Cryptogamie Algologie* 21(4): 315–354.

Riosmena-Rodríguez, R., W. J. Woelkerling, and M. S. Foster. 1999. Taxonomic reassessment of rhodolith-forming species of *Lithophyllum* (Corallinales, Rhodophyta) in the Gulf of California, Mexico. *Phycologia* 38(5): 401–417.

Ripa, P. 1990. Seasonal circulation in the Gulf of California. *Journal of Geophysical Research* 8: 559–564.

———. 1997. Towards a physical explanation of the seasonal dynamics and thermodynamics of the Gulf of California. *Journal of Physical Oceanography* 27: 597–614.

Ripa, P., and G. Velázquez. 1993. Modelo unidimensional de la marea en el Golfo de California. *Geofísica Internacional* 32: 41–56.

Rivera, M. G., R. Riosmena-Rodríguez, and M. S. Foster. 2004. Age and growth of *Lithothamnion muellerii* (Corallinales, Rhodophyta) in the southwestern Gulf of California, Mexico. *Ciencias Marinas* 30(1B): 235–249.

Rivera, M. G., and R. Scrosati. 2006. Population dynamics of *Sargassum lapazeanum* (Fucales, Phaeophyta) from the Gulf of California, Mexico. *Phycologia* 45(2): 178–189.

Roberts, N. C. 1989. *Baja California Plant Field Guide.* Natural History Publishing, La Jolla, CA.

Robertson, D. R. 2001. Population maintenance among tropical reef-fishes: Inferences from the biology of small-island endemics. *Proceedings of the National Academy of Sciences* 98: 5668–5670.

Robertson, D. R., and G. R. Allen. 2002. Shorefishes of the tropical eastern Pacific: An information system. CD-ROM Version 1.0.0. Smithsonian Tropical Research Institute, Balboa, Panama.

Robertson, D. R., J. S. Grove, and J. E. McCosker. 2004. Tropical transpacific shore fishes. *Pacific Science* 58(4): 507–565.

Robichaux, R. H. (ed.). 1999. *Ecology of Sonoran Desert Plants and Plant Communities.* University of Arizona Press, Tucson, AZ.

Robins, C. R. 1972. The state of knowledge of the coastal fish fauna of the Panamic region prior to the construction of an interoceanic sea-level canal. *Bulletin of the Biological Society of Washington* 2: 159–166.

Robinson, J., and D. A. Thomson. 1992. Status of the Pulmo Coral Reefs in the lower Gulf of California. *Environmental Conservation* 19: 261–264.

Robinson, M. K. 1973. *Atlas of Monthly Mean Sea Surface and Subsurface Temperatures in the Gulf of California, Mexico.* Memoir no. 5, San Diego Society of Natural History, San Diego, CA.

Robison, B. H. 1972. Distribution of the midwater fishes of the Gulf of California. *Copeia* 1972(3): 448–461.

Robles, A., and M. A. Carvajal. 2001. The sea and fishing. In P. Robles-Gil, E. Ezcurra, and E. Mellink (eds.), *The Gulf of California. A World Apart,* pp. 293–300. Agrupación Sierra Madre, México, D.F.

Robles, A., E. Ezcurra, and C. León. 1999. *The Sea of Cortés.* Editorial Jilguero/México Desconocido, México, D.F.

Robles Gil, P. 1996a. *Mexican Diversity of Cultures,* 2nd ed. (text by Víctor Manuel Toledo; foreword by José Sarukhán). CEMEX, México, D.F.

———. 1996b. *Mexican Diversity of Fauna,* 2nd ed. (text by Gerardo Ceballos and Fulvio Eccardi; foreword by Russell A. Mittermeier). CEMEX, México, D.F.

———. 1996c. *Mexican Diversity of Flora,* 2nd ed. (text by Rodolfo Dirzo; foreword by Peter Raven). CEMEX, México, D.F.

Robles Gil, P., E. Ezcurra, and E. Mellink (eds.). 2001. *The Gulf of California. A World Apart.* Agrupación Sierra Madre. México, D.F.

Rocha-Olivares, A., R. A. Leal-Navarro, C. Kimbrell, E. A. Lynn, and R. D. Vetter. 2003.

Microsatellite variation in the Mexican rockfish, *Sebastes macdonaldi*. *Scientia Marina* 67(4): 451–460.

Rocha-Olivares, A., R. H. Rosenblatt, and R. D. Vetter. 1999. Molecular evolution, systematics, and zoogeography of the rockfish subgenus *Sebastomus* (*Sebastes,* Scorpaenidae) based on mitochondrial cytochrome *b* and control region sequences. *Molecular Phylogenetics and Evolution* 11: 441–458.

Rocha-Olivares, A., and J. R. Sandoval-Castillo. 2003. Diversidad mitocondrial y estructura genética en poblaciones alopátricas del huachinango del Pacífico *Lutjanus peru* [Mitochondrial diversity and genetic structure in allopatric populations of the Pacific red snapper *Lutjanus peru*]. *Ciencias Marinas* 29(2): 197–209.

Roden, G. I. 1958. Oceanographic and meteorological aspects of the Gulf of California. *Pacific Science* 12(1): 21–45.

———. 1964. Oceanographic aspects of the Gulf of California. In T. H. Van Andel and G. G. Shor Jr. (eds.), *Marine Geology of the Gulf of California: A Symposium,* pp. 30–58. Memoir 3, American Association of Petroleum Geologists, Tulsa, OK.

Roden, G. I., and G. W. Groves. 1959. Recent oceanographic investigations in the Gulf of California. *Marine Research Journal* 18(1): 10–35.

Rodríguez, C. A., K. W. Flessa, and D. L. Dettman. 2001a. Effects of upstream diversion of Colorado River water on the estuarine bivalve mollusc *Mulinia coloradoensis*. *Conservation Biology* 15: 249–258.

Rodríguez, C. A., K. W. Flessa, M. A. Téllez-Duarte, D. L. Dettman, and G. E. Avila-Serrano. 2001b. Macrofaunal and isotopic estimates of the former extent of the Colorado River estuary, Upper Gulf of California, Mexico. *Journal of Arid Environments* 49: 183–193.

Rodríguez-R., J., L. A. Abitia-Cárdenas, J. De la Cruz-Agüero, and F. Galván-Magaña. 1992. Systematic list of marine fishes of Bahía Concepción, Baja California Sur, Mexico. *Ciencias Marinas* 18(1): 85–95.

Rodríguez-R., J., L. A. Abitia-C., F. Galván-M., J. Arvizu, and B. Aguilar-P. 1998. Ecology of fish communities from the soft bottoms of Bahía Concepción, México. *Archive of Fishery and Marine Research* 46: 61–76.

Rodríguez-R., J., L. A. Abitia-C., F. Galván-M., and H. Chávez. 1994. Composition, abundance and specific richness of fishes from Bahía Concepción, Baja California Sur, Mexico. *Ciencias Marinas* 20(3): 321–350.

Rodríguez-R., J., A. F. Muhlia-Melo, F. Galván-M., F. J. Gutiérrez-Sánchez, and V. Gracia-Lopez. 2005. Fish assemblages around Espiritu Santo Island and Espiritu Santo seamount in the lower Gulf of California, Mexico. *Bulletin of Marine Science* 77: 33–50.

Rohde, K., and M. Heap. 1996. Latitudinal ranges of teleost fish in the Atlantic and Indo-Pacific oceans. *American Naturalist* 147: 659–665.

Rojas-Bracho, L., and A. M. Jaramillo-Legorreta. 2002. Vaquita (*Phocoena sinus*). In W. F. Perrin, B. Würsig, and J. G. M. Thewissen (eds.), *Encyclopedia of Marine Mammals,* pp. 1277–1280. Academic Press, San Diego, CA.

Rojas-Bracho, L., R. R. Reeves, and A. Jaramillo-Legorreta. 2006. Conservation of the vaquita *Phocoena sinus*. *Mammal Review* 36(3): 179–216.

Rojas-Bracho, L., and B. L. Taylor. 1999. Risk factors affecting the vaquita (*Phocoena sinus*). *Marine Mammal Science* 15(4): 974–989.

Román-Rodríguez, M. J. 1998. Los sciaenidos en la reserva de la biosfera Alto Golfo de California. *Pesca y Conservación* 2: 7–8.

Román-Rodríguez, M. J., and M. G. Hammann. 1997. Age and growth of totoaba, *Totoaba macdonaldi* (Sciaenidae), in the upper Gulf of California. *Fishery Bulletin* 95: 620–628.

Romero-C., J. M. 1978. Composición y variabilidad de la fauna de acompañamiento del camarón en la zona norte del Golfo de California. Master's thesis, Escuela de Ciencias Marítimas y Alimentarias, Instituto Tecnológico y de Estudios Superiores de Monterrey–Campus Guaymas, Sonora.

Romo, S. F. 2004. Riqueza, distribución y taxonomía de cetáceos pertenecientes a las familias Kogiidae y Ziphiidae en Bahía de Banderas Nayarit-Jalisco, México. Bachelor's thesis, Instituto Tecnológico del Mar.

Rosales-Juárez, F. 1976. Composición y variabilidad de la fauna de acompañamiento del camarón en alta mar, frente a las costas de Sinaloa, México. In Instituto Nacional de la Pesca (ed.), *Memorias de la Reunión sobre los Recursos de la Pesca Costera en México*, 23–25 November, pp. 25–80. Veracruz, México.

Rosenblatt, R. H. 1959. A revisionary study of the blennioid family Tripterygiidae. Ph.D. dissertation, University of California, Los Angeles.

———. 1967. The zoogeographic relationships of the marine shore fishes of tropical America. *Studies in Tropical Oceanography* 5: 570–592.

Rosenblatt, R. H., J. E. McCosker, and I. Rubinoff. 1972. Indo-West Pacific fishes from the Gulf of Chiriqui, Panama. *Los Angeles County Museum Contributions in Science* 234: 1–18.

Rosenblatt, R. H., and T. D. Parr. 1969. The Pacific species of the clinid fish genus *Paraclinus*. *Copeia* 1969(1): 1–20.

Rosenblatt, R. H., and L. R. Taylor Jr. 1971. The Pacific species of the clinid fish tribe Starksiini. *Pacific Science* 25(3): 436–463.

Rosenblatt, R. H., and R. S. Waples. 1986. A genetic comparison of allopatric populations of shore fish species from the eastern and central Pacific Ocean. Dispersal or vicariance? *Copeia* 1986(2): 275–284.

Rosso, A. 2003. Bryozoan diversity in the Mediterranean Sea. *Biogeographia* 24: 227–250.

Rowell, K., K. W. Flessa, D. L. Dettman, and M. Román. 2005. The importance of Colorado River flow to nursery habitats of the Gulf corvina (*Cynoscion othonopterus*). *Canadian Journal of Fisheries and Aquatic Sciences* 62: 2874–2885.

Rowell, K., K. W. Flessa, D. L. Dettman, M. J. Román, L. R. Gerber, and L. T. Findley. In preparation. Pre- versus post-dam estuarine nursery habitat and life history of an endangered Mexican fish, *Totoaba macdonaldi*.

Rubio C., N. T., S. L. Mesnick, R. Vázquez-Juárez, J. Urbán R., C. A. J. Jodard, R. Payne, and A. E. Dizon. 2006. Genetic sex determination supports the Gulf of California as an important habitat for male and female sperm whales (*Physeter macrocephalus*). *Latin American Journal of Aquatic Mammals* 5(2): 125–128.

Ruíz-Cooley, R. I., D. Gendron, S. Aguinaga-García, S. L. Mesnick, and L. D. Carriquiry. 2004. Trophic relationships between sperm whales and the jumbo squid using stable isotopes of C and N. *Marine Ecology Progress Series* 277: 275–283.

Ruíz-Luna, A., and C. A. Berlanga-Robles. 1999. Modifications in coverage patterns and land use around the Huizache-Caimanero lagoon system, Sinaloa, Mexico: A multi-temporal analysis using Landsat images. *Estuarine Coastal and Shelf Science* 49: 37–44.

Rusnak, G. A., and R. L. Fisher. 1964. Structural history and evolution of Gulf of California. In T. H. van Andel and G. G. Shor Jr. (eds.), *Marine Geology of the Gulf of California: A Symposium,* pp. 144–156. Memoir 3, American Association of Petroleum Geologists, Tulsa, OK.

Russell, G. A., D. P. Middaugh, and M. J. Hemmer. 1987. Reproductive rhythmicity of the false grunion, *Colpichthys regis,* from Estero del Soldado, Mexico. *California Fish and Game* 73(3): 169–174.

Russell, P., and M. E. Johnson. 2000. Influence of seasonal winds on coastal carbonate dunes from the Recent and Plio-Pleistocene at Punta Chivato (Baja Californa Sur, Mexico). *Journal of Coastal Research* 16: 709–723.

Russell, S. M., and G. Monson. 1998. *The Birds of Sonora.* University of Arizona Press, Tucson, AZ.

Sáez-Arroyo, A., C. M. Roberts, J. Torre, and M. Cariño-Olvera. 2005a. Using fisher's anecdotes, naturalists' observations and grey literature to reassess marine species at risk: The case of the Gulf grouper in the Gulf of California, Mexico. *Fish and Fisheries* 6: 121–133.

Sáez-Arroyo, A., C. M. Roberts, J. Torre, M. Cariño-Olvera, and R. R. Enríquez-Andrade. 2005b. Rapidly shifting environmental baselines among fishers of the Gulf of California. *Proceedings of the Royal Society B* 272: 1957–1962.

SAGARPA (Agriculture, Livestock and Fisheries Ministry). 2004. Carta Nacional Pesquera. *Diario Oficial de la Federación,* March 15, México, D.F.

Sala, E. 2000. Manejo de peces de arrecife, conceptos y estrategias [Management of reef fishes, concepts and strategies]. In O. Aburto-Oropeza and C. A. Sánchez-Ortiz (eds.), *Recursos arrecifales del Golfo de California, estrategias de manejo para las especies marinas de ornato* [Reef resources of the Gulf of California, management strategies for the marine ornate species], pp. 107–111. Universidad Autónoma de Baja California Sur, La Paz.

Sala, E., O. Aburto-Oropeza, G. Paredes, I. Parra, J. C. Barrera, and P. K. Dayton. 2002. A general model for designing networks of marine reserves. *Science* 298: 1991–1993.

Sala, E., O. Aburto-Oropeza, G. Paredes, and G. Thompson. 2003. Spawning aggregations and reproductive behavior of reef fishes in the Gulf of California. *Bulletin of Marine Science* 72(1): 103–121.

Sala, E., O. Aburto-Oropeza, M. Reza, G. Paredes, and L. G. López-Lemus. 2004. Fishing down coastal food webs in the Gulf of California. *Fisheries* 29(3): 19–25.

Sale, P. F. 1980. The ecology of fishes on coral reefs. *Oceanography and Marine Biology Annual Review* 18: 367–421.

Sánchez-Ortiz, C., J. L. Arreola-Robles, O. Aburto-Oropeza, and M. Cortés-Hernández. 1997. Peces de arrecife en la región de La Paz, B.C.S. In J. Urbán-Ramírez and M. Ramírez-Rodríguez (eds.), *La Bahía de La Paz, investigación y conservación,* pp. 177–188. Universidad Autónoma de Baja California Sur, La Paz.

Sánchez-Velasco, L., B. Shirasago, and M. A. Cisneros-Mata. 2000. Spatial distribution of small pelagic fish larvae in the Gulf of California and its relation to the El Niño 1997–1998. *Journal of Plankton Research* 22(8): 1611–1618.

Sánchez-Velasco, L., J. E. Valdez-Holguín, B. Shirasago, M. A. Cisneros-Mata, and

A. Zárate. 2002. Changes in the spawning environment of *Sardinops caeruleus* in the Gulf of California during El Niño 1997–1998. *Estuarine, Coastal and Shelf Science* 54: 207–217.

Sanderson, I. T. 1951. *How to Know the American Mammals*. Little, Brown, Boston.

Sandoval-Castillo, J., A. Rocha-Olivares, C. Villavicencio-Garayzar, and E. Balart. 2004. Cryptic isolation of Gulf of California shovelnose guitarfish evidenced by mitochondrial DNA. *Marine Biology* 145: 983–988.

Santamaría-del-Ángel, E., S. Alvarez-Borrego, R. Millán-Nuñez, and F. E. Muller-Karger. 1999. Sobre el efecto de las surgencias de verano en la biomasa fitoplanctónica del Golfo de California. *Revista de la Sociedad Mexicana de Historia Natural* 49: 207–212.

Santamaría-del-Ángel, E., S. Alvarez-Borrego, and F. E. Muller-Karger. 1994a. Gulf of California biogeographic regions based on coastal zone color scanner imagery. *Journal of Geophysical Research* 99: 7411–7421.

———. 1994b. The 1982–1984 El Niño in the Gulf of California as seen in coastal zone color scanner imagery. *Journal of Geophysical Research* 99: 7423–7431.

Sarti, L., S. A. Eckert, P. Dutton, A. Barragan, and N. Garcia. 2000. The current situation of the leatherback population on the Pacific coast of Mexico and Central America, abundance and distribution of the nestings: An update. In H. J. Kalb and T. Wibbels (compilers), *Proceedings of the 19th Annual Symposium on Sea Turtle Biology and Conservation*, pp. 85–87. NOAA Technical Memorandum NMFS-SEFSC-443.

Sarti Martínez, A. L., A. R. Barragán, D. García-Muñoz, P. Huerta, and F. Vargas. 2007. Conservation and biology of the leatherback turtle in the Mexican Pacific. *Chelonian Conservation and Biology* 6(1): 70–78.

Sawlan, M. G., and J. G. Smith. 1984. Petrologic characteristics, age and tectonic setting of Neogene volcanic rocks in northern Baja California Sur, Mexico. In V. A. Frizzell Jr. (ed.), *Geology of the Baja California Peninsula*, pp. 219–236. Pacific Section, SEPM, Book 39, Society for Sedimentary Geology, Los Angeles, CA.

Scammon, C. M. 1968. *The Marine Mammals of the Northwestern Coast of America*. Dover, New York. [Originally published in 1874.]

Schaadt, C. P. 1989. Intraspecific comparisons of sexual and geographic variation in the growth of migratory and sedentary ospreys. Ph.D. dissertation, McGill University, Montreal.

Schaeffer, T. N., G. J. Smith, M. S. Foster, and A. Detomaso. 2002. Genetic differences between two growth-forms of *Lithophyllum margaritae* (Rhodophyta) in Baja California Sur, Mexico. *Journal of Phycology* 38(6): 1090–1098.

Schlanger, S. O., and C. J. Johnston. 1969. Algal banks near La Paz, Baja California—Modern analogs of source areas of transported shallow-water fossils in pre-alpine flysch deposits. *Palaeo* 6: 141–157.

Schmidt, N. 1990. Plate tectonics and the Gulf of California region. *Arizona Geology* 20(2): 1–4.

Schöne, B. R., K. W. Flessa, D. L. Dettman, and D. H. Goodwin. 2003. Upstream dams and downstream clams: Growth rates of bivalve mollusks unveil impact of river management on estuarine ecosystems (Colorado River Delta, Mexico). *Estuarine, Coastal Shelf Science* 58: 715–726.

Schwartzlose, R. A., and S. Alvarez-Borrego. 2002. The history of oceanography along the Mexican Pacific coast. In K. R. Benson and P. F. Rehbock (eds.), *Oceanographic History, the Pacific and Beyond*, pp. 167–173. University of Washington Press, Seattle, WA.

Schwartzlose, R. A., D. Alvarez-Millán, and P. Brueggeman. 1992. *Golfo de California: Bibliografía de las ciencias marinas* [Gulf of California: Bibliography of marine sciences]. Instituto de Investigaciones Oceanológicas, Universidad Autónoma de Baja California, Ensenada, Baja California, México.

Schwennicke, T., G. González-Barba, and N. DeAnda-Franco. 1996. Lower Miocene marine and fluvial beds at Rancho la Palma, Baja California Sur, Mexico. Departamento de Geología, Hermosillo, Sonora. *Boletín de la Universidad de Sonora* 13(1): 1–14.

Scoffin, T. P., D. R. Stoddart, A. W. Tudhope, and C. Woodroffe. 1985. Rhodoliths and coralliths of Muri Lagoon, Rarotonga, Cook Islands. *Coral Reefs* 4: 71–80.

Scott, M. D., and S. J. Chivers. 1990. Distribution and herd structure of bottlenose dolphins in the eastern tropical Pacific Ocean. In S. Leatherwood and R. R. Reeves (eds.), *The Bottlenose Dolphin,* pp. 387–402. Academic Press, San Diego, CA.

SDNHM (San Diego Natural History Museum). 1996. An interview with George Lindsay. Unpublished transcription of tape-recording conducted by Michael W. Hager, February 19, 1976. San Diego Natural History Museum, San Diego, CA.

Secretaría de Industria y Comercio. 1968. Programa Nacional de Marcado de Tortugas Marinas. Dirección General de Pesca e Industrias Conexas. Comisión Nacional Consultive de Pesca. Mexico City, Mexico.

Secretaría de Pesca. 1992. Legal Framework for Fisheries 1992. Secretaría de Pesca, México, D.F.

SEDUE. 1989. Información básica sobre las Áreas Naturales Protegidas de México. Subsecretaría de Ecología-SINAP, pp. 18–26, México, D.F.

SEMARNAP (Environment, Natural Resources and Fisheries Ministry). 2000. Carta Nacional Pesquera. *Diario Oficial de la Federación,* August 28, México, D.F.

Seminoff, J. A. 1994. Conservation of the marine turtles of Mexico: A survey of nesting beach conservation projects. Master's thesis, University of Arizona, Tucson.

———. 2000. The biology of the East Pacific green turtle (*Chelonia mydas agassizii*) at a warm temperate foraging area in the Gulf of California, Mexico. Ph.D. dissertation, University of Arizona, Tucson.

Seminoff, J. A., J. Alvarado, C. Delgado, J. L. Lopez, and G. Hoeffer. 2002a. First direct evidence of migration by an east Pacific green sea turtle from Michoacán, Mexico, to a foraging ground on the Sonoran Coast of the Gulf of California. *Southwestern Naturalist* 47: 314–316.

Seminoff, J. A., and P. H. Dutton. 2007. Leatherback sea turtles (*Dermochelys coriacea*) in the Gulf of California: Distribution, demography, and human interactions. *Chelonian Conservation and Biology* 6: 137–141.

Seminoff, J. A., T. T. Jones, A. Resendiz, W. J. Nichols, and M. Y. Chaloupka. 2003a. Monitoring green turtles (*Chelonia mydas*) at a coastal foraging area in Baja California, Mexico: Multiple indices describe population status. *Journal of the Marine Biological Association of the United Kingdom* 83: 1355–1362.

Seminoff, J. A., S. Karl, T. Swartz, and A. Resendiz. 2003b. Hybridization of the green turtle (*Chelonia mydas*) and the hawksbill turtle (*Eretmochelys imbricata*) in the Pacific Ocean: Indication of an absence of gender bias in the directionality of crosses. *Bulletin of Marine Science* 73: 643–652.

Seminoff, J. A., W. J. Nichols, A. Resendiz, and L. Brooks. 2003c. Occurrence of hawksbill turtles, *Eretmochelys imbricata* (Reptilia: Cheloniidae), near the Baja California Peninsula, Mexico. *Pacific Science* 57(1): 9–16.

Seminoff, J. A., A. Resendiz, and W. J. Nichols. 2002b. Diet of east Pacific green turtles (*Chelonia mydas*) in the central Gulf of California, Mexico. *Journal of Herpetology* 36(3): 447–453.

———. 2002c. Home range of the green turtle (*Chelonia mydas*) at a coastal foraging ground in the Gulf of California, Mexico. *Marine Ecology Progress Series* 242: 253–265.

Seminoff, J. A., A. Resendiz, W. J. Nichols, and T. T. Jones. 2002d. Growth rates of wild green turtles (*Chelonia mydas*) at a temperate foraging habitat in the Gulf of California, Mexico. *Copeia* 2002: 610–617.

Seminoff, J. A., A. Resendiz, B. Resendiz, and W. J. Nichols. 2004. Occurrence of loggerhead sea turtles (*Caretta caretta*) in the Gulf of California, Mexico: Evidence of life-history variation in the Pacific Ocean. *Herpetological Review* 35: 24–27.

Setchell, W. A. 1924. The marine algae. Expedition of the California Academy of Sciences to the Gulf of California in 1921. *Proceedings California Academy of Sciences, 4th series* 12: 695–949.

Setchell, W. A., and Gardner, N. L. 1924. Expedition of the California Academy of Sciences to the Gulf of California in 1921: The marine algae. *Proceedings California Academy of Sciences, 4th Series* 12: 695–949.

Sewell, A., M. Johnson, D. Backus, and J. Ledesma-Vazquez. 2007. Rhodolith detritus impounded by a coastal dune on Isla Coronados, Gulf of California. *Ciencias Marinas* 33(4): 483–494.

Shepard, F. P. 1950. *Submarine topography of the Gulf of California, Part 3 of the 1940 E. W. Scripps cruise to the Gulf of California.* Memoir 43, pp. 1–32, Geological Society of America, Boulder, CO.

Sheppard, P. R., A. C. Comrie, G. D. Packin, K. Angersbach, and M. K. Hughes. 2002. The climate of the U.S. Southwest. *Climate Research* 21: 219–238.

Sheridan, T. E. (ed.) 1999. *Empire of Sand: The Seri Indians and the Struggle for Spanish Sonora, 1645–1803.* University of Arizona Press, Tucson, AZ.

Shreve, F., and I. L. Wiggins. 1964. *Flora and Vegetation of the Sonoran Desert,* 2 vols. Stanford University Press, Stanford, CA.

Simian, M. E., and M. E. Johnson. 1997. Development and foundering of the Pliocene Santa Ines Archipelago in the Gulf of California: Baja California Sur, Mexico. In M. E. Johnson and J. Ledesma-Vázquez (eds.), *Pliocene Carbonates and Related Facies Flanking the Gulf of California, Baja California, Mexico,* pp. 25–38. Special Paper 318, Geological Society of America, Boulder, CO.

Simpson, J. H., A. J. Souza, and M. F. Lavín. 1994. Tidal mixing in the Gulf of California. In K. J. Beven, P. C. Chatwin, and J. H. Millbank (eds.), *Mixing and Transport in the Environment,* pp. 169–182. Wiley, London.

Smith, J. T. 1984. Miocene and Pliocene marine mollusks and preliminary correlations, Vizcaíno Peninsula to Arroyo la Purísima, northwestern Baja California Sur, Mexico. In V. A. Frizzell Jr. (ed.), *Geology of the Baja California Peninsula,* pp. 197–217. Pacific Section, SEPM, Book 39, Society for Sedimentary Geology, Los Angeles, CA.

———. 1991. Cenozoic marine mollusks and paleogeography of the Gulf of California. In J. T. Dauphin and R. T. Simoneit (eds.), *The Gulf and Peninsular Province of the*

Californias, pp. 447–480. Memoir 47, Association of Petroleum Geologists, Tulsa, OK.

Smith, J. T., J. G. Smith, J. C. Ingle, R. G. Gastil, M. C. Boehm, J. Roldán-Quintana, and R. E. Casey. 1985. Fossil and K/Ar age constraints on upper middle Miocene conglomerate, SW Isla Tiburón, Gulf of California. *Geological Society of America, Abstracts with Programs* 17(6): 409.

Snyder-Conn, E., and R. C. Brusca. 1977. Shrimp population dynamics and fishery impact in the northern Gulf of California. *Ciencias Marinas* 1(3): 54–67.

Sobel, J., and C. Dahlgren (eds.). 2004. *Marine Reserves: A Guide to Science, Design, and Use.* Island Press, Washington, DC.

Sokolov, V. A., and M. Wong-Rios. 1973. Investigaciones efectuadas sobre los peces pelágicos del Golfo de California (sardina, crinuda y anchoveta) en 1971. Instituto Nacional de Pesca/SI: 12. Informe Científico 2, México.

Soto-Mardones, L., S. G. Marinone, and A. Parés-Sierra. 1999. Time and spatial variability of sea surface temperature in the Gulf of California. *Ciencias Marinas* 25: 1–30.

Spitzer, P. R., R. W. Risebrough, J. W. Grier, and C. R. Sindelar Jr. 1977. Eggshell thickness-pollutant relationships among North American Ospreys. In J. C. Ogden (ed.), *Transactions of the North American Osprey Research Conference,* pp. 13–19. National Park Service, Department of the Interior, Washington, DC.

Spotila, J. R., R. D. Reina, A. C. Steyermark, P. T. Plotkin, and F. V. Paladino. 2000. Pacific leatherback turtles face extinction. *Nature* 405: 529–530.

Springer, V. G. 1959. Systematics and zoogeography of the clinid fishes of the subtribe Labrisomini Hubbs. *Institute of Marine Science, University of Texas* 5: 417–492.

Squires, D. 1959. Results of the Puritan-American Museum Natural History Expedition to Western Mexico. 7. Corals and coral reefs in the Gulf of California. *Bulletin of the American Museum of Natural History* 118(7): 367–432.

Squires, D., C. Janisse, J. A. Seminoff, and P. H. Dutton. In press. Fisheries mitigation in with Leatherback turtles in Baja California. In J. Bishop, S. Pagiola, and S. Wunder (eds.), *Market-Based Instruments for Biodiversity Conservation.* IUCN Press, Gland, Switzerland.

Squires, L. R., and R. A. Demetrion. 1992. Paleontology of the Eocene Bateque formation, Baja California Sur, Mexico. *Los Angeles County Museum Contributions in Science* 434.

Stager, K. E. 1957. The avifauna of the Tres Marias Islands, Mexico. *Auk* 74: 412–432.

Steinbeck, J. 1951. *The Log from the Sea of Cortez.* Viking Press, New York.

Steinbeck, J., and E. F. Ricketts. 1941. *The Sea of Cortez. A Leisurely Journal of Travel and Research.* Viking Press, New York.

Steindachner, F. 1877. Ichthyologische Beiträge. IV. *Anzeiger der Akademie der Wissenshaften in Wien* 72: 551–616.

Steller, D. L. 2003. Rhodoliths in the Gulf of California: Growth, demography, disturbance and effects on population dynamics of catarina scallops. Ph.D. dissertation, University of California, Santa Cruz.

Steller, D. L., and M. S. Foster. 1995. Environmental factors influencing distribution and morphology of rhodoliths in Bahía Concepción, B.C.S., México. *Journal of Experimental Marine Biology and Ecology* 194(2): 201–212.

Steller, D. L., M. Hernandez-Ayón, R. Riosmena-Rodríguez, and A. Cabello-Pasini. 2007a.

Effect of temperature on photosynthesis, growth and calcification rates of the free-living coralline alga *Lithophyllum margaritae*. *Ciencias Marinas* 33(4): 441–456.

Steller, D. L., R. Riosmena-Rodríguez, and M. S. Foster. 2007b. Sampling and monitoring rhodoliths. In R. Rigby, K. Iken, and Y. Shirayama (eds.), *Handbook for Sampling Coastal Seagrasses and Macroalgae Community Biodiversity*, pp. 93–97. Kyoto University Press, Kyoto, Japan.

Steller, D. L., R. Riosmena-Rodríguez, M. S. Foster, and C. A. Roberts. 2003. Rhodolith bed diversity in the Gulf of California: The importance of rhodolith structure and consequences of disturbance. *Aquatic Conservation: Marine and Freshwater Ecosystems* 13: S5–S20.

Steneck, R., and W. Adey. 1976. The role of environment in control of morphology in *Lithophyllum congestum*, a Caribbean algal ridge builder. *Botanica Marina* 19: 435–455.

Stephens, J. S., Jr. 1963. A revised classification of the blennioid fishes of the American family Chaenopsidae. *University of California Publications in Zoology* 68: 1–165.

Stephens, J. S., Jr., and V. G. Springer. 1971. *Neoclinus nudus*, a new scaleless clinid fish from Taiwan with a key to *Neoclinus*. *Proceedings of the Biological Society of Washington* 84(9): 65–72.

Stepien, C. A., R. H. Rosenblatt, and B. A. Bargmeyer. 2001. Phylogeography of the spotted sand bass, *Paralabrax maculatofasciatus:* Divergence of Gulf of California and Pacific coast populations. *Evolution* 55(9): 1852–1862.

Stewart, J. G. 1982. Anchor species and epiphytes in intertidal algal turf. *Pacific Science* 36: 45–59.

Stewart, R. H. 1985. *Methods of Satellite Oceanography*. University of California Press, Berkeley, CA.

Stock, J. M., and K. V. Hodges. 1989. Pre-Pliocene extension around the Gulf of California and the transfer of Baja California to the Pacific plate. *Tectonics* 8: 99–115.

Stoleson, S. H., R. S. Felger, G. Ceballos, C. Raish, M. Wilson, and A. Búrquez. 2005. Recent history of natural resource use and population growth in northern Mexico. In J.-L. E. Cartron, G. Ceballos, and R. S. Felger (eds.), *Biodiversity, Ecosystems, and Conservation in Northern Mexico*, pp. 52–86. Oxford University Press, New York.

Strand, S. W. 1977. Community structure among reef fish in the Gulf of California: The use of reef space and interspecific foraging associations. Ph.D. dissertation, University of California, Davis.

———. 1988. Following behavior and interspecific foraging associations among Gulf of California reef fishes. *Copeia* 1988(2): 351–357.

Streets, T. H. 1877. Ichthyology. In "Contributions to the Natural History of the Hawaiian and Fanning Islands and Lower California, Made in Connection with the United States North Pacific Surveying Expedition, 1873–75." *Bulletin of the United States National Museum* 7: 43–102.

Sverdrup, H. U. 1941. The Gulf of California: Preliminary discussion on the cruise of the E. W. Scripps in February and March 1939. *Proceedings of the 6th Pacific Science Congress*, vol. 3, pp. 161–166.

Swenson, J. E. 1979. The relationship between prey species ecology and dive success in ospreys. *Auk* 96: 408–412.

Sykes, G. 1937. *The Colorado Delta*. Carnegie Institution of Washington, American Geographical Society of New York, Baltimore, MD.

Sykes, L. 1968. Seismological evidence for transform faults, sea-floor spreading and continental drift. In R. A. Phinney (ed.), *History of the Earth's Crust, A NASA Symposium,* pp. 120–150. Princeton University Press, Princeton, NJ.

Tavera, J. J., A. F. González-Acosta, and J. De la Cruz-Agüero. 2005. First record of *Seriola peruana* (Actinopterygii: Carangidae) in the Gulf of California. *IMBA2-Biodiversity Records* (online).

Taylor, W. R. 1960. *Marine Algae of the Eastern Tropical and Subtropical Coasts of the Americas.* University of Michigan Press, Ann Arbor, MI.

Terry, A., G. Bucciarelli, and G. Bernardi. 2000. Restricted gene flow and incipient speciation in disjunct Pacific Ocean and Sea of Cortez populations of a reef fish species, *Girella nigricans. Evolution* 54: 652–659.

Tershy, B. R., D. Breese, and S. Alvarez-Borrego. 1991. Increase in cetacean and seabird numbers in the Canal de Ballenas during an El Niño–Southern Oscillation event. *Marine Ecology Progress Series* 69: 299–302.

Thayer, G. W., D. W. Engel, and K. A. Bjorndal. 1982. Evidence for short circuiting of the detritus cycle of seagrass beds by the green turtle, *Chelonia mydas. Journal of Experimental Marine Biology and Ecology* 62: 173–183.

Thomas, R. K. 1991. Papago land use west of the Papago Indian reservation, south of the Gila River, and the problem of Sand Papago Identity. In R. H. McGuire (ed.), *Ethnology of Northwest Mexico: A Sourcebook,* pp. 357–399. Spanish Borderlands Sourcebook no. 6, Garland, New York.

Thompson, R. W. 1968. *Tidal Flat Sedimentation on the Colorado River Delta, Northwestern Gulf of California.* Memoir 107, Geological Society of America, Boulder, CO.

Thomson, D. A., and W. H. Eger. 1966. *Guide to the Families of the Common Fishes of the Gulf of California.* University of Arizona Press, Tucson, AZ.

Thomson, D. A., L. T. Findley, and A. N. Kerstitch. 1979. *Reef Fishes of the Sea of Cortez: The Rocky-Shore Fishes of the Gulf of California.* Wiley, New York. [Revised edition published 2000 by University of Texas Press, Austin, TX.]

Thomson, D. A., and M. R. Gilligan. 1983. The rocky-shore fishes. In T. J. Case and M. L. Cody (eds.), *Island Biogeography in the Sea of Cortez,* pp. 98–129. University of California Press, Berkeley, CA.

———. 2002. Rocky-shore fishes. In T. Case, M. Cody, and E. Ezcurra (eds.), *A New Island Biogeography of the Sea of Cortés,* pp. 154–180. Oxford University Press, New York.

Thomson, D. A., and C. E. Lehner. 1976. Resilience of a rocky-intertidal fish community in physically unstable environment. *Journal of Experimental Marine Biology and Ecology* 22: 1–29.

Thomson, D. A., and N. McKibbin. 1976. *Gulf of California Fishwatcher's Guide.* Golden Puffer Press, Tucson, AZ.

———. 1978. *Peces del Golfo de California* (traducción de M. Mahieux). Centro de Investigaciones Científicas y Tecnológicas, Universidad de Sonora, Hermosillo, Sonora, México.

Thomson, D. A., and K. A. Muench. 1976. Influence of tides and waves on the spawning behavior of the Gulf of California grunion, *Leuresthes sardina* (Jenkins and Evermann). *Bulletin of the Southern California Academy of Sciences* 75: 198–203.

Thorade, H. 1909. Über die Kalifornische Meeresströmung. *Annals of Hydrography and Maritime Meteorology* 37: 17–34, 63–76.

Thunell, R., C. Pride, E. Tappa, and F. Muller-Karger. 1993. Varve formation in the Gulf of California: Insights from time series sediment trap sampling and remote sensing. *Quaternary Science Reviews* 12: 451–464.

Tiburcio-Pintos, G., P. Marquez-A., J. M. Sandez-C., and J. R. Guzman-P. 2006. First nesting report of black sea turtle (*Chelonia mydas agassizii*) in Baja California Sur, Mexico. In *Proceedings of the 24th Symposium on Sea Turtle Biology and Conservation,* April 2004, San Jose, Costa Rica.

Todd, E. S., and A. W. Ebeling. 1966. Aerial respiration in the longjaw mudsucker *Gillichthys mirabilis* (Teleostei: Gobiidae). *Biological Bulletin* 130: 265–288.

Topp, R. W. 1969. Interoceanic sea-level canal: Effects on the fish faunas. *Science* 165: 1324–1327.

Townsend, C. H. 1916. Scientific results of the expedition to the Gulf of California in charge of C. H. Townsend, by the U.S. Fisheries Steam-ship '*Albatross*' in 1911, Commander G. H. Burrage, U.S.N., Commanding. I. Voyage of the '*Albatross*' to the Gulf of California in 1911. *Bulletin of the American Museum of Natural History* 35: 399–476.

———. 1935. The distribution of certain whales as shown by the logbook records of American whale ships. *Zoologica* 19: 1–50.

Trujillo-Millán, O., J. De la Cruz-Agüero, and J. F. Elorduy-Garay. 2006. First reported records of the *Prionurus laticlavius* (Perciformes: Acanthuridae) from the Gulf of California. *Bulletin of Marine Science* 78: 393–395.

Turner, R. M., J. E. Bowers, and T. L. Burgess. 1995. *Sonoran Desert Plants: An Ecological Atlas.* University of Arizona Press, Tucson, AZ.

Underwood, A. J. 1997. *Experiments in Ecology.* Cambridge University Press, Cambridge, U.K.

Underwood, J. G., C. J. Hernandez-Camacho, D. Aurioles-Gamboa, and L. R. Gerber. 2008. Estimating sustainable bycatch rates for California sea lion populations in the Gulf of California. *Conservation Biology* 22(3): 701–710.

Upton, D. E., and R. W. Murphy. 1997. Phylogeny of the side-blotched lizards (Phrynosomatidae: *Uta*) based on mtDNA sequences: Support for a midpeninsular seaway in Baja California. *Molecular Phylogenetics and Evolution* 8: 104–113.

Urbán R., J. 1983. Taxonomía y distribución de los géneros *Tursiops, Delphinus* y *Stenella* en las aguas adyacentes a Sinaloa y Nayarit, México (Cetacea: Delphinidae). Tesis profesional, Facultad de Ciencias, Universidad Nacional Autónoma de México, México, D.F.

———. 1993. Varamiento y rescate de calderones de aletas cortas, *Globicephala macrorhynchus,* en la Bahía de La Paz, B.C.S. *Rev. Inv. Cient.* 1(No. especial SOMEMMA 1): 59–67.

———. 2000. Familia Balaenopteridae. In S. T. Alvarez-Castañeda and J. L. Patton (eds.), *Mamíferos del noroeste de México II,* pp. 661–683. Centro de Investigaciones Biológicas del Noroeste, S.C.

Urbán R., J., C. Alvarez F., M. Salinas Z., J. Jacobsen, K. C. Balcomb III, A. Jaramillo L., P. Ladrón de Guevara P., and A. Aguayo L. 1999. Population size of humpback whale, *Megaptera novaeangliae,* in waters off the Pacific coast of Mexico. *Fishery Bulletin* 97(4): 1017–1024.

Urbán R., J., and D. Aurioles G. 1992. First record of the pygmy beaked whale, *Mesoplodon peruvianus,* in the North Pacific. *Marine Mammal Science* 8(4): 420–425.

Urbán R., J., G. Cárdenas-Hinojosa, A. Gómez-Gallardo, U. González-Peral, and R. L. Brownell Jr. 2007. Mass stranding of Baird's beaked whales at San Jose Island, Gulf of California, Mexico. *Latin American Journal of Aquatic Mammals* 6(1): 83–88.

Urbán R., J., and S. Flores R. 1996. A note on Bryde's whales (*Balaenoptera edeni*) in the Gulf of California, Mexico. *Report of the International Whaling Commission* 46: 453–457.

Urbán R., J., A. Gómez-Gallardo U., and S. Ludwig. 2003a. Abundance and mortality of gray whales at Laguna San Ignacio, Mexico, during the 1997–98 El Niño and the 1998–99 La Niña. *Geofísica Internacional* 42(3): 439–446.

Urbán R., J., A. Gómez-Gallardo U., M. Palmeros R., and G. Velazquez Ch. 1997. Los mamíferos marinos de la Bahía de La Paz. In J. Urbán R. and M. Ramírez R. (eds.), *La Bahía de La Paz*, pp. 193–217. Investigación y Conservación. UABCS, CICIMAR, SCRIPPS.

Urbán R., J., U. González-Peral, G. Cárdenas-Hinojosa, and L. Rojas-Bracho. 2007. Informe para la Comisión para la Cooperación Ambiental del Plan de Acción de América del Norte para la Conservación de la Ballena Jorobada. Unpublished. Comisión para la Cooperación Ambiental del Tratado de Libre Comercio. UABCS.

Urbán R., J., A. Jaramillo L., A. Aguayo L., P. Ladrón de Guevara P., M. Salinas Z., C. Alvarez F., L. Medrano G., J. K. Jacobsen, K. C. Balcomb III, D. E. Claridge, J. Calambokidis, G. H. Steiger, J. Straley, O. von Ziegesar, J. M. Wite, S. Miszroch, M. E. Dahlheim, J. D. Darling, and C. S. Baker. 2000. Migratory destinations of humpback whales wintering in the Mexican Pacific. *Journal of Cetacean Research and Management* 2(2): 101–110.

Urbán R., J., and H. Pérez-Cortés M. 2000. Familia Ziphiidae. In S. T. Alvarez-Castañeda and J. L. Patton (eds.), *Mamíferos del noroeste de México II,* pp. 643–653. Centro de Investigaciones Biológicas del Noroeste, S.C.

Urbán R., J., S. Ramírez, and J. C. Salinas V. 1994. First record of the bottlenose whale *Hyperoodon* sp. in the Gulf of California. *Marine Mammal Science* 10: 471–473.

Urbán R., J., and L. Rojas B. 1999. Los programas de conservación de mamíferos marinos. In M. del C. Rodríguez H. and C. Hernández F. (eds.), *Océanos ¿Fuente inagotable de recursos?* pp. 541–573. Programa Universitario del Medio Ambiente, UNAM-SEMARNAP.

Urbán R., J., L. Rojas-Bracho, M. Guerrero-Ruíz, A. Jaramillo-Legorreta, and L. T. Findley. 2005. Cetacean diversity and conservation in the Gulf of California. In J. E. Cartron, G. Ceballos, and R. S. Felger (eds.), *Biodiversity, Ecosystems, and Conservation in Northern Mexico,* pp. 276–297. Oxford University Press, New York.

Urbán R., J., L. Rojas-Bracho, H. Pérez-Cortés, A. Gómez-Gallardo, S. L. Swartz, S. Ludwig, and R. L. Brownell Jr. 2003b. A review of gray whales on their wintering grounds in Mexican waters. *Journal of Cetacean Research and Management* 5(3): 281–295.

van Andel, T. H. 1964. Recent marine sediments of the Gulf of California. In T. H. van Andel and G. G. Shor Jr. (eds.), *Marine Geology of the Gulf of California: A Symposium,* pp. 216–310. Memoir 3, American Association of Petroleum Geologists, Tulsa, OK.

van Andel, T. H., and G. G. Shor Jr. 1964. *Marine Geology of the Gulf of California: A Symposium.* Memoir 3, American Association of Petroleum Geologists, Tulsa, OK.

van der Heiden, A. M. 1985. Taxonomía, biología y evaluación de la ictiofauna demersal del Golfo de California. In A. Yáñez-Arancibia (ed.), *Recursos pesqueros potenciales de*

México: La pesca acompañante del camarón, pp. 149–200. Programa Universitario de Alimentos, Instituto de Ciencias del Mar y Limnología, y el Instituto Nacional de la Pesca. Universidad Nacional Autónoma de México, México, D.F.

van der Heiden, A. M., and L. T. Findley. 1990. Lista de los peces marinos del sur de Sinaloa, México. Universidad Nacional Autónoma de México. *Anales del Instituto de Ciencias del Mar y Limnología* 15 [for 1988] (2): 209–223.

van der Heiden, A. M., and H. G. Plascencia-González. 2005. *Etropus ciadi,* a new endemic flatfish from the Gulf of California, Mexico (Pleuronectiformes: Paralichthyidae). *Copeia* 2005(3): 470–478.

van der Heiden, A. M., H. G. Plascencia-González, and S. Mussot-Pérez. 1986. Aportaciones al conocimiento de la ictiofauna demersal del Golfo de California. In *Memorias del I Intercambio Académico sobre las Investigaciones en el Mar de Cortés,* pp. 328–339. DICTUS-Universidad de Sonora/CONACyT, Hermosillo, Sonora, México.

van Gelder, R. G. 1960. Results of the Puritan-American Museum of Natural History Expedition to western Mexico, 10: Marine mammals from the coasts of Baja California and the Tres Marías Islands, Mexico. *American Museum Novitates* 1992: 1–27.

Van Rossem, A. J. 1932. The avifauna of Tiburón Island, Sonora, Mexico, with descriptions of four new races. *Transactions of the San Diego Society of Natural History* 7: 110–150.

Van Syoc, R. J. 1992. Living and fossil populations of a western Atlantic barnacle, *Balanus subalbidus* Henry, 1974, in the Gulf of California Region. *Proceedings of the San Diego Society of Natural History* 12: 9–27.

Van Voorhies, W. 1996. Bergmann size clines: A simple explanation for their occurrence in ectotherms. *Evolution* 50: 1259.

Varela-Romero, A. 1990. Aspectos tróficos de las mojarras (Pisces: Gerreidae) en tres sistemas costeros de Sonora. Tesis profesional [Bachelor's thesis], Universidad Autónoma de Baja California Sur, La Paz.

Vazquez-Elizondo, R. M. 2005. Revaluación taxonómica de *Lithophyllum bracchiatum* (Heydrich) ME. Lemoine (Rhodophyta: Corallinales) para el suroeste del Golfo de California. Tesis de Licenciatura, Universidad de Baja California Sur, México.

Vázquez-Hernández, S., A. L.Carreño, and A. Martín-Barajas, 1996. Stratigraphy and paleoenvironments of the Mio-Pliocene Imperial Formation in the eastern Laguna Salada area, Baja California, México. In P. L. Abbott and J. D. Cooper (eds.), *Field Conference Guide 1996,* pp. 373–380. Annual Meeting, American Association of Petroleum Geologists and Society for Economic Paleontology and Mineralogy. Pacific Section, AAPG, Guide Book 73; Pacific Section, SEPM, Book 80.

Velarde, E., and D. W. Anderson. 1994. Conservation and management of seabird islands in the Gulf of California. Setbacks and successes. In D. N. Nettleship, J. Burger, and M. Gachfeld, *Seabirds on Islands: Threats, Case Studies and Action Plans.* Birdlife Conservation Series no. 1, Bird Life International, Cambridge, U.K.

Velarde, E., J.-L. E. Cartron, H. Drummond, D. W. Anderson, F. Rebón Gallardo, E. Palacios, and C. Rodríguez. 2005. Nesting seabirds of the Gulf of California's offshore islands. In J.-L. E. Cartron, G. Ceballos, and R. S. Felger (eds.), *Biodiversity, Ecosystems, and Conservation in Northern Mexico,* pp. 452–470. Oxford University Press, New York.

Velarde, E., and E. Ezcurra. 2002. Breeding dynamics of Heermann's Gulls. In T. Case,

M. Cody, and E. Ezcurra (eds.), *A New Island Biogeography of the Sea of Cortés,* pp. 313–325. Oxford University Press, New York.

Velarde, E., E. Ezcurra, M. A. Cisneros-Mata, and M. F. Lavín. 2004. Seabird ecology, El Niño anomalies, and prediction of sardine fisheries in the Gulf of California. *Ecological Applications* 14(2): 607–615.

Venegas, Miguel. 1757. *Noticia de la California y su conquista,* 3 vol., Madrid. [Edited in 1979 as *Obras Californianas,* 5 tomes, Universidad Autónoma de Baja California, México, D.F.]

Verheij, E. 1993. *Marine plants on the reefs on the Spermonde Archipiélago, SW Sulawesi, Indonesia.* Rijksherbarium-Hortus Botanicus, Leiden, The Netherlands.

Victor, B. C., and G. M. Wellington. 2000. Endemism and the pelagic larval duration of reef fishes in the eastern Pacific Ocean. *Marine Ecology Progress Series* 205: 241–248.

Vidal, O. 1993. Aquatic mammal conservation in Latin America: Problems and perspectives. *Conservation Biology* 7: 788.

———. 1995. Population biology and incidental mortality of the vaquita, *Phocoena sinus. Reports of the International Whaling Commission, Special Issue* 16: 247–272.

Vidal, O., R. L. Brownell Jr., and L. T. Findley. 1999. Vaquita, *Phocoena sinus* Norris and McFarland, 1958. In S. H. Ridgway and R. Harrison (eds.), *Handbook of Marine Mammals,* vol. 6: *The Second Book of Dolphins and the Porpoises,* pp. 357–378. Academic Press, San Diego, CA.

Vidal, O., L. T. Findley, and S. Leatherwood. 1993. Annotated checklist of the marine mammals of the Gulf of California. *Proceedings of the San Diego Society of Natural History* 28: 1–16.

Vidal, O., L. T. Findley, P. J. Turk, and R. E. Boyer. 1987. Recent records of pygmy sperm whales in the Gulf of California, Mexico. *Marine Mammal Science* 3: 354–356.

Vidal, O., and J.-P. Gallo-Reynoso. 1996. Die-offs of marine mammals and sea birds in the Gulf of California, Mexico. *Marine Mammal Science* 12: 627–635.

Vidal, O., K. V. Waerebeek, and L. T. Findley. 1994. Cetaceans and gillnet fisheries in Mexico, Central America and the Wider Caribbean: A preliminary review. In W. F. Perrin, G. P. Donovan, and J. Barlow (eds.), "Gillnets and Cetaceans." *Reports of the International Whaling Commission, Special Issue* 15: 221–233.

Viesca-Lobatón, C. 2003. Cambios temporales en la estructura de la comunidad de peces de arrecifes rocosos en la parte sur-occidental del Golfo de California. Tesis de licenciatura [Bachelor's thesis], Universidad Autónoma de Baja California Sur, La Paz.

Viesca-Lobatón, C., E. F. Balart, A. González-Cabello, I. Mascareñas-Osorio, O. Aburto-Oropeza, H. Reyes-Bonilla, and E. Torreblanca. 2008. Los peces arrecifales. In G. Danemann and E. Ezcurra (eds.), *Bahía de los Ángeles: Recursos Naturales y Comunidad. Línea Base 2007.* Pronatura Noroeste, Instituto Nacional de Ecología, and San Diego Natural History Museum.

Villarreal, A. 1988. Distribución y abundancia de los peces del arrecife de Cabo Pulmo-Los Frailes, B.C.S. Tesis de licenciatura [Bachelor's thesis], Universidad Autónoma de Baja California Sur, La Paz.

Villavicencio-Garayzar, C. J. 1996. Observaciones sobre *Carcharhinus obscurus* (Pices: Carcharhinidae) en el Pacífico nororiental. *Revista de Biología Tropical* 44(1): 287–289.

Wagner, H. 1924. The voyage of Francisco de Ulloa. *California Historical Society Quarterly* 3(4): 307–383.

———. 1929. *Spanish Voyages to the Northwest Coast of America in the Sixteenth Century.* California Historical Society, San Franscisco.

———. 1930. Pearl fishing enterprises in the Gulf of California. *Hispanic American Historical Review* 10(2): 188–220.

Walker, B. W. 1960. The distribution and affinities of the marine fish fauna of the Gulf of California. *Systematic Zoology* 9(3–4): 123–133.

Walker, L. W. 1951. Sea birds of Isla Raza. *National Geographic* 99: 239–248.

———. 1965. Baja's island of birds. *Pacific Discovery* 18: 27–31.

———. 1981. Geographical variation in morphology and biology of bottlenose dolphins (*Tursiops*) in the eastern North Pacific. NOAA Administrative Report NMFS LJ-81-03C. [Available from the Southwest Fisheries Science Center, P.O. Box 271, La Jolla, CA.]

Wallerstein, B. R., and R. C. Brusca. 1982. Fish predation: A preliminary study of its role in the zoogeography and evolution of shallow-water idoteid isopods (Crustacea: Isopoda: Idoteidae). *Journal of Biogeography* 9: 135–150.

Warburton, K. 1978. Community structure, abundance and diversity of fish in a Mexican coastal lagoon system. *Estuarine and Coastal Marine Science* 7: 497–519.

Waters, M. R., and T. W. Stafford Jr. 2007. Refining the age of Clovis: Implications for the peopling of the Americas. *Science* 315: 1122–1126.

Watling, L., and E. A. Norse. 1998. Disturbance of the seabed by mobile fishing gear: A comparison to forest clearcutting. *Conservation Biology* 12(6): 1180–1197.

Watson, M. C., and W. R. Ferren Jr. 1991. A new species of *Suaeda* (Chenopodiaceae) from coastal northwestern Sonora, Mexico. *Madroño* 38: 30–36.

Weber-Van Bosse, A., and M. Foslie. 1904. The Corallinaceae of the Siboga Expedition. *Leiden Bulletin Report* 61: 1–110.

Wegener, A. 1928. *Die Entstehung der Kontinente und Ozeane* [English translation]. London, T. Murby.

Whitmore, R. C., R. C. Brusca, J. L. León de la Luz, P. González-Zamorano, R. Mendoza-Salgado, E. S. Amador-Silva, G. Holguin, F. Galván-Magaña, P. A. Hastings, J.-L. E. Cartron, R. S. Felger, J. A. Seminoff, and C. C. McIvor. 2005. The ecological importance of mangroves in Baja California Sur: Conservation implications for an endangered ecosystem. In J.-L. E. Cartron, G. Ceballos, and R. S. Felger (eds.), *Biodiversity, Ecosystems and Conservation in Northern Mexico,* pp. 298–333. Oxford University Press, New York.

Wicksten, M. K. 1983. A monograph on the shallow water caridean shrimps of the Gulf of California, Mexico. *Allan Hancock Monographs in Marine Biology* 13: 1–59.

———. 1987. A new species of hippolytid shrimp from the west coast of Mexico. *Bulletin, Southern California Academy of Science* 86(1): 27–33.

———. 1994. Taxonomic remarks on two species of the genus *Synalpheus* from the tropical eastern Pacific (Decapoda, Alpheidae). *Bulletin, Museum of Natural History, Paris, 4th series, section A* 16(1): 209–216.

———. 2000. The species of *Lysmata* (Caidea: Hippolytidae) from the eastern Pacific ocean. *Amphipacifica* 2(4): 3–22.

Wicksten, M. K., and M. E. Hendrickx. 2003. An updated checklist of benthic marine and brackish water shrimps (Decapoda: Penaoidea, Stenopodidea, Caridea) from the Eastern Tropical Pacific. In M. E. Hendrickx (ed.), *Contributions to the Study of East*

Pacific Crustaceans 2 [Contribuciones al estudio de los crustáceos del Pacífico Este 2], pp. 49–76. Instituto de Ciencias del Mar y Limnología, UNAM.

Wiemeyer, S. N., C. M. Bunck, and A. J. Krynitsky. 1988. Organochlorine pesticides, polychlorinated biphenyls, and mercury in Osprey eggs—1970–1979—and their relationships to shell thinning and productivity. *Archives of Environmental Contaminates and Toxicology* 17: 767–787.

Wiemeyer, S. N., P. R. Spitzer, W. C. Krantz, T. G. Lamont, and E. Cromartie. 1975. Effects of environmental pollutants on Connecticut and Maryland Ospreys. *Journal of Wildlife Management* 39: 124–139.

Wiemeyer, S. N., D. M. Swineford, P. R. Spitzer, and P. D. McLain. 1978. Organochlorine residues in New Jersey Osprey eggs. *Bulletin of Environmental Contaminates and Toxicology* 19: 56–63.

Wiggins, I. L. 1962. Investigations in the natural history of Baja California. *Proceedings of the California Academy of Sciences* 30(1): 1–45.

———. 1980. *Flora of Baja California*. Stanford University Press, Stanford, CA.

Wilen, J. 2005. Ensuring fisheries benefits for all generations. Disciplinary and personal perspectives on "the fisheries problem." University of British Columbia, Vancouver, http://oregonstate.edu/Dept/IIFET/NAAFE/Jim_Vancouver_with percent20 credit.ppt.

Williams, G. C., and O. Breedy. 2004. The Panamic Gorgonian genus *Pacifigorgia* (Octocorallia: Gorgoniidae) in the Galapagos Archipelago, with descriptions of three new species. *Proceedings of the California Academy of Sciences* 55(3): 55–88.

Willig, M. R., D. M. Kaufman, and R. D. Stevens. 2003. Latitudinal gradients of biodiversity: Pattern, process, scale and synthesis. *Annual Review of Ecology and Systematics* 34: 273–309.

Wilson, M. H. 2002. Checklist of fishes—Tropical eastern Pacific. CD-ROM Version 1.0.1. Smithsonian Tropical Research Institute, Balboa, Panama.

Winker, C. D., and S. M. Kidwell. 1996. Stratigraphy of a marine rift basin: Neogene of the western Salton Trough, California. In P. L. Abbott and J. D. Cooper (eds.), *Field Conference Guide 1996*, pp. 295–336. Annual Meeting, American Association of Petroleum Geologists and Society for Economic Paleontology and Mineralogy. Pacific Section, AAPG, Guide Book 73; Pacific Section, SEPM, Book 80.

Witzell, W. 1983. Synopsis of biological data on the hawksbill turtle *Eretmochelys imbricata* (Linnaeus, 1766). FAO Fisheries Synopsis, no. 137. Food and Agriculture Organization of the United Nations, Rome.

Woelkerling, W. J. 1988. *The Coralline Red Algae: An Analysis of the Genera and Subfamilies of Nongeniculate Corallinaceae*. London, British Museum of Natural History.

———. 1996a. Family Sporolithaceae. In H. B. S. Womersley (ed.), *The Marine Benthic Flora of Southern Australia–Part IIIB. Gracilariales, Rhodymeniales, Corallinales and Bonnemaisoniales*, pp. 153–158. Australian Biological Resources Study, Canberra, Australia.

———. 1996b. Subfamily Melobesioideae. In H. B. S. Womersley (ed.), *The Marine Benthic Flora of Southern Australia–Part IIIB. Gracilariales, Rhodymeniales, Corallinales and Bonnemaisoniales*, pp. 164–210. Australian Biological Resources Study, Canberra, Australia.

———. 1996c. Subfamily Choreonematoideae. In H. B. S. Womersley (ed.), *The Marine Benthic Flora of Southern Australia–Part IIIB. Gracilariales, Rhodymeniales, Corallinales and Bonnemaisoniales,* pp. 210–214. Australian Biological Resources Study, Canberra, Australia.

———. 1996d. Subfamily Lithophylloideae. In H. B. S. Womersley (ed.), *The Marine Benthic Flora of Southern Australia–Part IIIB. Gracilariales, Rhodymeniales, Corallinales and Bonnemaisoniales,* pp. 214–237. Australian Biological Resources Study, Canberra, Australia.

———. 1996e. Subfamily Mastophoroideae (excluding *Hydrolithon, Pneophyllum, Spongites* and *Neogoniolithon*). In H. B. S. Womersley (ed.), *The Marine Benthic Flora of Southern Australia–Part IIIB. Gracilariales, Rhodymeniales, Corallinales and Bonnemaisoniales,* pp. 237–255. Australian Biological Resources Study, Canberra, Australia.

Woelkerling, W. J., L. M. Irvine, and A. S. Harvey. 1993. Growth forms in non-geniculate coralline red algae (Corallinales, Rhodophyta). *Australian Systematic Botany* 6: 277–293.

Woo, H., E. Glenn, R. C. Brusca, and R. McCourt. 2004. Algae. In R. C. Brusca, E. Kimrey, and W. Moore (eds.), *A Seashore Guide to the Northern Gulf of California,* pp. 133–145. Arizona-Sonora Desert Museum Press, Tucson, AZ.

Woodring, W. P. 1966. The Panama land bridge as a sea barrier. *Proceedings of the American Philosophical Society* 110: 425–433.

World Wildlife Fund [WWF]. 2000. Camaronicultura, sociedad y ambiente en el Golfo de California. Internal report, WWF Washington office.

———. 2005a. Determinación de zonas de importancia para la pesca ribereña en el área marina frente a la costa entre Cabo Haro y Playa sur del Estero Tastiota, Sonora. Guaymas, Sonora, Internal report, WWF Gulf of California office.

———. 2005b. Integración del diagnóstico socioeconómico de las comunidades pesqueras de Santa Clara, Son. y San Felipe, B.C. en el Alto Golfo de California. Internal report, WWF Gulf of California office.

Wrobel, D., and C. Mills. 1998. *Pacific Coast Pelagic Invertebrates. A Guide to the Common Gelatinous Animals.* Sea Challengers and Monterey Bay Aquarium, Monterey, CA.

Wynne M. J., and J. N. Norris. 1976. The genus *Colpomenia* Derbés et Solier (Phaeophyta) in the Gulf of California. *Smithsonian Contributions to Botany* 35.

Yensen, N. P. 1980. Intertidal ants from the Gulf of California, Mexico. *Annals of the Entomological Society of America* 73(3): 266–269.

———. 2001. *Halófitas del Golfo de California y sus usos* [Halophytes of the Gulf of California and their uses]. Universidad de Sonora, Hermosillo.

Yensen, N. P., E. Glenn, and M. Fontes. 1983. Biogeographical distribution of salt marsh halophytes on the coasts of the Sonoran Desert. *Desert Plants* 5: 76–81.

Yépiz-Velázquez, L. M. 1990. Diversidad, distribución y abundancia de la ictiofauna en tres lagunas costeras de Sonora, México. Master's thesis, Centro de Investigación Científica y de Educación Superior de Ensenada, Baja California.

Yetman, D. 1996. *Sonora. An Intimate Geography.* University of New Mexico Press, Albuquerque, NM.

———. 2002. *Guarijíos of the Sierrra Madre: The Hidden People of Northwestern Mexico.* University of New Mexico Press, Albuquerque, NM.

———. 2007. *The Great Cacti.* University of Arizona Press, Tucson, AZ.

Young, R. H. 1982. The Guaymas shrimp bycatch program. *Gulf and Caribbean Fisheries Institute, Proceedings* (1981): 131–138.

Young, R. H., and J. M. Romero. 1979. Variability in the yield and composition of by-catch recovered from Gulf of California shrimping vessels. *Tropical Science (London)* 21(4): 249–264.

Zamora-Arroyo, F., P. Nagler, M. Briggs, D. Radtke, H. Rodriquez, J. Garcia, C. Valdes, A. Huete, and E. Glenn. 2001. Regeneration of native trees in response to flood releases from the United States into the delta of the Colorado River, Mexico. *Journal of Arid Environments* 49: 49–64.

Zamora-Arroyo, F., J. Pitt, S. Cornelius, E. Glenn, O. Hinojosa-Huerta, M. Moreno, J. García, P. Nagler, M. de la Garza, and I. Parra. 2005. *Conservation Priorities in the Colorado River Delta, Mexico and the United States.* Prepared and published by the Sonoran Institute, Environmental Defense, University of Arizona, Pronatura Noroeste Dirección de Conservación Sonora, Centro de Investigación en Alimentación y Desarrollo, and World Wildlife Fund–Gulf of California Program.

Zamorano, P., M. E. Hendrickx, and A. Toledano Granados. 2006. Distribution and ecology of deepwater mollusks from the continental slope, southeastern Gulf of California, Mexico. *Marine Biology* 150(5): 883–892.

Zanchi, A. 1994. The opening of the Gulf of California near Loreto, Baja California, Mexico: From basin and range extension to transtensional tectonics. *Journal of Structural Geology* 16: 1619–1639.

Zavala-González, A., and Mellink, E. 2000. Historical exploitation of the California Sea Lion, *Zalophus californianus* in Mexico. *Marine Fisheries Review* 62: 35–40.

Zavala-González, A., J. Urbán R., and C. Esquivel-Macías. 1994. A note on artisanal fisheries interactions with small cetaceans in Mexico. In W. F. Perrin, G. P. Donovan, and J. Barlow (eds.), "Gillnets and Cetaceans." *Reports of the International Whaling Commission, Special Issue* 15: 235–237.

Zeitzschel, B. 1969. Primary productivity in the Gulf of California. *Marine Biology* 3(3): 201–207.

Zink, R. M. 2002. Methods in comparative phylogeography, and their application to studying evolution in the North American aridlands. *Integrative and Comparative Biology* 42: 953–959.

Zwinger, A. H. 1983. *A Desert Country Near the Sea: A Natural History of the Cape Region of Baja California.* Harper & Row, New York.

About the Contributors

Saúl Alvarez-Borrego was born in Mazatlán, obtained his B.S. degree in Oceanology at the University of Baja California (UABC), and earned his M.S. and a Ph.D. at Oregon State University. Out of his more than 100 refereed contributions, 37 deal with physical, chemical, and biological oceanography of the Gulf of California. He was Director of UABC's School of Marine Sciences from 1973 to 1975 and was General Director of the national research and graduate institute, Centro de Investigación Científica y Educación Superior de Ensenada (CICESE), from 1975 to 1989. In 2005, Saúl was honored with the Baja California State Science and Technology Award and in 2006 with the Doctorate *Honoris Causa* by UABC.

Daniel W. Anderson is on the faculty at the University of California, Davis. His research involves studies of seabirds and other aquatic birds, raptors, and environmental contaminants and the distribution of organic and inorganic materials in birds. Dan's research area includes western North America, with a strong focus on Baja California and the Gulf of California, as well as the Klamath Basin and San Joaquin Valley in California. He is actively involved in the conservation and management of avian populations and their habitats, mainly in his study areas.

Richard C. Brusca is Senior Director of Conservation and Science at the Arizona-Sonora Desert Museum (Tucson), where he oversees all research and conservation programs, and Director of the ASDM Press. He is also a Research Scientist at the University of Arizona and an Adjunct Professor with Centro de Investigación en Alimentación y Desarrollo (CIAD, Hermosillo). His Ph.D. is from the University of Arizona. Rick is the author of more than 150 research publications and 15 books, including the largest-selling text on invertebrate zoology (*Invertebrates,* Sinauer Associates; in four languages) and the popular field guides *Common Intertidal Invertebrates of the Gulf of California* (UofA Press) and *A Seashore Guide to the Northern Gulf of California* (ASDM Press). He has been the recipient of more than a hundred research grants from the National Science Foundation, NOAA, National Geographic Society, National Park Service, The David and Lucile Packard Foundation, Charles Lindberg Fund, and other agencies and foundations. His research interests include natural history and conservation of the Sonora Desert and the Gulf of California, invertebrate zoology, and evolution within the Metazoa. Rick has served on panels and boards for many foundations and agencies, including the National Science Board, National Science Foundation, Smithsonian Institution, NOAA, PEW Program in Conservation and the Environment, Public Broadcasting Service, IUCN Species Survival Commission, U.S. Department of the Interior, and numerous nonprofit organizations. He has organized and conducted field expeditions throughout the world and on every continent, but he has maintained his research programs in the Sonoran Desert and the Gulf for more than thirty years. Rick is a Fellow in both the American Association for the Advancement of Science and the Linnean Society of London.

Roberto Carmona was born in México D.F., obtained his B.S. degree in Marine Biology at the Universidad Autónoma de Baja California Sur (UABCS; 1988), his M.Sc. in the Centro Interdisciplinario de Ciencias Marinas (fisheries, 1993), and a Ph.D. at the Universidad Autónoma de Baja California (Coastal Oceanography, 2007). He is a Titular Researcher in the UABCS and has published more than 100 research papers and trained over 25 graduate students. He has conducted studies on the ecology of waterbirds in northwestern Mexico for more than 20 years.

Ana Luisa Carreño obtained her B.S. in Biology and M.S. in Geology at UNAM, and she earned her Ph.D. in Paleontology (with a specialty in Micropaleontology) at the University of Paris. Her research examines the biostratigraphy of foraminiferans, ostracods, and calcareous nannoplankton in Neogene sequences related to the opening of the Gulf of California. She is also engaged in a large collaborative project with the Rio Grande do Sul University studying the recent and fossil marine and nonmarine ostracodes from Brazil.

Jean-Luc E. Cartron is Research Assistant Professor of Biology at the University of New Mexico and Director of the New Mexico office of the Drylands Institute. His research interests include raptor ecology and conservation, riparian ecosystem ecology, and macroecology. He is the senior editor or author of two books, *Biodiversity, Ecosystems, and Conservation in Northern Mexico* (Oxford University Press, 2005) and *A Field Guide to the Plants and Animals of the Middle Rio Grande Bosque* (University of New Mexico Press, 2008).

María de los Ángeles (Machángeles) Carvajal is associate founder of SuMar, a nonprofit organization focused on leadership strengthening as an agent for social change for sustainability in the Gulf of California region. For twenty years she has worked to establish and implement marine protected areas and best practices for wise use of coastal and marine resources in northwestern Mexico. Machángeles is co-founder and member of both Eco-Costas and ALCOSTA (Mexican Northwest Coastal Sustainable Development Alliance). She was a 2006 Donella Meadows Fellow and a 2008 Ashoka Fellow.

Miguel Á. Cisneros-Mata is a fisheries biologist who early on realized that good fisheries management required a deep understanding of marine ecology. Miguel holds M.S. (UABC, 1985) and Ph.D. degrees (U.C. Davis, 1995) in marine sciences. He has worked mainly for Mexico's National Fisheries Institute and is currently its Director-in-Chief. From 2004 to 2006 he worked for World Wildlife Fund–México's Gulf of California Program. His interests are fisheries modeling, policies, and strategies to conserve Gulf of California marine species and maintain fishers' livelihoods.

Exequiel Ezcurra has devoted his career to the study of northwestern Mexico. He has published more than 200 papers and books and has developed two museum exhibits and an award-winning film on the Gulf of California. He has been honored with a Conservation Biology Award and a Pew Fellowship in Marine Conservation; he was also Chair of the CITES

Convention and President of Mexico's National Institute of Ecology. Currently, he is the Director of the University of California Institute for Mexico and the United States (UC MEXUS) and Professor of Ecology at the University of California, Riverside.

Lloyd T. Findley is a research scientist at the Guaymas, Sonora, branch of the Centro de Investigación en Alimentación y Desarrollo (CIAD). He was raised in southern California, where two events during his high school years led him to the scientific study of fishes, especially those of the Gulf of California: weekend "skindiving" trips on the rocky shores of California; and a stint as a foreign exchange student to Los Mochis, Sinaloa, where he lived with a Mexican family and first saw Gulf fishes at nearby Topolobampo. Many trips to the Gulf followed, setting his mind to pursue higher studies on fishes and eventually to live and work in Mexico. His Ph.D. is from the University of Arizona, where he was Assistant Curator of Fishes and where eight years of intensive collecting and identifying of Gulf fishes led to collaborative writing and publication of *The Reef Fishes of the Sea of Cortez* (Thomson, Findley, and Kerstitch 1979; revised edition in 2000). Lloyd has taught at the University of Baja California in Ensenada, Baja California, and at the Marine Sciences School of the Instituto Tecnológico y de Estudios Superiores de Monterrey in Guaymas, Sonora—where, besides studying fishes, he researched the marine mammal fauna of the Gulf over a ten-year period. He has been awarded honorary membership in both the Mexican Society for the Study of Marine Mammals and the Mexican Society of Ichthyology, and he is a member of the Sistema Nacional de Investigadores in Mexico. Lloyd considers himself lucky to be studying what he enjoys and living on the warm desert shores of the Gulf in his adopted country of Mexico.

Michael S. Foster is Professor Emeritus at Moss Landing Marine Laboratories at San Jose State University. His research interests are the ecology of subtidal and intertidal reefs, algal assemblages in kelp forests, rhodolith beds, and ecology of rocky shores. He is a Fellow of the California Academy of Sciences and a Fulbright Scholar, and he has published more than 70 peer-reviewed works.

Philip A. Hastings spent the early years of his life exploring the bays and bayous near his home in northwest Florida. After receiving B.S. and M.S. degrees from the University of South Florida and University of West Florida, respectively, he entered graduate school at the University of Arizona. There, in 1980, he began his research on Gulf of California fishes, receiving a Ph.D. in Ecology and Evolutionary Biology. Phil remained at the University of Arizona as a Research Scientist and Curator of Fishes and Invertebrates until 1999, when he moved to the University of California San Diego and Scripps Institution of Oceanography, where he is now Professor and Curator of Marine Vertebrates. His research interests include the diversity, evolution, ecology, behavior, and conservation of marine fishes.

Michel E. Hendrickx received his Ph.D. from the Université Libre de Bruxelles, Belgium. He subsequently worked for two years at the Phuket Marine Biological Center in Thailand and then, in 1977, moved to a UNESCO position in Mexico as part of a marine sciences project between UNDP and the Universidad Nacional Autónoma de México (UNAM). In 1981, Dr. Hendrickx became a researcher for UNAM, where he has spent his time teaching and working on marine invertebrates. He is a member of the Sistema Nacional de Investigadores in Mexico. He is also curator of the Invertebrates Collection at UNAM's Mazatlán research station.

Charles J. Henny is a Research Zoologist at the Forest and Rangeland Ecosystem Science Center of the U.S. Geological Survey in Corvallis, Oregon. He received his B.S. and Ph.D. in Wildlife Ecology in 1970 from Oregon State University. After having spent four years at the Patuxent Wildlife Research Center in Laurel, Maryland, and two years at the Denver Wildlife Research Center (both of the U.S. Fish and Wildlife Service), he established the Pacific Northwest Field Station for the Patuxent Wildlife Research Center (1976). The field station became part of the U.S. Geological Survey in the early 1990s. Most of Charles's research, which includes about 200 published papers, has involved field studies of contaminant effects on wild populations. With most U.S. osprey populations greatly reduced in numbers by the 1970s, a study of the ospreys in northwestern Mexico was initiated in 1977 and then followed up in 1992/93 and again in 2006. Charles retired from the U.S. Geological Survey in January 2009 after 38 years of service.

Gustavo Hinojosa-Arango is Director of the School for Field Studies Wetland Studies Program for Mexico. He holds a Ph.D. in ecology from Queen's University of Belfast. Gustavo has a broad background in research and teaching on coastal ecology and conservation of the Baja California peninsula and Gulf of California. Much of his current research focuses on invertebrates associated with rhodolith communities.

Jorge Ledesma-Vázquez is professor of undergraduate and graduate studies in the Geology Department at the Universidad Autónoma de Baja California (UABC) and a member of Mexico's Sistema Nacional de Investigadores. Jorge has been an instructor for a U.S. National Science Foundation–sponsored Oceanography Short Course (with the University of San Diego); he co-teaches an extension course in Tectonics for San Diego State University; and he leads field trips in the Gulf of California in association with Williams College. In 2001 he was awarded the Academic Merit in Natural Sciences by UABC, and since 2003 he has had the honor of holding a Desirable Teacher Profile (Public Education Secretariat–México, SEP). He has published 36 scientific articles on the geology of the Baja California peninsula and the Gulf of California. Most of his work focuses on the stratigraphy, origin, and evolution of the region. Jorge has co-edited several volumes on the geology of the Gulf and the Baja California peninsula, and he is co-author of the field-trip guide *Caminos de Baja California; Geología y Biología para su viaje* (John Minch & Associates, 2003).

Richard McCourt received his Ph.D. from the University of Arizona for work elucidating the ecology and systematics of intertidal *Sargassum*. He continued his work on the molecular systematics of green algae with the late Dr. Robert Hoshaw. Rick's research has more recently focused on molecular phylogenetics of green algae and their evolutionary relatives, the green plants that invaded the land. He continues to be interested in the biogeography and evolution of algae in the Gulf of California. He taught at DePaul University in Chicago and is currently Associate Curator of Botany at the Academy of Natural Sciences in Philadelphia.

Rodrigo A. Medellín has worked on the ecology and conservation of mammals for thirty years. He is Professor of Ecology at the National Autonomous University of Mexico and advisor to the Mexican Federal Government on wildlife issues. Dr. Medellín is also the recipient of the Whitley Award for International Nature Conservation (from the hands of Her Royal Highness, Princess Anne of England), the National Nature Conservation Award (from President Vicente Fox), the Rolex Award for Enterprise, and the Volkswagen "Por Amor al Planeta" award.

Rafael Riosmena-Rodríguez has been a professor at the Departamento de Biología Marina at the Universidad Autónoma de Baja California Sur since 1989. His research interests in the Gulf of California include taxonomy and biogeography of marine plants in transitional zones, population and community ecology of marine plants, and global climate change and biodiversity conservation. Rafael is involved with numerous research projects throughout the Gulf and has published more than forty papers on the taxonomy of seaweeds, biogeography of seagrasses, and ecology of subtropical rhodoliths in the Gulf of California. He is also a guest researcher in the graduate program of CICIMAR (Mexico), Universidad de las Azores (Portugal), Universidad de Alicante (Spain), and Universidad de São Paulo (Brazil).

Alejandro Robles is Executive Director of Noroeste Sustentable (NOS). He has a B.S. degree in Biochemistry from the Instituto Tecnológico de Monterrey in Guaymas, Sonora (1982). He has dedicated his work to conservation and sustainable development issues in Mexico, especially those related to coastal and protected areas, governance, and fisheries. Completion of the LEAD (Leadership for the Environment and Development) Program–Mexico in 1998 positioned Alejandro to approach conservation and development problems from a multicultural and multistakeholder point of view. For the past fifteen years he has worked to coordinate the design of integrative conservation strategies and programs dealing with sustainable development on a regional scale in northwest Mexico, including designing and developing projects with an emphasis on community development, public policy formulation, economic alternatives, protected areas, conservation biology, education, and communication.

Jeffrey A. Seminoff leads the Marine Turtle Ecology and Assessment Program at the NOAA–Southwest Fisheries Science Center in La Jolla, California. He received his Ph.D. from the University of Arizona in 2000 for research conducted on the life history and conservation of green sea turtles in Bahía de los Ángeles, Baja California. He has conducted field research on sea turtles in Peru, Ecuador, El Salvador, Mexico, and the United States. Jeff serves as Adjunct Faculty at Indiana-Purdue University and the University of Florida, and he is a member of the IUCN Marine Turtle Specialist Group. Jeff lives with his wife, Jennifer, and young children Quin and Graeson, in San Diego along with an assortment of pets including George, a 30-kg tortoise.

Diana L. Steller is a research biologist, lecturer, and Diving Safety Officer at Moss Landing Marine Laboratories. She has been working and teaching in the Gulf of California for the past twenty years; her research interests include the ecology of temperate reefs and subtropical carbonate rhodolith beds, with an emphasis on macroalgal ecology, algal physiology, and species interactions. Diana is particularly interested in the role that macroalgae play as a substrate and food resource in subtidal communities and how algal dynamics influence community dynamics.

Albert M. van der Heiden obtained his Ph.D. in 1976 from the State University of Ghent, Belgium. In 1978 he joined UNESCO's program to investigate the fishes of the Gulf of California, working out of UNAM's marine research facility in Mazatlán, Sinaloa. In 1982 he became an Associate Researcher at UNAM and later, as a visiting scholar, he spent a sabbatical year at the Scripps Institution of Oceanography, where he focused his research on flatfishes. Currently, he is a Senior Research Scientist for the Centro de Investigación en Alimentación y Desarrollo (CIAD). He has recently been dedicating time to the conservation of the fast-dwindling tropical deciduous dry forest in northwestern Mexico.

Index

Italic page numbers indicate material in tables or figures.

Acanthaster planci (crown-of-thorns sea star), 90
Acanthemblemaria crockeri (browncheek blenny), 112
Acanthonyx petiveri (crustacean), 74
Acanthurus triostegus (convict surgeonfish), 103
Actinopterygii, *107*
"Adelita" (loggerhead turtle), 158
Agassiz, Alexander, 26
agriculture: drainage water from, 48, 73, 92, 183, 229; economic importance of, 223, 227; loss of wetlands to, 91; water needs of, 229
Agrupación Sierra Madre, 92
Alarcón Basin, 8, *12*, 19
Albatross cruise, 99–100, 117, 136, 144
ALCOSTA, 92, *220*, 227–228
algae: biogeography, 214–215; brown algae (Phaeophyta), 212–218; coralline red algae (Rhodophyta), 49–71, 213; *Gelidium robustum*, 87; *Gracilariopsis lemaneiformis*, 150; green algae (Chlorophyta), 150, 211, 212, 216–217; habitats of Northern Gulf, 215–217; history of exploration of, 213–214; kelps, 213, 214–215; overview, 210–213; reproduction, 212; *Sargassum*, 58, 212, 215–218; seasonality, 150, 215–217; substrates, 210–211, 216–217; turtle needs, 137–138, 150, 156, 165–166; *Ulva lactuca*, 150; zonation of, 217–218. *See also* chapters 3 and 10
alien species, 83–85, 222
allopatric speciation, 110–111
Alta California–Baja California Peninsula Range Batholith, 7
Alto Golfo (Upper Gulf), 93, 117, 238–245
amarilla, la. *See* loggerhead turtle (*Caretta caretta*)
Ambidexter symmetricus (crustacean), 74
ammonia concentration, 44
amphioxus (*Branchiostoma californiense*), 89
Anas acuta (northern pintail), 183
anchovies, 42, 109, 122
Anderson, Daniel, 233
Ángel de la Guarda Island. *See under* islands/islas
angelfishes, 113, 117
Anisotremus davidsonii (sargo), 109
Anita, Ramon, 142
Antennularia septata (cnidarian), 75
anthropogenic effects: on mammals, 205–208; on ospreys, 178, 184–187; from outboard motors, 222; on rhodolith beds, 67–71. *See also* artisanal fishing; commercial fishing; tourism
Aphrodita mexicana, *A. sonorae* (polychaete annelids), 81, 88
Aplysia spp. (Mollusca, Gastropoda—sea hares), 75, 77, 150
aquaculture industry: economic importance of, 121–129, 222, 223; estero-based, 73, 83–84, 88–90, 116; habitat loss, 2, 230
aquarium fish, harvesting of, 115, 117
Aratus pisonii (crustacean), 74
Arbacia incisa (echinoderm), 85
Archeolithothamnium (red alga), 55–56

Arctocephalus townsendi (Guadalupe fur seal), 189, 203, 205
Arizona-Sonora Desert Museum, 73, 92, 94, 230–231, 233
arribada, 160, 165
Arthron meleagris (spotted pufferfish), 90
articulate chiton (*Chiton articulatus*), 87
artisanal fishing: conservation efforts with, 2, 87–88, 93, 133, 225–226; environmental impact from, 73, 205–208; exclusive user rights, 131; history of, 123; number of jobs in, 124–125; percentage/volume of commercial catch, 126–127; regulation of, 129–130; status of, 119–121, 124, *131;* threat from, 73, 155, 157, 159, 205–208; tools used by, 119, 124; training of, 132. *See also* commercial fishing
Aruma histrio (slow goby), 111
Asociación Sudcaliforniana de Proteción al Medio Ambiente y a la Tortuga Marina (ASUPMATOMA), 161
Astrangia sanfelipensis, 81
Astrodictyum panamense (echinoderm), 86
Astrometis sertulifera (echinoderm), 85
Audubon Society, 231, 233
Axoclinus nigricaudus (Cortez triplefin), 109, *110*, 112

Bacescuella parvithecata (marine earthworm), 82
Bahía de La Paz: fishes, 101; hawksbills, 140; marine mammals, 192–196, 198–199, 201–204; ospreys, 174; rhodoliths, 54, 60–61; tourism at, 124; turtle bones, 140
Bahía de Los Ángeles: conservation in, 225, 249; osprey, 172–177, 180, 184, 186; rhodolith beds, 52; turtles, 144–145, 149, 151–152, *164*, 166
Bahía de San Rafael, 149, 198
Bahía Gonzaga, 145
Bahía Kino (Sonora), 145, 150, 151, 177
Bahía Magdalena, 66, 102, 152, *164*, 177, 200
Baird's beaked whale (*Berardius bairdii stejneger*), 193, 205
Baja California, 5n
Balaenoptera spp., 189–190, 201–202
Balistes polylepis (finescale triggerfish), 173
ballenas (whales), 3, 186–209. *See also* chapter 9
Ballenas Channel, 30, 32, 37, 41–42, 145, 202
Barbulifer mexicanus (saddlebanded goby), 111
barnacle bill blenny/borracho vacilón (*Hypsoblennius brevipinnis* Blenniidae), *98*
basaltic rocks, flow, and surfaces, 17, 76, 210, 216–217
Baseodiscus punnetti (nemertean), 75
Bechtel, Kenneth, 231
bekko, 157
Beltrán, Enrique, 231, 232
benthic food web, 1–3, 69, 76, *78*, 80, 87–88, 115, 138–139
benzene hexachloride contamination, 183
Berardius bairdii stejneger (Baird's beaked whale), 193, 205
Bergman size clines, 113
berrugata aleta amarilla/yellowfin croaker (*Umbrina roncador* Sciaenidae), 99
Betaeus longidactylus (crustacean), 75, 77
biogeography: effect on biodiversity, 4, 74–83, 101–108, 114, 214–215
biosphere reserves: Alto Golfo (Upper Gulf), 93, 117, 238–245; establishment of, 93, 117, 234, 237; fishing within, 87; and sea turtle conservation, 166; table of, *236*
bioturbation: and rhodoliths, 53, 66
black murex (*Hexaplex nigritus*), 86, 87
black turtle. *See Chelonia agassizii*
black zone (in intertidal region), 218
Blainville's beaked whale (*Mesoplodon densirostris*), 194
blennies, *98*, 100, 102, 109–111, 113; competition for space, 114; morphological

variation in, 112. *See also individual species*
blue whale (*Balaenoptera musculus*), 189, 190, 202
boat anchors, effects from, 90
borracho vacilón / barnacle bill blenny (*Hypsoblennius brevipinnis* Blenniidae), *98*
Bothidae, 174
bottlenosed dolphin (*Tursiops truncatus*), 192, 194–195, 205
bottlenose whale, 205
bottom trawling. *See* trawling
Boulder (Hoover) Dam, 28, 90, 229
Bourillón, Luis, 233
Branchiostoma californiense (amphioxus), 89
Brandtothura spp. (echinoderms), 85
brown algae (Phaeophyta). *See* algae
browncheek blenny (*Acanthemblemaria crockeri*), 112
Brusca, Richard, 73
Bryde's whale (*Balaenoptera edeni*), 189, 201–202
Bryzoa, 53, 64, 75, *80*, 82, 84–85, 211
bullseye electric ray (*Diplobatis ommanata*), 65
bullseye puffer (*Sphoeroides annulatus*), 173
Buston, Miguel Ángel Romo, 141
butterflyfishes, 117
by-catch, 1, 86–88, 93, 115, 123, *128*, 133, 136, 146, 155, 162, 166, 223, 225, 247

Cabo Corrientes, 7, 26, 29, 34, 170
Cabo Pulmo, 52, 90, 151, 157, *164;* Amigos de, *164*, 165; national park, 90, 235, *236*
Cabo San Lucas: biodiversity at, 102, 186; early research at, 25–26, 172; marine mammals at, 197, 201; ospreys at, 172, 186
Cabo Trough, 19–20
cachalotes (whales), 191–192. *See also* chapter 9
cadmium (Cd), 42–43
cahuama, la. *See* loggerhead turtle (*Caretta caretta*)
Calamus brachysomus (Pacific porgy), 173
calderones (whales), 197–198
Califanthura squamosissima (crustacean), 76
California Current Water (CCW), 40
California grunion (*Leuresthes tenuis*), 109–110, 113–114
Californian Province, 72, 75, 101–102
California sea lion (*Zalophus californianus*), 189, 203
Callinectes spp., 73, 77, 87
Campanularia castellata (cnidarian), 75
Canal de Infiernillo (Infiernillo Channel), 3, 142–143, 149, 151, *164*
Cantú, Antonio, 233
carbon: dissolved inorganic, 31; dissolved organic, 44–45; fixed, 43; Redfield's ratio, 43
carbonates: fossil deposits of, 67; from rhodoliths, 50, 54, 63, 66–67, 68; from sand and seashells (coquina), 216; shelf, 54; from tropical green algae, 212
carbon dioxide: rhodoliths as source of, 67; tidal mixing and, 31
Carcharodon carcharias (white shark), 97
Caretta caretta (loggerhead turtle), 137
carey, la. *See* hawksbill turtle (*Eretmochelys imbricata*)
Caribbean marine habitats, 103
Carmen Basin, *12*, 41
cartilaginous fishes, 107
CCW (California Current Water), 40
CEDO (Centro Intercultural de Estudios de Desiertos y Oceanos), 92
Central Gulf: defined, 74, 94–95, *106*; macroinvertebrates in, 81
Centrostephanus coronatus (echinoderm), 85
Cerebratulus lineolatus (nemertean), 75
CES (Centro Ecológico de Sonora), 240
chaenopsid blennies, 100
Chaetomorpha linum (green alga), 217
Chelonia agassizii (black turtle), 156

Chelonia mydas (green turtle), 66, 135–137, 150–151
Chelonidae, 152–153
Chicoreus erythrostomus (pink-mouth murex), 86
Chiton articulatus (mollusc), 87
Chloeia viridis (polychaete annelid), 74
chlorophyll. *See* photosynthesis
Chondrichthyes, 107
chuckwallas (*Sauromelas*), 168, 174
CIDESON (Centro de Investigación y Desarrollo de los Recursos Naturales de Sonora), 240
circulation, current, 33–40
Cirolana parva (crustacean), 74
CIRVA (International Committee for the Recovery of the Vaquita), 224
CITES (Convention on International Trade in Endangered Species), 3, 160, 161, 205
Cladophora microcladioides (green alga), 217
clingfishes, 100, 109–110
coastal/Mexican fishing bat (*Myotis vivesi mengaux*), 204
Coastal Zone Color Scanner, 45
COBI (Comunidad y Biodiversidad), 92
Cochimí people, 140, *141*, 143, 221
coldwater-adapted species, 102, 109
collapse: Aswan dam example, 28–29; of commercial shrimp trawling, 87–88, 225; El Niño–related, 123; leatherback turtle population, 155; sardine fisheries recovery from, 226
Coloradito, 3, *51*, 80, 81
Colorado River delta, 10, 23; biosphere reserve, 80, 93, 117, 233, 237, 239–241; dams, effect on, 29, 86, 90–91, 116, 229; effects of logging on, 229; people indigenous to, 140–141; pesticide contamination in, 183; restoration potential, 229, 235; seasonal changes in, 45–48; sediment discharge from, 91; and shrimp life cycle, 92; tidal range near, 31, 218; turtles of, 141–143, 155
Colorado River Sedimentary Province, 91
Colosio, Luis Donaldo, 240–241
Colpichthys hubbsi (delta silverside), 111
Colpichthys regis (false grunion), 114
Colpomenia (brown alga), 217
Comcáac (Seri) people, 142–143, 155, 157, 159, 221, 232
commercial fishing, 123; decreasing yields, 87–88, 127–128, 224–225; effect on ospreys, 184–185; effect on rhodolith beds, 68–69; effect on sea turtles, 146; effect on shrimp, 92; greenhouse gases, 225; invertebrate, 87–88; management tools, 133, 225–226; sardine recovery from collapse, 226; socioeconomics of, 124–129; support for conservation, 242–243; tools used in, 124. *See also* artisanal fishing
common bottlenosed dolphin (*Tursiops truncatus*), 192, 194–195, 205
common/harbor seal (*Phoca vitulina*), 203–204
common raven (*Corvus corax*), 175
community structure of reef fishes, 114
CONABIO (Comisión Nacional para el Conocimiento y Uso de la Biodiversidad), 4
CONANP, 93, *220*, 238
Conopea galeata (crustacean), 74
Conservación del Territorio Insular Mexicano (ISLA), 93
conservation: Alto Golfo (Upper Gulf), 240–242; biosphere reserves and turtle, 166; conceptual model, *132;* consumer-targeted, 94; cooperatives, 225–226; fish, 115–117; by gillnetters, 161; of Gulf fishes, 115–117; history of, 230–235; by indigenous groups, 143; invertebrate fauna, 85–92; of invertebrates, 85–92; Islas del Golfo, 93; lessons learned, 243–245; Loreto area, 69, 225; of mammals, 208–209; in Mexico, 2, 4, 92–94, 232–233; needs and objectives, 3, 94, 131–133,

223–224, 229–230, 246–250; regional, 245–249; of rhodolith beds, 70–71; sardines, 121, 130, 226; of sea turtles, 160–167; SLOSS debate, 4; table of organization acronyms, *220;* totoaba (*Totoaba macdonaldi*), 88, 93, 224, 233, 239–240; tourism and turtle, 147, 166; wetlands (estuaries, esteros, lagoons), 225, 229, 230, 237, 247
Conservation International, 233
Conservation International–Mexico, 73, 92–93, 233
consumer-targeted conservation, 94
contamination, 146, 169, 183, 208
continental shelves, 28, 101; species distribution and, 72, *78*, 82, 101–102, 108, 115
Convention on International Trade in Endangered Species (CITES), 3, 160, 161, 205
convict surgeonfish (*Acanthurus triostegus*), 103
coquina formations, 3, 76, 78, 80, 94, 210, 216–217
coral: *Astrangia sanfelipensis*, 81; diversity in Gulf, 81–82; in rhodolith beds, 50, 53, 64, 67; threats to, 81, 90
coralline red algae (Corallinales), 49–55, 213. *See also* chapter 3
cormorants (*Phalacrocorax* sp.), 174, 208
Corophium uenoi (crustacean), 75
Cortez garden eel (*Taeniconger digueti*), 65
Cortez/Gulf triplefin (*Axoclinus nigricaudus*), 109, *110*, 112
Cortez pipefish (*Syngnathus carinatus*), 111
Corvus corax (common raven), 175
Coryphaena hippurus (dorado/dolphinfish), 97
Cretaceous Period, 7, 21, 49
croakers, 109
Crocodilichthys gracilis (lizard triplefin blenny), 112
crown-of-thorns sea star (*Acanthaster planci*), 90
crustaceans, decapod, 85
cryptofauna: diversity of, 62–64
Cucapá people, 140–141, 221
current beds, 54, 60, 63
currents, 33–37; California Current, 104; as juvenile sea turtle habitat, 149; measuring, 37–38; modeling, 33–39; surface, 26, 30; types of, 26, 30. *See also* circulation; tides
Cuvier's beaked whale (*Ziphius cavirostris*), 192–193
cyanobacteria, 218
Cycloes bairdii (crustacean), 74

Dactyloscopidae, 102
dams: on Colorado River, 28–29, 91, 116, 229; Glen Canyon Dam, 45, 90; High Aswan Dam, 28–29; Hoover (Boulder) Dam, 28, 90, 229
damselfishes, 109, 113, 117
David and Lucile Packard Foundation, 93
Dawson, E. Yale, 214
DDT contamination, 183
delfines (dolphins), 194–196. *See also* chapter 9
Delphinus capensis (long-beaked common dolphin), 196–197, 205
Delphinus delphis (short-beaked common dolphin), 196, 205
delta, 91–92
delta clam (*Mulinia coloradoensis*), 92
delta silverside (*Colpichthys hubbsi*), 111
Dentalium spp. (molluscs), 75
Derby, George Horatio, 144
Dermochelys coriacea (leatherback turtle), 137, 152–155
development, housing and resort, 2, 73, 89, 116, *130*, 227
Diadumene leucolena (cnidarian), 75
Dictyota flabellate (brown alga), 217
dieldrin contamination, 183
Diopatra ornata (polychate annelid), 76
Diopatra splendidissima (polychate annelid), 76

Diplobatis ommanata (bullseye electric ray), 65
disjunct algae, 214
disjunct fishes, 108–109
disjunct invertebrates, 72, 75
dissolved organic carbon (DOC), 44
divers, effects from, 90
diversity (biodiversity): decrease from trawling, 88, 115–116; previous and future, 222; of reef fish communities, 114
dolphinfish/dorado (*Coryphaena hippurus*), 97
Dosidicus gigas (jumbo, or Humboldt squid), 119–122, 123, 125, 127–131, 133, 191
drift nets, 146, 155, 159, 161, 166
Ducrue, Norberto, 143–144

eared grebe (*Podiceps nigricollis*), 174
earthworms, 82
east Pacific barrier, 102–103
east Pacific green turtle. *See Chelonia agassizii and C. mydas*
East Pacific Rise, 7–8, 9, 20
Echeverría Álvarez, Luis, 231, 232
Echinometra vanbrunti (echinoderm), 85
economic importance: of aquaculture industry, 121–129, 222, 223
education: of seafood consumers, 94; and turtle conservation, 166
eelgrass (*Zostera marina*), 150
eels, 65, 109
egg poaching, 155
ejido (communal land), 232
El Desemboque, 142, 180, 185
electrocution from power poles, 185, 187
elephant seals, 189, 204, 205
El Niño, 25–26, 90, 104, 108, 123; Southern Oscillation (ENSO), 40, 60, *61*, 150
Emblemaria (chaenopsids), 111
Encope grandis (echinoderm), 75
Encope micropora (echinoderm), 86
endangered species act (Mexico), 3, 89, 205
Endangered Species Act (U.S.), *138*, 160, 161
endemic fishes, 109–112
endemic invertebrates, 77–83
ENDESU (Espacios Naturales y Desarrollo Sustentable), 93
ENSO. *See under* El Niño
entanglement net fishing. *See* gillnetting
environmental laws. *See* Mexico, environmental laws and decrees
Epinephelus labriformis (flag cabrilla), 112
epiphytes, 217
Equatorial Surface Water (ESW), 40
Eretmochelys imbricata (hawksbill turtle), 137, 157
Escalera Náutica (Nautical Stairway), 147, 226–227, 246, 248
Eschrichtius robustus (gray whale), 189, 191, 200, 205
estero, defined, 83, 210
estuaries. *See* wetlands (estuaries, esteros, lagoons)
ESW (Equatorial Surface Water), 40
Eubalaena japonica, 199–200, 205
Eucidaris thouarsii (echinoderm), 85
Eulerian method, 37
Eurythoe complanata (polychaete annelid), 74
Excirolana mayana (crustacean), 74
E. W. Scripps cruises, 26, 34
exotic species, 83–85, 222, 233

false grunion (*Colpichthys regis*), 114
false killer whale (*Pseudorca crassidens*), 197–198, 209
FAO Code of Conduct for Responsible Fisheries, 132–133
Farallon Basin, *12*, 41
Farallón Island, 157
Farallon Plate, 7–8, *13*
farmland. *See* agriculture
faunal regions, 105, *106*
Federation of Independent Seafood Harvesters (FISH), 161

Feresa attenuata (pigmy killer whale), 209
fertilization mechanisms, natural, 28
Findley, Lloyd, 73
finescale triggerfish (*Balistes polylepis*), 173
fin whale (*Balaenoptera physalus*), 190, 202
fisheries. *See* commercial fishing
fishes: associated with rhodoliths, 65–66; biogeographic origins of Gulf, 108–112; conservation issues, 115–117; diversity in Gulf, 65, 83, 100–108; Gulf endemic, 109–112; morphological variation within Gulf, 112–113; number of currently valid species, *98;* reproductive behavior of, 113–115; studies/documentation of Gulf, 97–101, 113–115
fish meal, 130
flag cabrilla (*Epinephelus labriformis*), 112
flatback turtle (*Natator depressus*), 152
foca común (harbor seal), 203–204
FONATUR (National Fund for the Promotion of Tourism), 227
foraging associations, 114
Forgotten Peninsula (J. W. Krutch), 231
fossil deposits: rhodolith beds, 49–50, 53, 56–57, 66–67; studies of, 10, 16
Fossothuria rigida (echinoderm), 85
Fox, Vicente, 226, 227–228
Fregata magnificens (magnificent frigatebird), 175
freshwater resources, 91, 228–229
Fungina (corals), 81

Gardner, Nathanial Lyon, 213
Garman, Samuel, 98
GCW (Gulf of California Water), 40
GEF (Global Environmental Facility), 238
Gelidium crinale (red alga), 217
Gelidium robustum (red alga), 87
General Law of Wildlife (Mexico), 208–209
geographical names, 5n
Geograpsus lividus (crustacean), 74
geological/geophysical definition, 105
geology of Gulf: effect on fish diversity, 101–112; evolution of, 7–10, 22–23, 105; opening of Gulf, 11–19; and phylogeography, 10–11, 20–22, 41, 101–108, 221; rhodolith sediment production, 66–67; typography of, 26–27. *See also* chapter 1
Gerreidae, 174
giant aphroditid polychaetes (*Aphrodita mexicana*, *A. sonorae*), 81, 88
giant Mexican limpet (*Patella mexicana*), 87
giant sea cucumber (*Isostichopus fuscus*), 87
gigantothermy, 153
Gilbert, Charles Henry, 99
Gill, Theodore, 99
gillnets/gillnetting, 124, 222, 239; damage to fish, 116, 222; damage to marine mammals, 199, 205, 240; damage to ospreys, 184–185; legal status of, 93, 242; support for conservation by users of, 161; uses of, 124
ginkgo-toothed beaked whale (*Mesoplodon ginkgodens*), 194
Girella nigricans (opaleye), 109
Girella simplicidens (Gulf opaleye), 174
glacial events, 74, 75, 109
Glen Canyon Dam, 45, 90
Globicephala macrorynchus, 198–199
Glycera tesselata (polychate annelid), 76
Gnathophyllum panamense (crustacean), 77, 85
gobies, 109–111, 113
gooseneck barnacle (*Lepas anatifera*), 158
Gortari, Carlos Salinas de, 161
Gracilaria subsecundata (red alga), 217
Gracilariopsis lemaneiformis (red alga), 150
Grampus griseus (Risso's dolphin), 197
granitic surfaces, 76, 216
Grayson, Andrew Jackson, 172
gray whale (*Eschrichtius robustus*), 3, 189, 191, 200, 205
grebes, 208
green algae (Chlorophyta). *See* algae

green turtle (*Chelonia mydas*), 66, 136–137, *138*, 143, 150–151, 155–156
groupers, 3n, 113
Grupo Tortuguero, 162–164, 166
Guadalupe fur seal (*Arctocephalus townsendi*), 189, 203, 205
Guaymas Basin, 8, 9, 19, 37–44, 129
Gulf/Cortez triplefin (*Axoclinus nigricaudus*), 109, *110*, 112
Gulf corvina, 116
Gulf Extensional Province, 8, 9
Gulf grunion (*Leuresthes sardina*), 110, 114
Gulf of California, name, 5n
Gulf of California Geological Province, 8, 11, *12*
Gulf of California harbor porpoise (*Phocoena sinus*), 88, 93, 199, 205, 224, 239–240
Gulf of California Water (GCW), 40
Gulf opaleye (*Girella simplicidens*), 174
Günther, Albert, 97–98
Gymnomuraena zebra (zebra moray), 103
gyres, 34–36, 38–39, 149

habitats: critical, 3; degradation, effect on sea turtles, 146–147; loss to urbanization, 91
hagfishes, 107
hake, 125
harbor/common seal (*Phoca vitulina*), 203–204
Hastings, Philip, 73
Haustellum elensis (mollusc), 86
hawksbill turtle (*Eretmochelys imbricata*), 137, *138*, 157
heat exchange, effect on circulation, 33–34
Heliaster kubiniji (echinoderm), 85
Hemigrapsus spp. (crustaceans), 75
Hendrickx, Michel, 73
Henricia aspera (echinoderm), 75
herbivorous fishes, 113
hermatypic corals, 81
Hexaplex nigritus (mollusc, black murex), 86, 87
Hexaplex princeps (mollusc), 86
Hia C-ed O'odham (Sand Papago) people, 140, 142, 241
High Aswan Dam, 28–29
high seas fishes, 102
hookah-diving, 150
Hoover (Boulder) Dam, 28, 90, 229
hot spots, geographic, 87
Huaycuras people, 140, *141*
Humboldt squid (jumbo squid). *See Dosidicus gigas*
humpback whale (*Megaptera novaeangliae*), 189, 191, 200–201, 205
hurricanes: and rhodolith beds, 54
hydrothermal vents, 9, 221
Hypsoblennius brevipinnis (Blenniidae), *98*
Hypsoblennius jenkinsi (mussel blenny), 109

incidental taking. *See* by-catch
indigenous peoples, 93; Cochimí, 140, *141*, 143, 221; Comcáac (Seri), 142–143, 159, 221, 232; Cucapá, 140–141, 221; Huaycuras, 140, *141;* O'odham (Hia C-ed O'odham/Sand Papago), 140, 142, 241; and sea turtles, 139–143; Yaqui, 143, 221
Indopacetus pacíficus (Logman's beaked whale), 193
industrial fisheries. *See* commercial fishing
Infiernillo Channel (Canal de Infiernillo), 3, 142–143, 149, 151, *164*
International Committee for the Recovery of the Vaquita (CIRVA), 224
International Water Treaty, 229
intertidal marine earthworm (*Bacescuella parvithecata*), 82
intertidal zone, 72–73, 76–77, *78*, 82–83, 85–86, 94, 114
invertebrate fauna: comparisons to other regions, 83–85; conservation of, 85–92; origin of, 74–76; patterns of diversity, 76–83. *See also* chapter 4
Ishige foliacea (fish), 214
ISLA (Conservación del Territorio Insular Mexicano), 93
islands/islas, 9–10; Ángel de la Guarda,

27–28, 41–42, 145, 172, 177, 180; Bota, 184; Cholludo (Roca Foca), 172, 180, 182; Coronado, 63, 67; El Requeson, 61; Espíritu Santo, 70, 81, 180–181, 184, 186, 197; Guadalupe, 203, *236*, 238; Isabel, 184, 201, *236;* Lobos, 203; Marietas, 53, 170, *236;* Partida, 180, 235; Rasa, 174, 219, 230–233, 235; Revillagigedos, 71; Salsipuedes, 172; San Benito, 203; San Esteban, 172, 182, 193; San José, 193; San Lorenzo, 172, 235; San Lorenzo Norte, 172; San Luís, 172, 177, 180; San Pedro Mártir, 70, 192, 235, *236*, 238; Tiburón, 11, 27–28, 172, 179, 182, 232; Tres Marías, 169, 172, 173, 201, 238; Turner, 180, 182; Ventana, 184. *See also* Midriff Islands
Islas del Cinturón. *See* Midriff Islands
Islas del Golfo conservation program, 93
Isostichopus fuscus (echinoderm), 85, 86, 87
IUCN Red List, 222

Japan, protection of sea turtles, 158–159
javelina, la. *See* loggerhead turtle (*Caretta caretta*)
jawfish, 174
jellyfish, 82
Jordan, David Starr, 98–100
jumbo squid. *See Dosidicus gigas*
jumbo squid fisheries, 123, 125, 130, 133

kelps, 213, 214–215
Kemp's Ridley (*Lepidochelys kempii*), 152
keystone predators, turtles as, 139
killer whale (*Orcinus orca*), 198, 209
Kogia breviceps (pygmy sperm whale), 192
Krutch, Joseph Wood, 231

labrisomid blennies, 100, 102, 110, 113
Lagenorynchus obliquidens (Pacific white-sided dolphin), 197
la golfina (olive Ridley turtle), 137, *138*, 159–160
lagoons. *See* wetlands (estuaries, esteros, lagoons)
Lagrangian method, 37–38
Laguna San Ignacio (San Ignacio Lagoon), 164, 186, 200, 225
laminated diatomaceous sediments, 41–42
La Paz area. *See* Bahía de La Paz
La Paz Fault, *12*, 20, *21*
Larus livens, 175
larval fishes, 114–115
laws. *See* Mexico, environmental laws and decrees
leatherback turtle (*Dermochelys coriacea*), 137, *138*, 152–155
Lepas anatifera (pelagic goose barnacle), 158
Lepidochelys kempii (Kemp's Ridley), 152
Lepidochelys olivacea (olive Ridley), 137
Leptoseris (corals), 81
Leuresthes sardina (Gulf grunion), 110, 114
Leuresthes tenuis (California grunion), 110
Ley de la Vida Silvestre, 3
Ley Federal sobre Metrología y Normalización, 3
Ley General del Equilibrio Ecológico y la Protección al Ambiente, 3
Ley General de Vida Silvestre, 208–209
Ley Orgánica de Administración Pública Federal, 3
Linckia columbiae (echinoderm), 85
Lindbergh, Charles, 231
Lindsay, George, 231, 233
Lithoamnion (red alga), 55–56
Lithophyllum margaritae (red alga), 53, 55, *56*, 58, 60–61, 63
Lithothamnion australe (red alga), 55
Lithothamnion muellerii (= *Lithothamnium crassiusculum*) (red alga), 55, *57*, 58, 60, 63
lizard triplefin (*Crocodilichthys gracilis*), 112
lobo fino de Guadalupe, 203
lobo marino de California, 203
loggerhead turtle (*Caretta caretta*), 137, *138*, 158–159
logging, *130*, 229

Logman's beaked whale (*Indopacetus pacificus*), 193
longbeaked common dolphin (*Delphinus capensis*), 196–197, 205
long-line fishing, 116, 155, 159
longnose hawkfish (*Oxycirrhites typus*), 103
loons, 208
López Mateos, Adolfo, 232
Loreto area, *51;* conservation at, 70, 225; marine mammals, 197, 199, 202; marine park, 117, 166, 236; pictographs, petroglyphs, *141;* turtles, 141, 145, 149, 164
Los Cabos block, *12*, 17, *19*, 20, *21*
Lovenia cordiformis (echinoderm), 86
Lower Colorado River channel, 90–91
Luidia spp. (echinoderms), 85
Luria isabellamexicana (mollusc), 86
Lutjanus peru, 112
Lytechinus pictus (echinoderm), 85

MAB (Man and the Biosphere) program, 234, 237–238
macroalgae, 52, 64–65, 218
Macrofauna Golfo Project, 72–74, 84, 94, 105–107
macroinvertebrate biodiversity. *See* chapter 1
magnificent frigatebird (*Fregata magnificens*), 175
mammals: diversity and abundance, reasons for, 31–32, 190–191; species accounts, 191–204; status of, *206–207;* threats to, 205–208; whaling, 189
mangroves, 83, 89, 116, 176, 185, 230, 247
María Magdalena Rise, 8, 19
marina development, 89, 116, 222, 226
marine algae, defined, 211
marine algae pastures, 165
marine protected areas, 70, 117, 165–166, 225
Marisla, 93
Mar Vermejo (Vermillion Sea), 5n
Marsupenaeus japonicus (crustacean), 85
Mediterranean Sea, 83–85
Megaptera novaeangliae (humpback whale), 189, 191, 200–201, 205
melon-headed whale (*Peponocephala electra*), 197, 209
Mesophyllum engelhartii (red alga), 55, *57*, 58
Mesoplodon spp., 193–194
mesoplodontes (whales), 193–194
mesotrophic sites, 52
metals, trace, 42–43
Metapenaeus monoceros (crustacean), 85
Mexican/coastal fishing bat (*Myotis vivesi mengaux*), 204
Mexican Directive for Species under Some Threat, *69*
Mexican rockfish (*Sebastes macdonaldi*), 109
Mexican Whale Refuge, 209
Mexico: conservation movement, 2, 4, 92–94, 232–233; economic importance of Gulf fishes to, 115; geographical place names, 5n; national fisheries program, 86
Mexico, environmental laws and decrees: on Gulf islands, 231, 232, 234, 237; on industrial shrimping, 88; on land/water rights of Seri people, 232; Ley de Pesca (Fishing Law), 208; Ley Federal sobre Metrología y Normalización, 3; Ley General del Equilibrio Ecológico y la Protección al Ambiente (General Law of Ecological Balance and Environmental Protection), 3, 208, 235–237; Ley General de Vida Silvestre (General Law of Wildlife), 3, 208–209; Ley Orgánica de Administración Pública Federal, 3; on marine mammals, 208; National Fisheries Chart, 133; National Fishing Law, 132–133; NOM-059-ECOL-2001, 3, 189, 205; NOM-131-ECOL-1998, 209; NOM-ECOL 095, *68;* Ordenamiento Ecológico Marino (Marine

Habitat Use Plan), 227–228; Reglamente Interior de la Secretaría de Medio Ambiente y Recursos Naturales (SEMARNAT), 93, 208–209, 234–235; on sea turtles, 146, 160–162, 208
microalgae, 65
microallopatric speciation, 111
Microporella cribosa (bryozoan), 75
Midriff Islands region, 28, *171;* biodiversity, 80; chemical water properties in, 42, 45; currents in, 31–32, 36–37; marine mammals in, 196, 197, 202, 204; ospreys in, 173–174, 177–180, 186; turtles in, 145, 149
minke whale (*Balaenoptera acutorostrata*), 189, 201
Miocene Period, 8, 9, 14–19
Mirounga angustirostris, 189, 204, 205
Mithrodia bradleyi (echinoderm), 85
mojarras (Gerreidae), 174
Molgula occidentalis (ascidian), 75
molluscs: distribution of, 64, 67, 73, 78–85; in Mediterranean, 84; threats to, 29, 73, 89, 124
monitoring programs, 166
monsoons, 30
Monterey Bay Aquarium, 94
Moorish idol (*Zanclus cornutus*), 103
moray eels, 103
mountains, 29
mudflats, 39, 216. *See also* wetlands (estuaries, esteros, lagoons)
Mugil spp. (mullets), 168, 173–175
Mulinia coloradoensis (delta clam), 29, 92
Mullets. *See Mugil* spp.
murciélago pescador (bat). *See* Mexican/coastal fishing bat
mussel blenny (*Hypsoblennius jenkinsi*), 109
Myrichthys maculosus, 65
Myxini, 107

NAFTA, 242
narcotrafficking, 208
Natator depressus, 152
National Fisheries Chart (Mexico), 133
National Fishing Law (Mexico), 132–133
National Marine Fisheries Service (U.S.), 161
Naturalia, 93
Nature Conservancy, 93, 233
Nautical Stairway (Escalera Náutica), 147, 226–227, 246, 248
needlefish (*Strongylura* and *Tylosurus* spp.), 173–175
Neogene Period, 10–11
Neogoniolithon trichotomum, 55, *57*, 58, 61, 63
Nereis riisi (polychaete annelid), 74
nets. *See* drift nets; gillnetting
New World silversides, 109, 113–114
NGOs (nongovernmental organizations), 92–93, 162–164, 233, 235
Nichols, Wallace J., 162
Nidorellia armata (echinoderm), 77, 85
Nile River, 28
Niparajá, 93
nitrate, 28, *32*, 42–44, 48
nitrogen (N), 44
NK-19 contamination, 208
nongovernmental organizations. *See* NGOs (nongovernmental organizations)
Norma Oficial Mexicana (government standards): NOM-059-ECOL-2001, 3, 189, 205; NOM-131-ECOL-1998, 209; NOM-ECOL 095, *68*
North American Plate, 8, *13*, 14, 22
northern disjunct fishes, 108–109
northern elephant seal (*Mirounga angustirostris*), 189, 204, 205
northern fraildisc clingfish (*Pherallodiscus funebris*), 110
Northern Gulf: invertebrate species diversity, 75–78, 80–82; defined, 74, 94–95, *106*
northern pintail (*Anas acuta*), 183
North Pacific right whale (*Eubalaena japonica*), 199–200, 205
NOS, 226, 249

oceanographic definition, 105
Ocopode occidentalis (crustacean), 77
octopus (*Octopus bimaculatus*), 87
Odontaster crassus (echinoderm), 75
Oligocene Period, 7–8, 13–14
Oliva porphyria (mollusc), 86
olive Ridley turtle (*Lepidochelys olivacea*), 137, *138*, 159–160
O'odham (Sand Papago) people, 140, 142, 241
opaleye (*Girella nigricans*), 109
Ophioblennius steindachneri, 112
Ophioderma spp. (echinoderms), 85
Ophiocoma spp. (echinoderms), 85
Ophiodromus pugettensis (polychaete annelid), 75
Ophiopholis longispina (echinoderm), 75
Opistognathus ("jawfish"), 174
orcas, 197–198, 209
Orcinus orca (killer whale), 198
Ordenamiento Ecológico, 235
Ordenamiento Ecológico Marino (Marine Habitat Use Plan), 228
Oreaster occidentalis (echinoderm), 85
organochlorine pesticide contamination, 183
osprey (*Pandion haliaetus*): contamination threat to, 183; diet and foraging, 173–175; dispersal, 177; distribution, 169–171; future research needs, 185–187; history of research, 171–173; management and protection needs, 187; nesting, 175–177, 178–180; regional population levels and trends, 177–178; variation and recent local declines, 180–182. *See also* chapter 8
outboard motors, 222
overfishing, 116–117, 224–226
Oxycirrhites typus (longnose hawkfish), 103
oxygen: concentrations, *32*, 41, 42, 215; effect on species diversity, 215; fish tolerance of low, 114; minima, 25, 41; vertical distribution of, *32*
Pachycerianthus aestuari (cnidarian), 75
Pachygrapsus crassipes (crustacean), 77
Pacific Bottom Water (PBW), 40
Pacific hake, 42
Pacific Intermediate Water (PIW), 40
Pacific mackerel (*Scomber japonicus*), 42, 174
Pacific Ocean: coastal fishes of tropical eastern, 100; east Pacific barrier, 102–103; water exchange with, 28, 37, 43–44
Pacific Plate, *13*, 22, 102
Pacific porgy. *See Calamus brachysomus*
Pacific red snapper (*Lutjanus peru*), 112
Pacific sierra (*Scomberomorus sierra*), 174
Pacific white-sided dolphin (*Lagenorynchus obliquidens*), 197
Padina durvellei (brown alga), 216
Panama Canal, 103
Panamic fanged blenny (*Ophioblennius steindachneri*), 112
Panamic Region, 101–102
Pandion haliaetus. *See* osprey (*Pandion haliaetus*)
pangas, 119, 124
pantropical spotted dolphin (*Stenella attenuata*), 195, 205
Paralabrax maculatofasciatus, 173–174
Paranthura squamosissima (crustacean), 76
particle movement, 33–40
particulate organic matter (POM), 44
patch reefs, 81
Patella mexicana (mollusc), 87
Pavona (corals), 81
PBW (Pacific Bottom Water), 40
pelagic species, ecosystems, and food webs, 1–2, 29, 44, 76, *78*, 80, 82, 94, 102, 110, 116, 122–129, *147*, 153, 158, 159, *164*, 173–175, 197
pelagic red crab. *See Pleuroncodes planipes*
pelicans, 208
penca, 157
Pentaceraster cumingi (echinoderm), 77
Peponocephala electra, 197, 209
Percnon gibbesi (crustacean), 75

perica, la. *See* loggerhead turtle (*Caretta caretta*)
Pericu, 140
Pescadero Basin, *12*, 41
pesticide contamination, 183
petroglyphs, *141*
pH, 31, *32*, 41, 42
Phalacrocorax sp., 174
Pharia pyramidata (echinoderm), 85
Phascolosoma perlucens (sipunculan), 74
Pherallodiscus funebris (northern fraildisc clingfish), 110
Pherallodiscus varius (northern fraildisc clingfish), 110
Phoca vitulina (harbor seal), 203–204
Phocoena sinus (vaquita porpoise), 88, 93, 199, 205, 224, 239–240
phosphate, 28, *32*, 41, 42–43, 48
photosynthesis: in algae, 210–213; chlorophyll concentration values, 30, 39, 45; measuring, 44; nutrients needed for, 28; pigment concentration values, 31, 38
Phyllonotus erythrostomus (pink-mouth murex), 87
phylogeography of Gulf, 10–11, 20–22, 41, 74–76, 101–108, 221
Physeter macrocephalus, 189, 191–192, 205
physiographic regions, *122*
phytoplankton: ENSO and, 40; nutrient fluxes and, 43–45; upwelling effect on, 30
pictographs, *141*
pigmy. *See* pygmy
Pilumnus reticulates (crustacean), 74
pink-mouth murex (*Phyllonotus erythrostomus*), 87
PIW (Pacific Intermediate Water), 40
place names, 5n
plate tectonics, 7–10
Plumularia reversa (cnidarian), 75
Pleistocene Period, 21–22, 49
Pleurobranchus areolatum (mollusc), 75
Pleuroncodes planipes (pelagic red crab), 158
Pliocene Period, 10, 19, 22–23
poaching, 165
Podiceps nigricollis, 174
Polydora nuchalis (polychaete annelid), 75
Porites (corals), 81
power poles, concrete, 185, 187
precipitation, 29
pre-fledging cormorants (*Phalacrocorax* sp.), 174
Procuraduría Federal de Protección al Ambiente: (PROFEPA), 93, 162, 165, 243
Proesteros, 93
Pronatura, 93
ProPeninsula, 93
Proto-Gulf, 8–9, *18*
Psammocora (corals), 81
Pseudorca crassidens, 197–198, 209
Pseudostylochus burchami (Platyhelminthes), 75
Pteropurpura macroptera (mollusc), 75
Ptilosarcus undulatus, 88, 150
Puerto Peñasco: black zone near, 218; commercial fishing in, 88, 125, 145, 241–242; conservation efforts at, 240–242; coquina formations, 3, 78, 216; marine mammals at, 192, 199; osprey at, 177; tourism in, 88–89, 124
pufferfish. *See Arthron meleagris*
Punta Borrascosa (Sonora), 78
pygmy beaked whale (*Mesoplodon peruvianus*), 193–194
pygmy killer whale (*Feresa attenuata*), 209
pygmy sperm whale (*Kogia breviceps*), 192

rainfall, 29, 221
Ramírez Ruíz, Jesús, 233
rays, 65, 86, 107
red algae (Rhodophyta). *See* algae; rhodoliths
Redfield's ratio, 43
redrump blenny (*Xenomedea rhodopyga*), 112
reef fishes, 3, 100, 114–115
refugia, 2–3, 76, 80, 89, 94

Refugio Ballenero Mexicano, 209
Reglamento Interior de la Secretaría de Medio Ambiente y Recursos Naturales (SEMARNAT), 3
research needs, 2, 93–94
rhodoliths, 49–71; anatomical features of, *59;* anthropogenic disturbance and conservation, 67–71; coralline red algae, 49–52, 66, 213; cryptofauna and diversity, 62–63; distribution of beds, *51*, 52–55, 217; diversity of communities of, 61–66; effect of light on, 54, 60; effect of temperature on, 54; fishes associated with, 65–66; harvesting of, 50; morphology and growth of, 56, 58–61; overview, 49–50; reproduction, 60; seasonal changes in growth, 60–61, 63–65; sediment production and fossil deposits, 66–67; taxonomic diversity of, 55–59; taxonomic key to, *58*; vulnerabilities of, 50; water motion and, 53, 60, 67–68; wave beds vs. current beds, 54. *See also* chapter 3
Risso's dolphin (*Grampus griseus*), 197
rivers: biodiversity of, 29, 86; dams, effect of, 29, 86, 90–91, 228–229; effects of loss of, 90–92; Nile River, 28; nutritional input to Gulf from, 28. *See also* Colorado River delta
Rocinela signata (crustacean), 74
rockfishes, 103, 109, 111, 123–125, 173
rorcual tropical, 201–202
rough-toothed dolphin (*Steno bredanensis*), 194
Roy Chapman Andrews Fund, 231, 233

saddlebanded goby (*Barbulifer mexicanus*), 111
Salinas de Gortari, Carlos, 161, 241
salinity: changes in Colorado River delta, 29, 116; in irrigated farmland, 48, 73, 229; seasonal changes in, 48; tolerance in fishes, 114; vertical distribution of, *32*, 40–43
Salsipuedes Basin. *See* Ballenas Channel
Salton Trough, 8–10, 11, *12*, 16–17, 23
San Andreas Fault, 8, 11
sanctuaries, gray whales, 3
sanddabs (Bothidae), 174
Sand Papago (O'odham) people, 140, 142, 241
sand stargazers, 100, 102
San Felipe: commercial fishing at, 88, 241–242; conservation in, 240–242; coquina formations in, 3, 80, 81; geology, 17; location of, *12;* marine mammals at, 192, 199, 201; ospreys in, 172, 177; tourism in, 88–89, 124; turtles in, 145, 151; water properties, 48
San Ignacio Lagoon (Laguna San Ignacio), 164, 186, 200, 225
sardines (*Sardinops* spp.): catch volume vs. value, 127–129; commercial fishing of, 1, 120, 123, 125, 127; competition with anchovies, 42; conservation needs, 121, 130, 226; effect of dams on, 28; effect of Gulf circulation on, 38; El Niño–related collapse of, 123, 226; as important prey animal, 122
Sargassum, 212, 215–218
sargo (*Anisotremus davidsonii*), 109
Sauromelas (chuckwallas), 174
scallops, harvesting of, 50, 68
Scarus spp. (parrotfishes), 90
scavenger species, 69, 86, 88
Scomber japonicus (Pacific mackerel), 174
Scomberomorus sierra (Pacific sierra), 174
sea basses, 113, 173
sea fans, 82
seagrass beds, 165
sea hares (*Aplysia* spp.), 150
sea lettuce (*Ulva* spp.), 150, 212
sea mice. *See Aphrodita*
Sea of Cortez. *See* Gulf of California
sea pen (*Ptilosarcus undulatus*), 88, 150
seasonality: air–water heat exchange and, 33; and algae, 65, 150, 215–217; Central Gulf, 77; Colorado River, 45–48,

90; effect on sea turtles, 149–150; and El Niño events, 40, 45, 150; and fish diversity, 108–109, 114; and Gulf temperatures, 45, 104, 108, 215; and invertebrate diversity, 72, 77; mapping of, *27;* and marine mammal migration, 191, 197, 200–202; Northern Gulf, 45–48, 76–77; and nutrient values, 42, 45; and osprey migration, 170; and rainfall, 29; and rhodolith growth, 60–61, 63–65; and salinity, 48; shrimp season, 120–127; Southern Gulf, 77; and surface circulation, 34–39; tides and, 31–32, 34–40, 104; and whale strandings, 193; and wind-induced upwelling, 24, 26, 30, 31

sea star. *See* crown-of-thorns sea star

sea turtles. *See* turtles

seaweeds. *See* algae

Sebastes macdonaldi (Mexican rockfish), 109

Sebastiinae (rockfishes), 103, 109, 111, 123–125, 173; *S. macdonaldi*, 109

sedimentation: from Colorado River, 10; effects of dams on, 28, 91; effects of trawling on, 68; geological history of, 10–17, 20, 22–23; habitat loss from, 222; laminated diatomaceous, 41–42; and nutrient flux, 43–48, 239; patterns, 41–42; from rhodoliths, 50, 53–54, 62–64, 66–68

sei whale (*Balaenoptera borealis*), 202

selenium contamination, 183

SEMARNAT, 93, 208–209, 234–235

Seri (Comcáac) people, 142–143, 159, 221, 232

Setchell, William Albert, 213

sharks: distribution of, 97, 107, 113; effects of long lines and gillnets on, 116; fishing of, 123–124, 146, 205

sharpnose puffers, 113

shooting of ospreys, 185

short-beaked common dolphin (*Delphinus delphis*), 196, 205

short-finned pilot whale (*Globicephala macrorynchus*), 198–199, 209

shrimp farming: and competition with trawling, 88; and demand for fish meal, 130; environmental impact of, 3, 88–89, 116, 123, 130, 230; history of, 123; inland relocation possibility, 73

shrimps (Penaeidae), 87, 92

shrimp season, 123, 124–125

shrimp trawling. *See* trawling

Sicyonia laevigata (crustacean), 74

silversides, 109, 111, 113–114

Sipunculus nudus (sipunculan), 76

sister-species pairs, 75

skates, 86

SLOSS debate, 4

slow goby (*Aruma histrio*), 111

snappers, 113

Sociedad Cooperativa de la Producción Pesquera Seri, 232

southern fraildisc clingfish (*Pherallodiscus varius*), 110

Southern Gulf: biodiversity in, 96, 108, 111, 214; definition, 74, 94–95, *106*; geological evolution of, 11–12, 16, 19; gyres in, 38; macroinvertebrates in, 81; marine mammals in, 193–195, 202; seasonality and, 77, 104; water properties of, 34–36, 38

Southern Oscillation (ENSO), 40, 60, *61*, 150

Southwest Seafood Watch Cards, 94

souvenir-hunting by tourists, 2, 89

speciation, 110–111, 115

sperm whale (*Physeter macrocephalus*), 189, 191–192, 205

Sphoeroides annulatus (bullseye puffer), 173

spinner dolphin (*Stenella longirostris*), 195

Spirobranchus giganteus (polychaete annelid), 74

sponges, 82, 83, 84, 157

Sporolithon (red alga), 56

spotted pufferfish (*Arthron meleagris*), 90

spotted sand bass (*Paralabrax maculatofasciatus*), 173–174
Spyridia filamentosa (brown alga), 217
SSW (Subtropical Subsurface Water), 40
Steindachner, Franz, 98
Stenella attenuata (pantropical spotted dolphin), 195
Stenella coeruleoalba (striped dolphin), 195–196
Steno bredanensis (rough-toothed dolphin), 194
storms, tropical, 29–30
strike-slip faulting, 13–14, 21, 26–27
striped dolphin (*Stenella coeruleoalba*), 195–196
striped mullet (*Mugil cephalus*), 173
Strongylura, 173
Strongylura exilis (needlefish), 174
Subtropical Subsurface Water (SSW), 40
surgeonfishes, 103
sustainable fisheries, 93–94
swimming crabs, 73, 77, 87
swordfish fishery regulations, 161
Syllis elongata (polychate, annelid), 76
sympatric speciation, 111
Syngnathus carinatus (Cortez pipefish), 111

Taeniconger digueti (Cortez garden eel), 65
taxonomic key: to rhodoliths, *58*
taxonomy: of rhodoliths, 55–59
TEDs (turtle excluder devices), 161–162, 166
temperature: air–water heat exchange, 33–34; glacial events and, 75, 109; North vs. South Gulf, 108; offshore surface, 39; and sea turtle harvesting, 150–151; thermal gradients, 103–104, 109, 215–218; tolerance in fishes, 114
thermohaline circulation, 28, 33, 40, 122
Tiburón Basin, 37, 42
Tiburón Island. *See under* islands/islas
tides: contrasted with circulation, 33; early research on, 25–26; habitats created by, 211, 216–218; impact of freshwater flows on, 91; increasing due to dams, 91; modeling of, 35–37; sea level and, 39–40; and tidal mixing, 31–32
tiger snake eel (*Myrichthys maculosus* Curvier), 65
Tohono O'odham people, 140, 142, 241
tonina (bottlenose dolphin), 192, 194–195, 205
topography of Gulf, 7–10, 26–28
Torre, Jorge, 73
tortoiseshell, 140–141, 157, 165
totoaba (*Totoaba macdonaldi*): commercial fishing of, 3, 116, 123, 222, 224, 239; conservation of, 88, 93, 224, 233, 239–240; and loss of prey, 88
tourism: economic importance of, 89, 115, 122, 223; ecotourism, 136, 166, 223, 248–249; effects of camping and hiking, 184, 187; hand-collecting, 2, 89; promise and threats of, 89, 147, 198, 226–228; restaurants, 89; sport fishing, 115, 124, 226; and turtle conservation, 147, 166; whale-watching tours, 225
Townsend, Charles, 144
trace metals, 42–43
trans-Panamic seaway, 103
trawling: conservation needs, 247; damage from, 1–2, 87–90, 133; effect on invertebrates, 73, 80, 81, 87–88; effect on rhodolith beds, 68–69; effect on sea turtles, 146, 161–162; effect on soft-bottom fishes, 115; increase in, 224–225; loss of diversity from, 86, 88, 115, 222–223; regulation of, 93–94, 242–243; research needs, 2; technology of, 124, 130, 222–223; use of inappropriate technologies, 130
tres aletas colinegra (*Axoclinus nigricaudus*), 109, *110*, 112
triplefin blennies (Tripterygiidae), 100, 113, 115
tube blennies (Chaenopsidae), 110, 113, 115

tuna, 127, 130
tunicates, 82, 83
tursión, 192, 194–195
Tursiops truncatus (bottlenosed dolphin), 192, 194–195, 205
turtle excluder devices (TEDs), 161–162, 166
turtles, sea: arribada, 160, 165; commercial harvesting of, 123, 144–147; current status of, 146–147, 224; environmental laws on, 93; foraging needs of, 149–151; future conservation needs, 165–167; and indigenous peoples, 139–143; life history of, 147–149, *147;* nesting of, *148*, 151–160; research and monitoring sites, *164;* species accounts, 152–160
Tylosurus (needlefish), 173

Ulloa, Francisco de, 5n, 140, 143
Ulva lactuca (green alga), 150, 212
Umbrina roncador (Sciaenidae, yellowfin croaker), *99*
UNCED (United Nations Conference on Environment and Development), 242
Upeneus moluccensis (fish), 85
upper Gulf: seasonal changes in, 45–48
Upper Gulf of California and Colorado River Delta Biosphere Reserve, 93, 117
upwelling: and biodiversity, 80, 215, 221; and circulation, 34, 38–39; as fertilization mechanism, 28, 122; meteorological aspects of, 29–30; in Midriff Islands region, 26, 80; seasonal, 104, 149–150
urbanization, 2, 28, 91, 116, *130*, 184
Urechis caupo (echiuran), 75
U.S. Hydrographic Office, 34
utility poles, osprey nests on, 178

van der Heiden, Albert, 73
vaquita porpoise (*Phocoena sinus*), 88, 93, 199, 205, 224, 239–240
Velarde, Enriqueta, 233
Vermillion Sea (Mar Vermejo), 5n
vertical distribution of physical, chemical properties, *32*, 40–43
Villa, Bernardo, 231, 233

Walker, Lewis Wayne, 230–231
water exchange (with Pacific Ocean), 28, 37, 43–44
wave beds, 54, 63
wetlands (estuaries, esteros, lagoons): biodiversity of, 39–40, 72, 83, 89, 91, 101; conservation of, 225, 229, 230, 237, 247; effect of agriculture drainage on, 48, 183, 229; effect of dams on, 29, 91, 92, 229; effect of inshore fishing on, 222; effect of sea level on, 39–40, 91; effect of tidal currents on, 32; *estero*, defined, 83, 210; estero-based aquaculture, 73, 116; fishes in, 116, 173; fragmentation of, 221; and habitat lost, 89, 91, 228–230; lagoons, coastal, 28, 116; marine algae in, 216; mudflats, 39, 216; ospreys in, 173, 182; pesticide contamination in, 183; research needs, 93; and shrimp life cycle, 92; and shrimp lifecycle, 92; species distribution in, 83; and tidally driven current beds, 54; turtles in, 156. *See also* Colorado River delta
whales: conservation laws on, 209; gray, 3; history of whaling, 189; species accounts, 191–194; strandings, 193
white shark (*Carcharodon carcharias*), 97
Wildcoast, 93
wildlife protection areas, *236*, 237–238
wind-induced upwelling. *See* upwelling
Winkler method, 41
World Wildlife Fund, 93
wrasses, 113, 117

Xántus, John, 25, 99, 172
Xenomedea rhodopyga (redrump blenny), 112

Yaqui people, 143, 221
yellowfin croaker/berrugata aleta amarilla (*Umbrina roncador* Sciaenidae), *99*
yellowfooted gull (*Larus livens*), 175

Zalophus californianus (California sea lion), 189, 203
Zanclus cornutus (Moorish idol), 103
Zavala, Alfredo, 233
zebra moray (*Gymnomuraena zebra*), 103
Zedillo, Ernesto, 234
Zífido de baird, 193
Zífido de cuvier, 192–193
Zífido de logman, 193
Ziphius cavirostris (Cuvier's beaked whale), 192–193
Zostera marina (eelgrass), 143, 150